Molecular Chaperones in the Cell

EDITED BY

Peter Lund

School of Biosciences
University of Birmingham
Birmingham B15 2TT

OXFORD

UNIVERSITY PRESS

OXFORD

UNIVERSITY PRESS

Great Clarendon Street, Oxford OX2 6DP

Oxford University Press is a department of the University of Oxford
It furthers the University's objective of excellence in research, scholarship,
and education by publishing worldwide in

Oxford New York

Athens Auckland Bangkok Bogotá Buenos Aires Cape Town
Chennai Dar es Salaam Delhi Florence Hong Kong Istanbul Karachi
Kolkata Kuala Lumpur Madrid Melbourne Mexico City Mumbai Nairobi
Paris São Paulo Shanghai Singapore Taipei Tokyo Toronto Warsaw

with associated companies in
Berlin Ibadan

Oxford is a registered trade mark of Oxford University Press
in the UK and in certain other countries

Published in the United States
by Oxford University Press Inc., New York

A catalogue record for this book is available from the British Library

Library of Congress Cataloging in Publication Data
Molecular chaperones in the cell / edited by Peter Lund.
(Frontiers in molecular biology)
Includes bibliographical references and index.
1. Molecular chaperones. I. Lund, Peter A. II. Series.

QP552.M64 M646 2001   572'.633–dc21   2001016278
ISBN 0 19 963868 3 (Hbk.)   1 0 0 2 5 0 8 3 0 7
ISBN 0 19 963867 5 (Pbk.)

Typeset by Footnote Graphics, Warminster, Wilts

Printed in Great Britain on acid free paper by
Bookcraft (Bath) Ltd,
Midsomer Norton, Avon

Molecular Chaperones in the Cell

Frontiers in Molecular Biology

SERIES EDITORS

B. D. Hames

*Department of Biochemistry
and Molecular Biology
University of Leeds, Leeds LS2 9JT, UK*

D. M. Glover

*Department of Genetics,
University of Cambridge, UK*

TITLES IN THE SERIES

Preface

A search through the scientific literature using 'chaperone' as a key word is informative. Prior to 1987, there is no mention of the word in the context of the stress response or protein folding. In 1987 and 1988, papers which refer to 'molecular chaperones' and which are now regarded as classics, begin to appear—demonstrating, for example, sequence similarity between proteins involved in bacteriophage assembly and the import of proteins into chloroplasts. Then as the 90s begin, there is an explosion in the number of papers in what is clearly a newly emerging field of biology. The numbers of papers published with the word 'chaperone' in the title or abstract are in the hundreds by 1992, and in 1998 pass the 1000 mark. The number published in 2000 was the highest yet; indeed, there has yet to be a year in which the numbers have dropped. If only our investments all performed so well.

The extraordinary growth in interest in molecular chaperones is the result of two quite different phenomena. First, the advent of the 'chaperone hypothesis'—that is, that many proteins require other proteins to assist them in order to reach their final correctly folded and active conformation—appeared at first sight to violate the widely held opinion that proteins contain within their amino-acid sequence all the information that is required for them to fold correctly. This notion acted on many in the protein folding community like a red rag to a bull, and also brought many others, who had not previously had a strong interest in the process of protein folding, into that area of research. Molecular chaperones were the gatecrashers that livened up the protein folding party. In due course, it was realized that the discovery of molecular chaperones did not in fact require anyone to set aside any long-cherished thermodynamic principles about protein folding, but it did compel researchers to begin thinking about protein folding in the context where it is naturally takes place—within the cell. And this in turn linked with the second reason that people started to get excited about molecular chaperones, which is that it has become increasingly obvious that large numbers of cellular events require diverse molecular chaperones to be present in order to take place at all.

Much elegant work has been done over the last decade in understanding the structure and the function of the major molecular chaperones, to the extent that in some cases we now have a remarkable degree of insight into how these proteins work, although in no case is the picture complete. More recently, the emphasis in research has begun to shift more to understanding the precise roles of molecular chaperones in the many different processes with which they are involved inside the cell, and this aspect of molecular chaperone biology is the focus of the present book. The decision to look at chaperones mainly from a process point of view, rather than to review them in terms of their sequence homologies or mechanisms of action, has been a very deliberate one. Part of the reason for this is that there are now so many examples known of proteins which have a molecular chaperone function, that a book reviewing each and every one of them would be prohibitively long. More importantly, it would also be rather disjointed. By presenting chaperones more in the context of the processes that they are involved with, I hope that readers will gain a fuller understanding of the roles of molecular chaperones in the context of cellular processes, while still learning something about how the chaperones display the properties that they do. I hope that readers will also come to appreciate what a broad group of proteins the molecular chaperones represent, and also how at the edges the definitions of what is, and is not, a molecular chaperone become rather fuzzy: some chaperones can also act as proteases, for example, and some protein folding events (particularly those that take place in oxidizing environments such as the endoplasmic reticulum or the bacterial periplasm) require a mixture of covalent and non-covalent interactions between substrate proteins and their cognate chaperones.

The book begins by looking at prokaryotic systems, where much of the preliminary work on chaperones was done, and where the level of our understanding is greatest. It is becoming clear that networks of chaperones operate within cells, and these are better understood in *E. coli* that in any other organism. Mogk, Deuerling, and Bukau give a description of the key chaperones in *E. coli*, and show how their modes of action overlap under normal growth conditions and under the stressed conditions, such as heat shock, where many chaperones are strongly induced. In the next chapter, Harms, Luirink, and Oudega go on to consider the special case of protein secretion and show how all the steps in protein secretion from initial recognition of signal sequences, through targeting to and transport across the cytoplasmic membrane, modification in the periplasm, and insertion into or out of the outer membrane, all have different chaperones associated with them. As might be expected, the same is true of the movement of proteins into and around plastids and mitochondria from the cytoplasm in eukaryotic cells, and Voos and Pfanner cover this topic in the third chapter. In the eukaryotic cytosol itself there are many different molecular chaperones and these are discussed in the next few chapters. Willison and Grantham look at chaperones in actin and tubulin assembly, in particular the role played by the CCT protein complex, distant cousin to the Hsp60s of prokaryotes and plastids that started the chaperone ball rolling in the classic papers mentioned above. The pictures of the CCT complex together with bound actin, shown in this chapter, are an exciting example of how powerful new imaging techniques have contributed to the growth of our understanding of how molecular chaperones interact with their substrates. Pfund, Yan, and Craig then examine the multitude of roles played by the Hsp70 proteins and their various cofactors in the eukaryotic cytosol particularly under normal growth conditions, while Morimoto and Song go on to examine the roles of different chaperones, focusing again on Hsp70 chaperones and their cofactors, in stressed eukaryotic cells. The story here is a complex one, and it is becoming evident that the many different roles played by the numerous Hsp70 proteins in eukaryotes are to some extent co-ordinated by the cofactors with which they interact and which are discussed here. The chapter by Smith then discusses the role that different chaperones, in particular Hsp90, play in the central signal transduction pathways of the eukaryotic cell. Pelletier, Bergeron, and Thomas then go on to consider the chaperone pathways that exist in the cell's own protein folding organelle: the endoplasmic reticulum, where proteins must be folded with the correct disulphide bonds and other modifications to be delivered to the secretory pathways. Two chapters then link prokaryotic and eukaryotic aspects of chaperone biology (although many parallels are already discussed in the chapters above). Maurizi looks at the relationships between chaperones and proteases, which is critical for the whole area of what might be called protein quality control in both normally grown and stressed cells, and Lund considers the routes whereby the expression of chaperone genes can respond to greater demands for chaperone capacity in the cell, for example the presence of increased concentration of unfolded or partially folded proteins. Finally, Kan and Radford discuss the phenomenon of protein misfolding and disease, an area which is starting to receive much attention because of its obvious potential applications, and point out the possible links between this and chaperone biology. This is undoubtedly an area where we can expect a good deal of research effort over the next few years and it seems fitting to close this collection of reviews with a survey of a field where so much work remains to be done and where the potential consequences for our understanding of ageing and many degenerative diseases is so profound.

It is difficult, in a field as fast moving as this, to keep up to date with developments in the literature and in the many labs around the world. Moreover, because of limitations of space there are some topics which have only been lightly touched upon in the present collection: the newly proposed role of Hsp90 chaperones as 'molecular capacitors' in masking much genetic variation is one obvious area. However, given the considerable effort which the authors of all the chapters have expended to present a picture which summarizes the current state of our knowledge in all the different areas above, but which also looks to the future and to where the field is likely to move, I like to think that this book will be useful for some years to come. It only remains for me to thank the authors for their efforts in summarizing the enormous body of knowledge in this field into bite-size chunks, and for their patience in the process of turning their hard work into this book.

Birmingham   P.L.
January 2001

Contents

3 The role of chaperone proteins in the import and assembly of proteins in mitochondria and chloroplasts   61

WOLFGANG VOOS AND NIKOLAUS PFANNER

4 The roles of the cytosolic chaperone, CCT, in normal eukaryotic cell growth   90

KEITH R. WILLISON AND JULIE GRANTHAM

9 The function of chaperones and proteases in protein quality control and intracellular protein degradation   205

MICHAEL MAURIZI

10 Regulation of expression of molecular chaperones   235

PETER LUND

11 Partial unfolding as a precursor to amyloidosis: a discussion of the occurrence, role, and implications   257

NEIL M. KAD AND SHEENA E. RADFORD

Contributors

JOHN J. M. BERGERON
Department of Anatomy and Cell Biology, McGill University, Montreal, Quebec, H3A 2B2, Canada.

BERND BUKAU
Universität Freiburg, Institut für Biochemie und Molekularbiologie Hermann-Herder-Str.7, 79104 Freiburg, Germany.

ELIZABETH A. CRAIG
Department of Biomolecular Chemistry, University of Wisconsin–Madison, 1300 University Avenue, Madison, WI 53706, USA.

ELKE DEUERLING
Universität Freiburg, Institut für Biochemie und Molekularbiologie Hermann-Herder-Str.7, 79104 Freiburg, Germany.

JULIE GRANTHAM
Institute of Cancer Research, Chester Beatty Laboratories, 237 Fulham Road, London SW3 6JB, UK.

NELLIE HARMS
Department of Microbiology, Institute of Molecular Biological Sciences, BioCentrum Amsterdam, De Boeleaan 1087, 1081 HV Amsterdam, The Netherlands.

NEIL M. KAD
School of Biochemistry and Molecular Biology, University of Leeds, Leeds LS2 9JT, UK.

JOEN LUIRINK
Department of Microbiology, Institute of Molecular Biological Sciences, BioCentrum Amsterdam, De Boeleaan 1087, 1081 HV Amsterdam, The Netherlands.

PETER LUND
School of BioSciences, University of Birmingham, Birmingham B15 2TT, UK.

MICHAEL R. MAURIZI
Laboratory of Cell Biology, National Cancer Institute, Bethesda, MD 20892, USA.

AXEL MOGK
Universität Freiburg, Institut für Biochemie und Molekularbiologie Hermann-Herder-Str.7, 79104 Freiburg, Germany.

RICHARD I. MORIMOTO
Department of Biochemistry, Molecular Biology and Cell Biology, Rice Institute for Biomedical Research Northwestern University, Evanston, IL 60208, USA.

BAUKE OUDEGA
Department of Microbiology, Institute of Molecular Biological Sciences, BioCentrum Amsterdam, De Boeleaan 1087, 1081 HV Amsterdam, The Netherlands.

MARC F. PELLETIER
Genetics Group, Biotechnology Research Institute, National Research Council of Canada, 6100 avenue Royalmount, Montreal, Quebec H4P 2R2, Canada, and Department of Biology, McGill University, Montreal, Quebec, H3A 2B2 Canada.

NIKOLAUS PFANNER
Universität Freiburg, Institut für Biochemie und Molekularbiologie Hermann-Herder-Str.7 79104 Freiburg, Germany.

CHRISTINE PFUND
Department of Biomolecular Chemistry, University of Wisconsin–Madison, 1300 University Avenue, Madison, WI 53706, USA.

SHEENA E. RADFORD
School of Biochemistry and Molecular Biology, University of Leeds, Leeds LS2 9JT, UK.

DAVID F. SMITH
Dept of Research, Mayo Clinic Scottsdale, Scottsdale, AZ 85259, USA.

JAEWHAN SONG
Department of Biochemistry, Molecular Biology and Cell Biology, Rice Institute for Biomedical Research Northwestern University, Evanston, IL 60208, USA.

DAVID Y. THOMAS
Department of Biology, McGill University, Montreal, Quebec, H3A 2B2, Canada.

WOLFGANG VOOS
Universität Freiburg, Institut für Biochemie und Molekularbiologie Hermann-Herder-Str.7 79104 Freiburg, Germany.

KEITH R. WILLISON
Institute of Cancer Research, Chester Beatty Laboratories, 237 Fulham Road, London SW3 6JB, UK.

WEI YAN
Department of Biomolecular Chemistry, University of Wisconsin–Madison, 1300 University Avenue, Madison, WI 53706, USA, and Department of Microbiology, University of Washington, Box 357242, Seattle, WA 98195, USA.

Abbreviations

AAA	ATPase associated with different cellular activities
ACP	acyl phosphatase
AFM	atomic force microscopy
ANS	1-anilinonapthalene-8-sulfonic acid
APP	amyloid precursor protein
AR	androgen receptor
ARE	androgen receptor element
CAT	chloramphenicol acetyl transferase
CCT	chaperone containing TCP-1
CFTR	cystic fibrosis transmembrane conductance regulator
CIRCE	controlling inverted repeat of chaperone expression
CPY	carboxypeptidase Y
CS	citrate synthase
DTT	dithiothreitol
EM	electron microscopy
ER	endoplasmic reticulum
ERAD	endoplasmic reticulum associated degradation
ESI-MS	electron spray ionization mass spectrometry
FAP	familial amyloid polyneuropathy
FKBP	FK506 binding protein
FTIR	Fourier transform infrared spectroscopy
GAP	glucosaminoglycan
GFP	green fluorescent protein
GR	glucocorticoid receptor
HCMV	human cytomegalovirus
HGF	hepatocyte growth factor
HIV	human immunodeficiency virus
HLA	human leucocyte antigen
HMW	high molecular weight
HRI	haem regulated eIFα kinase
HSE	heat shock element
HSF	heat shock transcription factor
Hsp	heat shock protein
IAP	import associated protein
IEM	chloroplast inner membrane
IM	inner membrane
IMP	inner membrane protein
IMS	intermembrane space

LPS	lipopolysaccharide
mAb	monoclonal antibody
MAP	microtubule associated protein
MDH	malate dehydrogenase
MHC	major histocompatibility complex
MSF	mitochondrial import stimulation factor
mtHsp70	mitochondrial Hsp70 (Ssc1p)
NBD	nucleotide binding domain
NMR	nuclear magnetic resonance
NSF	N-ethylmaleimide sensitive factor
OEM	chloroplast outer membrane
OM	outer membrane
OR	oestrogen receptor
OMP	outer membrane protein
PDGF	platelet derived growth factor
PDI	protein disulfide isomerase
PDZ	domain of approximately 90 residues found in a number of proteins associated with receptors, membrane channels, and signal transduction proteins
PPIase	peptidyl prolyl *cis*/*trans* isomerase
PR	progesterone receptor
RAR	retinoic acid receptor
RCM-La	reduced and carboxymethylated lactalbumin
RNC	ribosome nascent chain complex
ROSE	repression of heat shock gene expression
RUBISCO	ribulose *bis*-phosphate carboxylase
SAP	serum amyloid component
SAPK	stress activated protein kinase
SBA	soybean agglutinin
SDS-PAGE	sodium dodecyl sulphate polyacrylamide gel electrophoresis
SRP	signal recognition particle
SSD	sensor and substrate discrimination
TCP-1	t-complex polypeptide 1
TF	trigger factor
TicX	translocase of inner chloroplast membrane, subunit of X kDa
TimX	translocase of inner mitochondrial membrane, subunit of X kDa
TocX	translocase of outer chloroplast membrane, subunit of X kDa
TomX	translocase of outer mitochondrial membrane, subunit of X kDa
TPR	tetratricopeptide repeat
TTR	transthyretin
UBC	ubiquitin conjugating enzyme
UGGT	UDP-glucose:glycoprotein glucosyl transferase
UPR	unfolded protein response
VSV	vesicular stomatitis virus

1 | Cellular functions of cytosolic *E. coli* chaperones

AXEL MOGK, BERND BUKAU, and ELKE DEUERLING

1. Introduction

The folding of newly synthesized polypeptide chains into their unique three-dimensional structures is of fundamental importance in biology. Protein folding mechanisms were first studied *in vitro* using purified, chemically denatured polypeptides as substrates. In 1973 Anfinsen demonstrated the correct (re)folding of RNase upon removal of denaturant, suggesting that all the information needed for proper (re)folding was contained in the primary sequence of the protein (1). However, the concept of spontaneous protein folding cannot be extrapolated to living cells. A major difference between the idealized conditions used in *in vitro* refolding studies and those encountered within cells is that the intracellular environment is highly crowded due to the presence of high concentrations of soluble and insoluble macromolecules (*E. coli* cytosol: 170 mg proteins/ml) (2) . Theoretical considerations (3) predict that such crowding leads to excluded volume effects which strongly affect biochemical rates, for example by increasing protein association constants thereby increasing the propensity of intermolecular interactions including aggregation. Consistent with this theory are results of recent *in vitro* experiments with unfolded reduced lysozyme (4). The yields of correctly folded enzyme decreased dramatically, and aggregation increased concomitantly, in the presence of different crowding agents. This effect was not due to decreased folding rates, and did not occur under oxidizing conditions which preserve native disulfide bonds in lysozyme and increase the folding rate by 2–3 orders of magnitude. For lysozyme, its folding rate thus determines whether the folding yield is sensitive to crowding agents. Molecular crowding may therefore endanger especially the slow folding process of multidomain proteins by intramolecular misfolding and intermolecular aggregation (5). However, an earlier study reported that crowding agents nonspecifically protect other proteins from heat denaturation probably by decelerating association rates between folding proteins or, alternatively, by unspecific hydrophobic or electrostatic interactions between folding and protecting proteins (6). Taken together, macromolecular crowding has to be considered as an important, though poorly understood, parameter influencing protein folding.

Table 1 Major cytosolic chaperon families of *E. coli*

Chaperone family	E. coli member	Interacting protein (co-chaperone)	Eukaryotic homolog	Action	Null mutant phenotype
AAA+ (Hsp100/Clp)	ClpA	ClpP		ATP-dependent proteolysis	No phenotype
	ClpB		Hsp104, Hsp78	ATP-dependent disaggregation of protein aggregates	Impaired thermotolerance
	ClpX	ClpP		ATP-dependent proteolysis	No phenotype
	ClpY (HslU)	ClpQ (HslV)		ATP-dependent proteolysis	No phenotype
	FtsH		Afg3p, Rcalp	ATP-dependent proteolysis	lethal
	Lon		Lon	ATP-dependent proteolysis	mucoid growth
Hsp90	HtpG		Hsp82, etc.	ATP-dependent chaperone	Reduced growth rate at 44 °C
Hsp70	DnaK	DnaJ, CbpA, DjlA	Hsc70, Hsp72, Bip, etc	ATP-dependent chaperone	Temperature-sensitive growth (39 °C)
	HscA	HscB	Ssq1p	Unknown	Slow growth
	Hsc62	(YbeS, YbeV)		Unknown	No phenotype
Hsp60	GroEL	GroES	CCT, TriC	ATP-dependent chaperone	lethal
sHSP	IbpA, IbpB		Hsp25, etc	ATP-independent chaperone	No phenotype
Hsp33	Hsp33			Redox-regulated ATP-independent chaperone	Slightly sensitive towards high temperature H_2O_2
Trigger factor (TF)	Trigger factor			PPIase, ATP-independent chaperone	No phenotype

Living cells have developed systems for preventing the misfolding and aggregation of newly synthesized proteins and for controlling the folding status of polypeptides. These functions are fulfilled by molecular chaperones. Chaperones prevent inappropriate inter- and intra-molecular interactions by binding to hydrophobic patches of non-native proteins, thereby influencing the partitioning between productive and unproductive folding pathways. Importantly, chaperones do not form part of the final structures of the folded proteins. Instead substrates are released from the chaperones, thereby providing non-native proteins with a new opportunity for productive folding. Some chaperones perform an ATP-dependent mode of substrate interaction (Hsp100, Hsp90, Hsp70, Hsp60), while others (TF, sHsps) do not (7). Finally, depending on their abilities to either prevent protein aggregation or to support proper protein (re)folding, 'holder' and 'folder' chaperones are distinguished. Holder and folder chaperones may functionally cooperate in a folding network. Table

1 summarizes the major cytosolic chaperone families of *E. coli* and their eukaryotic homologues.

This review first describes the basic features of major cytosolic *E. coli* chaperones. The roles of individual chaperones in the processes of *de novo* folding, protein secretion, and protein protection during stress are discussed in the following sections. Emphasis is given to other excellent reviews describing chaperone functions in the individual processes (8–12).

2. Major cytosolic *E. coli* chaperones

2.1 The Hsp100/Clp chaperones

Members of the Hsp100/Clp family are highly conserved among prokaryotes and eukaryotes. These chaperones are discussed in more detail in Chapter 9, and will only be briefly surveyed here. This family has been classified according to their structural features and sequence similarities into two groups. Proteins of the first group (ClpA, ClpB, ClpC, ClpD, and ClpE), contain two nucleotide-binding domains (NBDs), whereas the second group of smaller proteins has only a single NBD (ClpX and ClpY/HslU). Each class is further subdivided according to specific signature motifs or the length of the interdomain region separating the two NBDs (13). Recent sequence analysis and structural determination revealed striking similarities between Hsp100/Clp proteins and members of the AAA (ATPases associated with different cellular activities) protein family, thereby defining a new AAA+ superfamily also including the *E. coli* proteases Lon and FtsH (14–16).

All Clp ATPases analysed so far oligomerize to hexameric ring-like structures. Many of the Clp proteins (ClpA, ClpX, HslU. and ClpC) act as the ATPase subunit of an ATP-dependent protease by associating with either the ClpP proteolytic subunit or, in the case of HslU, with the peptidase HslV. ATPase subunits confer substrate specificity to the proteolytic subunits (17, 18). Degradation of protein substrates by ClpP is strictly dependent on the associated Clp ATPase. Through *in vitro* studies of the ATPase components, it has been demonstrated that ClpA and ClpX have chaperone-like activity.

Purified ClpA activates the bacteriophage P1 replication protein, RepA, for DNA binding (19) but in the presence of ClpP, RepA is degraded in a ClpA and ATP-dependent manner. ClpA has therefore been proposed to destabilize protein structure, allowing passage of substrates destined for degradation through a central channel into the ClpP proteolytic chamber (17, 20). A global unfolding activity of ClpA towards the stable monomeric green fluorescent protein GFP was recently demonstrated (21). ClpX stimulates the replication of bacteriophage Mu by promoting the dissociation of the MuA protein from DNA (22, 23), an activity which also requires ATP hydrolysis. *In vivo* Mu replication is impaired in a *clpX* mutant but not in a *clpP* mutant, confirming that the biologically important reaction dependent on ClpX is related to its chaperone activity rather than its activity as part of the ClpXP protease (24).

Two close relatives of ClpA, the *E. coli* ClpB and the *Saccharomyces cerevisiae* homologue Hsp104, have not been implicated in proteolysis and do not associate with proteolytic subunits *in vitro*. They instead exhibit chaperone-like activity on a variety of substrates and are necessary for thermotolerance *in vivo* (25–27). ClpB and Hsp104 act to resolubilize protein aggregates. However, this disaggregating activity only works in cooperation with the corresponding Hsp70 system of *E. coli* or *S. cerevisiae*, respectively (28–32). For the *E. coli* disaggregation system it was shown that sub-stoichiometric amounts of the DnaK system (the major Hsp70 of *E. coli*) together with ClpB could solubilize and reactivate large amounts of a wide array of both aggregated natural and model substrates (29, 31). Experiments using SO_4^{2-} as a specific inhibitor of ClpB revealed that in a first phase of the disaggregation process ClpB is strictly essential. During this phase large aggregates are converted to smaller particles. It is unlikely that a complex comprising ClpB and DnaK is formed during this process since optimal disaggregating and refolding yields of malate dehydrogenase aggregates were observed in the presence of sub-stoichiometric amounts of hexameric ClpB with respect to monomeric DnaK. In the second phase, which is independent of ClpB, smaller aggregates are completely resolubilized and refolded into native proteins by the DnaK system. Thus ClpB and the DnaK system act in a sequential manner. ClpB was shown to interact with protein aggregates, thereby increasing the hydrophobic exposure of aggregate surfaces. It was therefore proposed that ClpB or Hsp104 alter the structure of these particles in a 'crowbar mechanism', thereby creating new binding sites for the Hsp70 systems (30, 31).

2.2   The HtpG (Hsp90) chaperone

Members of the Hsp90 family are present in the cytoplasm of eubacteria, yeast, and higher eukaryotes. Homologues also exist in the ER, mitochondria, and chloroplasts, while a homologue has not yet been found in Archaea. Biophysical studies characterized Hsp90 as a dimeric protein of elongated shape (33) with the dimerization site located to the C-terminal region. Hsp90 proteins consist of two domains separated by a charged region. The crystal structure of the 25-kDa N-terminal domain contains a binding site for ATP, which also represents the high affinity binding site for geldanamycin (GA), an anti-tumour drug that specifically inhibits Hsp90 activity (34, 35). Dissection of Hsp90 into N- and C-terminal domains surprisingly revealed that both domains could suppress the aggregation of non-native proteins (36, 37). These results strongly suggest that Hsp90 contains two chaperone sites, which contribute independently to its chaperone activity. Both chaperone sites differ in substrate specificity and nucleotide dependence. While the N-terminal site interacts with unfolded proteins and peptides in an ATP- and GA-dependent way, the C-terminal domain appears to act as an ATP-independent general chaperone (37). Several *in vitro* experiments showed that Hsp90 can act as a holder chaperone. Bovine Hsp90 as well as *E. coli* HtpG are able to suppress the aggregation of thermally denatured citrate synthase (38). Human Hsp90 was shown to hold denatured β-galactosidase in a refolding competent state for subsequent refolding by the Hsp70 system (39).

   S. cerevisiae and *Drosophila melanogaster* Hsp90 are essential proteins at all temperatures (40, 41). Despite being essential for viability, loss of Hsp90 function in yeast, carrying a conditional temperature-sensitive hsp90 allele, does not affect protein folding or refolding in a general manner (42). Thus, Hsp90 is not essential for protein folding in a general sense, although functional redundancy may exist between Hsp90 and other chaperone systems in general protein folding. Hsp90 appears to be a dedicated chaperone for proteins involved in signal transduction, such as steroid-hormone receptors and cell-cycle kinases (43, 44). The *in vivo* function of the pro-karyotic homologue, HtpG, remains enigmatic. *htpG* null mutants of *E. coli* are viable at high temperatures, however they exhibit a reduced growth rate above 44 °C. Loss of HtpG function did not affect the development of thermotolerance (45, 46). A specialized chaperone function of HtpG, as proposed for eukaryotic Hsp90, may also hold true in prokaryotes, since *E. coli htpG* null mutants exhibit no increased protein aggregation even at high temperatures (29, 46). Interestingly, heat-shock induction of the *Bacillus subtilis htpG* gene is regulated by a separate mechanism, allowing an independent regulation of *clpC*, *dnaK*, and *groEL* expression, pointing to a specialized function of *htpG* (47, 48).

2.3 The DnaK (Hsp70) chaperone system

Hsp70 homologues are widespread in prokaryotes as well as in eukaryotes where they occur in the cytosol, mitochondria, chloroplasts, and the endoplasmic reticulum. In *E. coli* three Hsp70 proteins are present: DnaK, the major Hsp70 protein, HscA, and Hsc62. Hsp70 chaperones exert their ATP-dependent chaperone activities as monomers. The striking characteristic of these chaperones is their binding to short, linear segments of unfolded proteins or peptides containing hydrophobic residues (49, 50). The binding motif was characterized by screening of DnaK binding sites in peptide libraries and consists of a core of up to four or five hydrophobic residues enriched particularly in Leu. Flanking regions are enriched in basic residues, while acidic residues are disfavored (49).

DnaK (Hsp70) proteins are built of three domains, an N-terminal ATPase domain, a central substrate binding domain, and a C-terminal domain with potential regulatory function. ATP binding and hydrolysis allow DnaK to shift between low and high affinity states for substrates, respectively. ATP-controlled binding of protein substrates is linked to different activities of DnaK. DnaK can efficiently prevent the aggregation of unfolded proteins (29, 51–53), thereby exhibiting a 'buffer' function. Refolding activities of DnaK rely in parts on its ability to suppress protein aggregation, thereby abolishing off-pathways of protein folding. However, prevention of aggregation is necessary but not sufficient for refolding of several proteins (51, 52, 54). It is therefore tempting to speculate that binding of a misfolded polypeptide to DnaK leads to conformational changes within the bound protein. Finally, DnaK has a disaggregation activity towards protein aggregates, although this activity seems to be limited to certain protein substrates (55–57). Importantly, ATP hydrolysis and therefore binding and release of unfolded polypeptides from DnaK is regulated by a set of co-chaperones.

DnaJ (Hsp40) co-chaperones are multidomain proteins which share a highly conserved signature motif of approximately 78 amino acids, the J-domain (58–61). These co-chaperones stimulate the low ATPase activity of the Hsp70 partner proteins. Importantly *E. coli* DnaJ is a chaperone in its own right as well as an activator of the DnaK ATPase, functions which are crucial for the functional cycle of the DnaK system. DnaJ was found to associate with substrates of the DnaK system with a kinetic rate that is fast enough to prevent their aggregation (51, 52, 62). In addition DnaJ can present a polypeptide substrate to the ATP-bound form of DnaK, which has an open peptide-binding cleft, thereby allowing transfer of the substrate to DnaK and the subsequent stimulation of the ATPase activity of DnaK via the J-domain of DnaJ. This mechanism tightly couples substrate binding and ATP hydrolysis and guarantees stable binding of substrates to DnaK–ADP (63). Substrate specificity of DnaJ was analysed by screening of DnaJ binding to peptide libraries, revealing an overall affinity towards hydrophobic and aromatic residues (S. Rüdiger and B. Bukau, unpublished results). DnaJ and DnaK thus share overall similar and overlapping binding specificities. However, in contrast to DnaK, DnaJ seems to interact only with amino acid side chains, but not with the backbone of polypeptides (S. Rüdiger and B. Bukau, unpublished results). This important difference may allow DnaJ to interact very quickly with protein substrates and enables the co-chaperone to scan the surface of polypeptides for the presence of hydrophobic patches. Exposure of hydrophobic residues in a misfolded protein would therefore allow binding of the 'scanning factor' DnaJ, followed by subsequent transfer of the bound substrate to DnaK. In prokaryotes an additional co-chaperone, GrpE, catalyses the exchange of DnaK-bound ADP for ATP, thereby facilitating polypeptide release and DnaK reactivation.

The *E. coli* DnaK system was shown to be involved in a large variety of cellular processes, including folding of newly synthesized polypeptides (64, 65), protection and rescue of thermolabile proteins (29, 52, 53, 66), protein secretion (67), proteolysis (68, 69), regulation of the heat-shock response (70–73) and λ, F, and P1 replication (74–77). *dnaK* null mutants grow slowly at 30 °C and can not form colonies at lower (15 °C) or higher (39 °C) temperatures, unless a suppressor mutation partly restores the heat shock regulation defects in these mutants (78). In addition *dnaK* null mutants have defects in chromosome segregation, cell division, and motility and are impaired in the development of thermotolerance (79–82). Mutant strains of the co-chaperone DnaJ also exhibit a temperature-sensitive growth phenotype, although less severe than Δ*dnaK* mutants (83). This can be explained by functional redundancy between DnaJ and CbpA, a second general DnaJ-protein of *E. coli*. The *cbpA* gene was shown to function as a multicopy suppressor of *dnaJ* mutations. *cbpA* insertional mutants showed no noticeable phenotype, particularly with regard to temperature sensitivity. However, Δ*cbpA* Δ*dnaJ* double mutants have similar severe phenotypes to Δ*dnaK* cells (84, 85).

In *E. coli* four additional DnaJ-homologues DjlA, HscB, YbeS, and YbeV have been identified. The membrane-associated DjlA homologue interacts with DnaK and targets the Hsp70 protein to membrane-associated substrates that are involved in the signal transduction pathway regulating capsular polysaccharide synthesis (86, see

Chapter 4). Thus DjlA does not function as a general co-chaperone supporting DnaK in protein folding or aggregation prevention, but confers substrate specificity to the Hsp70 protein.

HscB interacts with HscA (Hsc66), the second Hsp70 homologue of *E. coli*, which is at least five times less abundant than DnaK (66). It is suggested that HscA–HscB and DnaK–DnaJ–GrpE comprise separate molecular chaperone systems with distinct cellular functions, since HscA/HscB are unable to complement missing DnaK functions in Δ*dnaK* mutants (66, 87). Moreover Δ*hscA*/Δ*dnaK* double knockout mutants are viable and have no additional phenotype over the sum of the single mutants. Recent studies suggest a function of HscA/HscB in the assembly of Fe-S clusters in Fe-S-proteins. *E. coli hscA* and *hscB* genes exist in an operon and are part of a gene cluster containing *isc* (iron sulfur cluster assembly) genes (88). Isc gene products are believed to possess a 'housekeeping' function in the biosynthesis and/or repair of the standard Fe-S protein equipment of a bacterial cell. Co-expression of this gene cluster together with recombinant ferredoxins in *E. coli* indeed increased the formation of the mature Fe-S proteins (89). Interestingly, Ssq1p, the mitochondrial homologue of HscA in *S. cerevisiae*, was shown to be involved in iron metabolism. Ssq1p was required for normal activity of several mitochondrial Fe-S proteins (90, 91).

Hsc62 (F556) represents the third Hsp70 homologue of *E. coli*. Although detailed genetic and biochemical analysis of this Hsp70 protein are still missing, Hsc62 obviously does not function as a general chaperone system, since *hsc62* knockout mutants do not exhibit a strong phenotype and Hsc62 overproduction cannot rescue the temperature-sensitive growth phenotype of Δ*dnaK* mutants (C. Kluck, M. Mayer and B. Bukau, unpublished results). Two recently identified DnaJ homologues, YbeS and YbeV, are encoded by corresponding genes located in close vicinity to *hsc62* and are therefore proposed to interact with Hsc62 (92). Thus, *E. coli* seems to have three gene clusters composed of Hsp70 and Hsp40 homologues: the DnaK/DnaJ, the HscA/HscB, and the Hsc62/YbeS/YbeV systems.

2.4 The GroEL (Hsp60) chaperone system

The GroEL system is a highly sophisticated chaperone machinery found in bacteria, mitochondria, and chloroplasts (93). GroEL exhibits a complex overall structure composed of two heptameric rings of the large subunit GroEL (57 kDa) stacking back to back and forming a 14-subunit hollow cylinder with two identical binding sites for non-native proteins. The smaller co-chaperone GroES (10 kDa) forms a seven-membered dome-shaped single ring. The interaction between GroEL and GroES is necessary for certain proteins *in vitro* to fold under otherwise non-permissive conditions. Non-native substrates bind through hydrophobic interactions to the apical domains of the unoccupied ring of a GroEL-GroES asymmetric complex. ATP and GroES then bind to the same ring occupied by the substrate, forming a cis ternary complex in which the polypeptide is encapsulated within the GroEL-GroES structure. Binding of GroES induces large conformational changes in GroEL. The apical substrate binding domains of GroEL, which hold the substrate by multiple interactions, show strong upward and outward motions upon GroES binding, thereby greatly expanding the distance between the polypeptide binding regions (93–95). Substrates bound to multiple apical domains may therefore be subjected to stretching forces ('mechanical stress'), which can induce their unfolding as shown for the model substrates hen lysozyme and rhodanese (96, 97). In addition binding of GroES roughly doubles the volume of the central cavity as well as obscuring GroEL's hydrophobic polypeptide recognition regions. As a consequence the substrate polypeptide is encapsulated within a relatively polar environment that favors folding (98). Thus the central cavity provides an environment equivalent to infinite dilution of the substrate, thereby preventing nonproductive aggregation. The combination of both principles may ensure proper folding especially of kinetically trapped folding intermediates, caught in a stable non-native conformation. Nucleotide binding to GroEL is responsible for the transition from the high- to the low-affinity state of the chaperone for its substrates and thereby controlling their dissociation or rebinding. ATP hydrolysis therefore provides a timer, giving the substrate at least 6 seconds to fold. ATP hydrolysis primes GroEL to release GroES, allowing the substrate to exit into the bulk solution. Substrates that have not reached their native states are then either recaptured by GroEL or another chaperone system or eventually targeted for proteolysis or aggregation.

GroEL and GroES are the only chaperones which are essential for growth of *E. coli* under all conditions (99). The GroEL system was shown to be involved in many cellular processes, including the *de novo* folding of newly synthesized polypeptides (100,101), conformational maintenance of preexisting proteins (102), proteolysis (103), secretion (104, 105), and λ and T4 morphogenesis (106, 107).

2.5   The IbpA / IbpB (small Hsps) chaperones

Members of the family of small Heat shock proteins (sHsps) are ubiquitous and abundant proteins exhibiting a subunit molecular mass of 15–40 kDa. The sequence homology in this group is much lower than within the other Hsp families and is mainly restricted to the C-terminal part of sHsps, the so called 'α-crystallin' domain (108, 109). sHsps form large oligomeric complexes up to mega-Dalton-range with an average of 16–32 subunits (110, 111). Recently the crystal structure of Hsp16.5 from *Methanococcus jannaschii* was solved, revealing a hollow football-like structure of the oligomer (112).

sHsps perform their chaperone functions *in vitro* independent of ATP. sHsps bind non-native proteins with high promiscuity, thereby inhibiting aggregation and holding the substrates in a recoverable form (109,113–116). The substrate binding site of sHsps is not clearly defined, but from structural models and electron microscopic data it was proposed that substrates are bound to the outer surface of the oligomeric complexes (113,115). Since sHsps bind substrates very efficiently and stably with stoichiometries of up to one molecule per subunit, they are well suited to provide an energy-independent reservoir for non-native proteins under stress conditions. In a recent study Hsp26 from *S. cerevisiae* was shown to be a temperature-regulated chaperone (117). At heat shock temperatures oligomers of Hsp26 dissociated into dimeric, active species. This dissociation was shown to be a prerequisite for efficient chaperone activity, obviously by allowing the formed Hsp26 dimers to bind a larger amount of non-native substrates (117).

In *E. coli*, the sHsp representatives IbpA and IbpB were initially found associated with inclusion bodies formed during the overproduction of heterologous proteins (118). In addition IbpA/IbpB were recovered from heat-shocked cells in a fraction containing aggregated proteins (29, 119), pointing to a stress-related protein repair function. *In vitro* IbpB was shown to prevent heat-denaturation of malate dehydrogenase (MDH). This prevention of MDH aggregation was the prerequisite for the subsequent refolding of the enzyme by the DnaK and GroEL systems (114). However IbpB was unable to prevent heat-induced aggregation of thermolabile *E. coli* proteins in cell extracts and Δ*ibpAB* null mutants do not exhibit a temperature-sensitive growth-phenotype and show no increased protein aggregation after temperature upshift (29, 46). Thus IbpA/IbpB do not constitute a protein repair function that is essential for viability or prevention of heat-induced protein aggregation. It is however not excluded that *E. coli* sHSPs may become more important if other chaperone systems, like DnaK, become saturated by substrates.

2.6 Trigger factor (TF)

TF is a highly abundant and conserved cytosolic protein of eubacteria even found in *Mycoplasma genitalium*, a bacterium believed to be free from genetic redundancy. So far, no homologues have been found in archaea or eukaryotes. The role of TF in protein secretion is discussed extensively in Chapter 2, and only the salient points are covered here. TF displays PPIase- and chaperone-activities *in vitro* and catalyses the proline-limited refolding of RNaseT1 with a much higher efficiency than all other PPIases tested so far, probably due to the cooperation of its catalytic and chaperone activities (120). Unfolded proteins like RCM-La (reduced and carboxymethylated lactalbumin) competitively inhibit the RNase T1 refolding activity of full-length, but not the activity of the isolated PPIase domain towards peptide substrates *in vitro*.

In *E. coli*, TF binds to the large subunits of ribosomes (121). This association is salt-sensitive when ribosomes are not active and converts to a high-salt resistant association when ribosomes synthesize nascent polypeptides (122). *In vitro*, TF is the major protein that crosslinks to virtually all nascent chains of cytosolic and secreted proteins tested so far (122, 123). It crosslinks to ribosome associated short chains of 57 residues in length and thus may be positioned very close to the exit site of ribosomes (124). After puromycin induced release of nascent chains from ribosomes, TF fails to crosslink to the released polypeptides (122). These findings suggest a cycling scenario, where TF first loosely associates with ribosomes; when ribosomes become active and synthesize nascent chains the low affinity binding to ribosomes is converted to a high affinity, salt-resistant ribosome association. Subsequently, TF may leave the ribosome in association with the nascent chain which is followed by a rapid release of TF and recycling of it back to the ribosome for a new round of action (see Fig. 2). So far, it is unknown what features of the unfolded nascent chains are recognized and bound by TF. However, modelling the PPIase domain of TF based on the human FKBP structure reveals a gallery of aromatic residues lining the substrate binding pocket of TF (E. Deuerling and B. Bukau, unpublished results). Therefore, TF may recognize preferentially aromatic residues in unfolded protein substrates.

Together, these findings suggest an important role for TF in assisting the co-translational folding of nascent polypeptide chains. Surprisingly, complete deletion of the TF gene *tig* revealed that TF is not essential (65).

3. Housekeeping activities of cytosolic *E. coli* chaperones

Molecular chaperones are involved in a variety of different cellular processes and chaperone activity is sometimes responsible for controlling specialized pathways. The guiding of newly synthesized proteins and the protection of pre-existing proteins under stress conditions represents the general, housekeeping activities of cytosolic chaperones. The contributions of individual chaperones in the general processes of protein *de novo* folding, protein secretion, and prevention of protein aggregation under stress conditions are described in the following sections and are illustrated in Fig. 1.

3.1 *De novo* folding of proteins in the *E. coli* cytosol

In living cells folding of newly synthesized polypeptides is linked to the vectorial process of translation. A nascent polypeptide chain may not be able to fold completely during synthesis on a ribosome, as sequences required to attain the native conformation are buried in the ribosome or not yet synthesized. Hydrophobic residues that are buried in the final structure may be transiently exposed during synthesis and consequently the growing polypeptide may be particularly aggregation-prone, especially with regard to the macromolecular crowding of the cell and the high concentration (30–50 µM) of folding chains in the *E. coli* cytosol. The molecular events leading to misfolding are poorly understood, but conceivably include interference of folding of a domain by incorrect interactions with adjacent domains. Aggregation may occur between regular folding intermediates or between kinetically trapped misfolded intermediates, both exposing hydrophobic regions at their surfaces. Folding of many proteins is therefore guided by molecular chaperones *in vivo*.

3.1.1 Co-translational folding of nascent polypeptides at the ribosome

Several studies showed that proteins can fold co-translationally in cell free translation systems, involving formation of intermediates in which only N-terminal domains have attained native-like structures (125). Recently a particularly elegant study investigated the kinetics of protein folding in the cytosol of living cells, using the Semliki Forest virus capsid protein (C protein) as a model protein (126). The C pro-

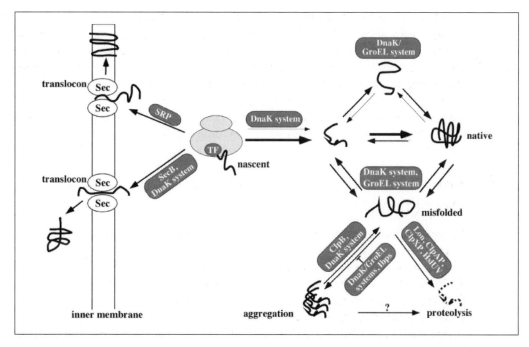

Fig. 1 Involvement of cytosolic *E. coli* chaperones and proteases in cellular processes. Depicted are the major cellular events where chaperones and proteases participate to promote correct protein folding and protein sorting, or ensuring repair or removal of aggregated and misfolded proteins. Chaperones have been implicated to act during protein *de novo* folding, protein secretion, repair of misfolded proteins, prevention of protein aggregation, and resolubilization of protein aggregates. Proteases are predominantly responsible for the degradation of non-native proteins.

tein corresponds to the N-terminal of five proteins which are translated as a poly-protein precursor. It contains a chymotrypsin-like domain which, once folded to its native structure, acts in cis to cleave itself from the precursor. In both, hamster cells and *E. coli*, this domain folds rapidly during translation, well before termination of synthesis of the polyprotein. These results establish the principle, that protein domains can fold co-translationally in the cytosols of prokaryotic and mammalian cells. The domain-wise co-translational folding of proteins may be beneficial to the folding process because it minimizes the possibility of incorrect interactions between differ-ent folding domains. In support of this possibility are the findings that the co-translational stepwise folding of model proteins in cell-free translation systems is more efficient and rapid than the folding of the corresponding denatured full length proteins (127–129). It is important to note that there is no general built-in code for a vectorial, N- to C-terminal, deciphering of the folding information *in vivo*. This was shown by random circular permutation experiments in which N-terminal sequences of various lengths were genetically moved to the C-terminus of a model protein (130). In many cases such circular permutation did not affect the folding efficiency.

The cellular strategy to avoid misfolding and aggregation of cytosolic proteins during co- and post-translational folding relies on assistance of this process by a large arsenal of molecular chaperones (Table 1). In *E. coli* the ribosome associated TF is a perfect candidate to await a nascent polypeptide for chaperoning as soon as it emerges from the ribosome (Fig. 2). TF is present in large amounts in *E. coli* (20000 copies per cell) with a twofold molar excess over ribosomes (64, 121). Therefore, most probably all ribosomes are associated with TF, and TF may interact with every nascent chain to assist co-translational folding or to suppress premature or incorrect folding events during synthesis. Cells without TF showed no defects in *de novo* folding of reporter enzymes or bulk proteins. DnaK, the major Hsp70 of *E. coli*, was also proposed to play a role in folding of newly made polypeptides (51). *In vivo*, however, Δ*dnaK* cells are viable between 30 °C and 37 °C and the absence of DnaK function caused only a very mild increase in levels of aggregated cytosolic proteins at 30 °C (29, 66). However, two recent studies showed that DnaK becomes essential in mutant cells lacking TF, suggesting overlapping or cooperative action of both chaperones (64, 65). *E. coli* cells which were depleted of the DnaK system in the absence of TF showed strongly decreased folding yields of reporter enzyme luciferase and a dramatic increase in aggregation of more than 40 cytosolic proteins, particularly of higher molecular weight proteins (65). A physical interaction between DnaK and newly synthesized proteins was demonstrated by co-immunoprecipitation experiments (64, 65). In wild type *E. coli* cells about 5–18% of newly made proteins interact transiently with DnaK (see Fig. 2) and this level increased two- to threefold (26–36%) in the absence of TF. A fraction of these proteins are nascent, indicating that DnaK interacts co- and post-translationally. The interaction is transient for many proteins (≤ 2 min), but long-lasting for others, suggesting considerable differences in the number of DnaK ATPase cycles involved in these interactions. The fraction of newly made polypeptides interacting with DnaK may be significantly greater, considering the instability of Hsp70-substrate complexes during immunoprecipitation (64, 65). The profile of polypeptides bound to DnaK extended from small chains of less than 14 kDa to polypeptides larger than 90 kDa, although smaller proteins seemed to be preferentially excluded from DnaK interaction. Besides the assistance in folding of newly made polypeptides, DnaK may alternatively act co- and post-translationally to repair misfolded proteins which may accumulate in Δtig mutants (see below). Taken together these data show that the ribosome associated TF cooperates with the cytosolic Hsp70 chaperone DnaK in folding of newly made polypeptides. The molecular basis for the functional cooperation between TF and the

Fig. 2 Model for folding of newly synthesized cytosolic proteins in the *E. coli* cytosol. Ribosome bound TF associates with emerging polypeptides and may migrate with nascent chains for a short period, then TF dissociates and recycles back to the ribosome. The majority of proteins (relative estimation in % of total) may fold subsequently spontaneously without additional chaperone action. A subpopulation of newly synthesized polypeptides transit through additional chaperone systems. The DnaK system (acting together with its co-chaperones DnaJ and GrpE) may either associate co- and/or post-translationally. This scenario was estimated for about 9–18% of newly synthesized proteins. For 10–15% of the newly made polypeptides the GroEL system (acting together with its co-chaperone GroES) assists folding post-translationally. Both chaperone systems may share a subset of substrates that may either be shuttled between both systems or may require a sequential mode of chaperone action.

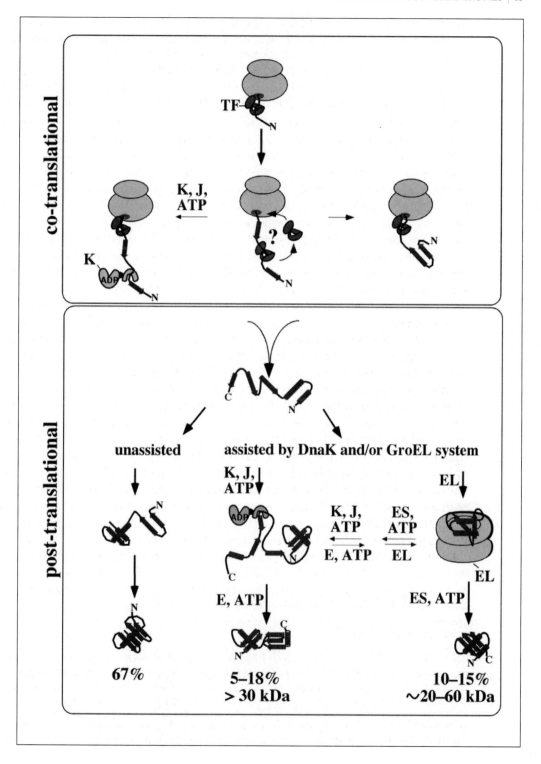

DnaK system is still not understood. DnaK may mechanistically replace TF in its absence, although the lack of ribosome binding and prolyl isomerase activities argues against this possibility.

3.1.2 Post-translational folding in the cytosol

Besides TF and DnaK the GroEL system functions to ensure proper folding of newly made polypeptides, however, exclusively in a post-translational manner (see Fig. 2) (100, 101). Recent studies analysed the quantitative contributions of GroEL to protein folding in *E. coli* by co-immunoprecipitation, showing that 10–15% of all cytoplasmic proteins, corresponding to approximately 300 newly translated polypeptides, interact with the GroEL system during *de novo* folding (101, 102). The majority of identified GroEL substrates were of intermediate size between 20 and 60 kDa. About half of these proteins were still detectable on GroEL 10 min after *de novo* synthesis, albeit in reduced amounts. It has been proposed that many of these proteins have a stringent chaperonin requirement for folding and that they fold slowly during multiple GroEL binding and release cycles, as demonstrated for rhodanese (101). A significant fraction of higher molecular weight proteins (> 60 kDa) was also observed to bind GroEL for more than 2 hours. It was speculated that the majority of this protein class represents pre-existing proteins, which may be structurally unstable and therefore may use the chaperonin for conformational maintenance (102). Interestingly, these larger proteins exceed the size limitation of the GroEL–GroES cage (55 kDa) and cannot be enclosed in the GroEL cavity by GroES (98). It is therefore unclear whether GroEL supports their folding to the native structure.

GroEL substrates, identified by co-immunoprecipitation and subsequent mass spectrometry, preferentially contain αβ-domains (102). As a common occurrence in such structural motifs, the β-sheets expose a hydrophobic surface packed against the hydrophobic surfaces of α-helices. These β-sheets and the corresponding hydrophobic surfaces of the α-helices would provide good binding surfaces to mediate high-affinity interactions with the apical domains of GroEL. Interestingly, several stringent model substrates of GroEL used for *in vitro* studies, including ornithine transcarbamylase, malate dehydrogenase, rhodanese, and RUBISCO, belong to this category of αβ-proteins. For αβ-domains, the formation of the β-sheet is expected to be the most difficult step in the folding process. Unlike the formation of α-helices, the assembly of β-sheets requires the formation of a large number of specific long-range contacts in the correct orientation. Thus αβ-proteins are expected to exhibit relatively slow folding rates and misfolding or kinetic trapping may occur through improper packing of helices and sheets within one domain, between domains within one molecule, or between an αβ-domain and another molecule. Typical GroEL substrates are proposed to consist of preferentially two or more αβ-domains. The estimated number of proteins with multiple αβ-domains in the *E. coli* cytoplasm is between 200 and 600, and GroEL may assist folding and conformational maintenance of at least a subset of these proteins (102). It remains to be determined, however, whether the identified GroEL substrates require GroEL for *de novo* folding *in vivo*. So far only two studies have investigated the consequences of missing GroEL function on protein folding *in vivo* (100, 131). Horwich and co-workers examined the role of GroEL by production of a temperature-sensitive lethal mutation in the essential *groEL* gene. After shift to non-permissive temperature a defined group of cytoplasmic proteins, including citrate synthase, ketoglutarate dehydrogenase, and polynucleotide phosphorylase, were translated but failed to reach the native state. Approximately 30% of protein species were found to be affected by loss of GroEL function; however, indirect effects cannot be excluded since protein translation was also reduced under non-permissive conditions (100). McLennan and Masters provided genetic evidence for a role of GroEL in generating active dihydropicolinate synthase (DapA), an essential protein involved in cell wall synthesis. *E. coli* cells deprived of GroEL lysed because of defects in cell wall synthesis due to a decreased stability of newly synthesized DapA. The instability of DapA is probably a consequence of the failure of GroEL to assist its folding. Addition of purified DAP or the presence of *dapA* on a multicopy plasmid delayed lysis by several hours but did not restore viability upon GroEL depletion (131). Thus additional essential *E. coli* proteins must depend on the GroEL system during *de novo* folding.

One important and still unsolved question is the functional interplay between TF, DnaK, and GroEL. As mention above, DnaK is essential in cells lacking TF, suggesting overlapping or cooperative action of both chaperones; however, the mechanism is still unknown (64, 65). The relationship between DnaK and GroEL is also very complex. At present three potential models exist describing the events that occur during *de novo* protein folding. It is important to notice that these models share two common features. First, TF is always the first chaperone encountering nascent chains. Second, after interaction of the nascent chains with ribosome-associated TF, the majority (about 2/3) of newly made polypeptides can fold spontaneously without additional chaperone assistance to native states whereas about 1/3 of polypeptides needs the additional action of chaperones (see Fig. 2). The three models are distinct concerning the mode of DnaK and GroEL action. The first model suggests a sequential mode of chaperone action, where the association of polypeptides with DnaK is a prerequisite for further interaction with the GroEL system to gain native conformation (51, 64). The second model favors the idea that both chaperones may act as a network, shuttling the substrate back and forth between the two systems depending on the availability and specificity of both systems (54). Both models are mainly based on *in vitro* studies, showing that the holder function of DnaK can be followed by the folder function of GroEL, and vice versa (51, 54). The final model suggests that GroEL and DnaK have quite distinct substrate populations with only limited overlap. This model is supported by the finding that DnaK substrates, identified by analysing aggregated proteins found in Δ*dnaK* at 42 °C ranges from 30 to over 150 kDa, with a notable enrichment of proteins larger than 90 kDa. This size distribution differs considerably from that of GroEL, where the majority of substrates are 20–60 kDa in size. Also very striking is the finding that only 9% of the 52 identified GroEL substrates were found to be DnaK substrates *in vivo* (29, 102). Additional experiments are definitely required to prove which model holds true *in vivo*.

3.2   Protein quality control in the *E. coli* cytosol

As pointed out in the previous section, molecular chaperones assist *de novo* folding of proteins, thereby controlling and monitoring their folding state. Binding of newly synthesized proteins to DnaK or GroEL is in general transient; however, some substrates show prolonged association, indicating multiple rounds of binding and release (64, 65, 102). Thus DnaK and GroEL may exhibit repair functions towards structurally unstable proteins, thereby ensuring their conformational maintenance. Besides molecular chaperones ATP-dependent proteases like Lon, ClpAP, ClpXP, and HslUV are also involved in this quality control of cytoplasmic proteins (see Fig. 1). ClpAP and ClpXP were shown to degrade abnormal polypeptides, marked with a C-terminal peptide tag. Addition of the peptide tag to nascent polypeptides translated from damaged mRNAs without in-frame stop codons is mediated by SsrA RNA and guarantees removal of non-native proteins (132, 133). The quality control is in addition a crucial function during heat shock (see below).

Although the relation of chaperone and protease function in protein quality control is not well understood, a number of observations have suggested that the DnaK and GroEL systems participate in protein degradation *in vivo*, although a coupled unfolding-degradation reaction requiring both chaperones and ATP-dependent proteases has not been demonstrated *in vitro* (134).

Mutations in specific chaperones (DnaK/DnaJ/GrpE and GroEL/GroES) generally result in lowering the amount of protein degradation in the cell although, in specific instances, they can lead to increased degradation of individual substrates (68, 69, 103, 135, 136). As mutations in *dnaJ*, *dnaK*, and *grpE* cause constitutive expression of the heat shock genes in *E. coli* and consequently lead to increased synthesis of proteases, the decreased degradation generally seen in these chaperone mutants is very striking. The bulk of the available evidence supports an indirect role of chaperones in proteolysis, reflecting the ability of chaperones to maintain abnormal proteins in a soluble state. Thus chaperones assist degradation by stabilizing a population of soluble, misfolded, non-native forms of polypeptides by binding and release. This pool include non-native forms arising from newly synthesized proteins and/or from heat or chemical denaturation. The fates of these misfolded proteins, and therefore quality control, would reflect the relative affinities of non-native proteins for proteases or chaperones and the relative rates of degradation, aggregation and folding. If the native form of a protein is reached either spontaneously or via chaperone binding and release, the protein would no longer be recognized by chaperones or proteases as an appropriate target. Alternatively, some of the non-native proteins might have higher affinity for proteases than for chaperones, thereby directing the polypeptide to degradation. Aggregation of misfolded proteins especially during stress conditions would render them inaccessible to proteases. Indeed ATP-dependent proteases were found to be trapped in the aggregated protein fraction of Δ*dnaK* mutant cells after heat shock (29). Thus this model of kinetic partitioning of substrates can explain why enhanced aggregation of proteins in chaperone mutant cells is coupled to decreased degradation (136).

3.3 Involvement of cytosolic *E. coli* chaperones in protein secretion

A large number of proteins synthesized in the *E. coli* cytosol have to be translocated into and across the plasma membrane. This process is mediated by specific targeting sequences at the amino terminus of the proteins and by a specific protein machinery comprising a set of integral and peripheral membrane proteins, as well as molecular chaperones in the cytosol. As will be discussed in Chapter 2, polytopic membrane proteins are guided by the bacterial SRP particles, while periplasmic and outer membrane proteins require SecB (see below) and SecA interactions for translocation across the plasma membrane (see Fig. 1) (137–139). SRP specificity results from the preferential binding of the more hydrophobic signal sequences of integral membrane proteins by SRP at the ribosome (124, 140). Interestingly, a recent study suggests TF is involved in the sorting pathway of membrane and secretory proteins. TF abolished *in vitro* SRP binding to signal sequences of SecB-dependent secretory proteins, thereby restricting SRP to its high affinity targets. It is therefore proposed that the decision for the different targeting pathways is taken early during protein synthesis at the ribosome, mediated by SRP and TF (141). However TF-depleted *E. coli* cells show no mistargeting or reduced targeting of SecB-dependent secretory proteins (142), pointing to a possible redundancy in protein sorting.

The tetrameric SecB chaperone is devoted to maintaining precursor proteins in a translocation-competent form, and pilots preproteins to the membrane-associated receptor, SecA, which provides the link to the translocon complex (143, 144). Disruption of *secB* was demonstrated to inhibit the translocation of a number of outer membrane proteins at an early stage of the translocation process (145). SecB appears to bind to longer nascent chains or precursors of a subset of outer membrane and periplasmic proteins at a late co-translational or post-translational stage (146). Signal sequences are in general dispensable for association of SecB with substrates (147–150). Randall and co-workers therefore postulated a kinetic partitioning model, according to which SecB does not bind specifically to the signal sequence, but to various segments of the precursor polypeptide. The role of the signal sequence in this model is to reduce the folding rate of newly synthesized precursor to allow binding of SecB (151–153). However, *in vitro* binding studies indicate that the association of SecB with substrate proteins is much faster than their rate of folding, raising the possibility that the selectivity of SecB *in vivo* is determined by its affinity for structural elements in nascent polypeptides (154, 155). SecB was shown to bind preferentially stretches of polypeptides enriched in basic and aromatic residues (150, 156, 157). This suggests that SecB has binding pockets or surfaces that are specific for aromatic residues. The aromatic side chains of high affinity SecB-binding regions were shown to occur in general within the core regions of folded proteins (150). Thus the nature of this substrate-binding motif allows SecB to bind preferentially to unfolded conformers of protein substrates and forms a basis for its function as a chaperone.

Loss of SecB function was shown to be complemented at least in parts by the DnaK system (67). Moreover the DnaK system was shown to be required for viability and protein secretion in *secB* null mutant strains (67). Thus the DnaK system plays a major role in secretion in *secB*-deficient strains, while in wild type *secB⁺* cells mutations in *dnaK*, *dnaJ*, and *grpE* do not alter the secretion of SecB-dependent proteins. Overlapping functions of SecB and DnaK can be explained by their ability to associate with newly synthesized polypeptides and to distinguish between native and non-native conformations of a protein substrate. The SecB-binding motif shares an overall similarity with the motif recognized by the DnaK chaperone (49, 150). Therefore SecB and DnaK have the potential to interact with similar sets of proteins.

Besides its involvement in secretion of SecB-dependent secretory proteins, the DnaK system was also demonstrated to participate in export of SecB-independent precursor proteins, including β-galactosidase hybrid proteins, alkaline phosphatase, ribose-binding protein, and β-lactamase (67, 158, 159). Secretion of β-lactamase was also shown to depend on the GroEL system (104, 105). However, loss of DnaK or GroEL function in *E. coli* wild type cells did not lead to general secretion defects or aggregation of precursor proteins (29, 100), suggesting functional redundancy in the protein secretion machinery of *E. coli*. Interestingly, accumulation of protein precursors induces the heat shock response of *E. coli*, while the loss of heat shock chaperones resulted in an increased production of SecB (160, 161). Thereby these regulatory mechanisms ensure sufficient amounts of chaperones for assisting the translocation of precursor proteins. Together, these results suggest that many preproteins can utilize multiple, functionally redundant chaperones, including the DnaK system, to facilitate their export (see Fig. 1).

3.4   Protection of thermolabile *E. coli* proteins during heat stress

Misfolding and aggregation of proteins are major damaging consequences of stress such as heat shock (162–164). The transient induction of heat shock proteins, including molecular chaperones and ATP-dependent proteases, represents an important protective and homeostatic mechanism to cope with the physiological and environmental stress at the cellular level. This heat shock response is highly conserved in prokaryotes and eukaryotes. Despite differences in the regulatory mechanisms of eubacteria, two features of the heat shock response are shared by all prokaryotes investigated so far. First, non-native proteins, generated by temperature-upshift, serve as induction signals, since addition of puromycin, amino acid analogues, or overproduction of aggregation-prone proteins lead to heat shock induction (52, 164, 165). Second, molecular chaperones serve as cellular thermometers by sensing the amount of misfolded proteins in the cytosol (166–168). This is discussed further in Chapter 10.

The capacity of molecular chaperones to function as holder chaperones during heat stress was mainly defined by their ability to suppress aggregation of chemically denatured model proteins *in vitro*. However, the contributions of individual chaperones to the holding and folding network *in vivo* and the identity of the stress sensitive cellular proteins remained unknown. Moreover bacterial cells were shown to have only a limited chaperone capacity to prevent protein aggregation under stress conditions (166, 168, 169). However, the ability of chaperones in prokaryotes to resolubilize aggregated proteins that escaped the protective function of holder chaperones remained enigmatic. Recent studies now provide strong evidence that the DnaK system has central, dual protective roles for thermolabile proteins by preventing their aggregation and, cooperatively with ClpB, mediating their disaggregation (29, 31).

The DnaK system was shown to efficiently prevent aggregation of a wide variety of heat-denatured proteins both in *E. coli* cell extracts and *in vivo*, in contrast to other chaperone systems tested, including GroEL/GroES, HtpG, ClpB, and IbpB (29, 46, 66). Moreover only Δ*dnaK* mutants exhibit strongly increased aggregation of approximately 10% of the amount of soluble cytosolic proteins upon heat shock to 42 °C. The missing function of the DnaK system cannot be replaced *in vivo* by other chaperones, even though their levels are two- to threefold increased in Δ*dnaK* mutant cells as compared to wild type due to regulatory defects (78). This interpretation is further supported by the findings that overproduction of GroEL/GroES, ClpB, HtpG or IbpA/B do not rescue the temperature-sensitive growth phenotype of Δ*dnaK* mutant cells at 42 °C and only partially suppress protein aggregation at high temperatures (29). In addition DnaK was found to be the most abundant cytosolic chaperone in *E. coli*: DnaK is at least eightfold more abundant as an active species than any other major cytosolic chaperone. Thus its high cellular concentration combined with its promiscuous and efficient substrate binding capacity qualifies DnaK as the central holder chaperone in the *E. coli* cytosol (29).

The number of aggregation prone *E. coli* proteins is surprisingly high: 150–200 species (corresponding to 15–25% of detected cytosolic proteins) were recovered from the insoluble cell fraction of Δ*dnaK* mutant cells at high temperatures (29). Since most of these proteins are also found aggregated in *E. coli* cell extracts at 45 °C, but are prevented from aggregation by DnaK/DnaJ, they represent the major thermolabile *in vivo* substrates of the DnaK system. Temperature-dependent association of many of these proteins with DnaK was demonstrated by co-immunoprecipitation with DnaK-specific antiserum. Identification by mass spectrometry revealed that all aggregated proteins are cytoplasmic and participate in various cellular processes including metabolism, cell wall synthesis, cell division, and in particular transcription and translation. Many of these proteins have essential functions, and aggregation of any of them may cause temperature sensitivity of Δ*dnaK* mutants. The thermosensitivity of key proteins of transcription (RNA polymerase) and translation (elongation factor) may have protective roles in heat treated cells, since reduced levels of these proteins may slow down protein synthesis during stress when the availability of chaperones is limited. Analysis of thermolabile proteins revealed that large proteins are strongly enriched: 80% of the large (> 90 kDa) but only 18% of small (< 30 kDa) cytosolic proteins were thermolabile (29). It is tempting to speculate that the appearance of large, multidomain proteins in evolution was accompanied by the appearance of a powerful DnaK chaperone system. Three features of large proteins may contribute to their aggregation propensity. First, misfolded conformers of large proteins may statistically expose more hydrophobic surface patches than smaller proteins. Since such patches are considered to be involved in, or even to trigger, intermolecular aggregation (170, 171), there is a higher probability for larger proteins to aggregate. Second, large proteins are composed of more domains than small proteins. In many proteins the interactions between domains are flexible and subject to regulation. Interdomain surface contacts may be particularly vulnerable to heat and, if exposing hydrophobic surfaces, initiate aggregation. Third, the rates of refolding of unfolded conformers of large proteins may be slower than that of small proteins which consequently favours competing aggregation reactions. Analysis of native structures of thermolabile and thermostable proteins revealed that, in general, thermosensitive proteins expose more patches of hydrophobic side chains, than thermostable proteins do. Although the surface exposure of most of these hydrophobic sites in the native structures is insufficient for DnaK association, they may become more surface exposed upon thermal unfolding of the proteins and trigger aggregation or DnaK binding.

Severe stress conditions lead to transient aggregation of thermolabile proteins even in *E. coli* wild type cells (29, 172), probably resulting from limitations in chaperone capacity. Removal of aggregated proteins is abolished or retarded in *dnaK* and *clpB* mutants, while the disaggregation process is largely unaffected in other chaperone gene mutants tested (29, 173, 174). In addition, overproduction of the DnaK system together with ClpB is necessary and sufficient to allow resolubilization of large amounts of protein aggregates, formed by temperature-upshift in Δ*dnaK* or Δ*rpoH* mutant cells, which lack major chaperones and ATP-dependent proteases. Other major cytosolic chaperones remain inefficient in protein disaggregation even if overproduced to large amounts (29; T. Tomoyasu and B. Bukau, unpublished results). Cooperation of DnaK and ClpB is further demonstrated by their capacity to disaggregate and refold several denatured model substrates *in vitro*, including MDH, firefly luciferase, and a large number of heat-aggregated proteins in *E. coli* cell extracts (29, 31, 32). The DnaK system on its own has been reported to dissolve aggregates of RNA polymerase and DnaA (55–57), but is inefficient in disaggregating other substrates, including heat-aggregated firefly luciferase or MDH (52, 114) and therefore seems to be limited in this process.

Thus prokaryotes possess a highly efficient bi-chaperone system, composed of the DnaK system and ClpB, to dissolve a wide variety of aggregated proteins both in cell extracts and *in vivo*. These observations extend the pioneering findings of Lindquist and co-workers, who identified a protein disaggregating activity for the yeast homologues, Hsp104 and Ssa1/Ydj1 (26, 27, 30). These findings also explain why in several bacteria the heat-shock inducible *clpB*, *dnaK*, and *dnaJ* genes are organized in an operon (175, 176), or are at least regulated by a common, specific mechanism (177).

The disaggregating activity of the bi-chaperone system is directly linked to the development of thermotolerance in living cells, since expression of *dnaK*, *dnaJ*, *grpE*, and *clpB* is necessary and sufficient to restore thermotolerance in *E. coli* Δ*rpoH* mutant cells (T. Tomoyasu and B. Bukau, unpublished results). In agreement with this finding, knockout mutations of *clpB* in *E. coli* and *Helicobacter pylori*, and of *hsp104 in S. cerevisiae*, prevent the cells acquiring thermotolerance and surviving severe stress conditions (25, 26, 46, 178–180). Thermotolerance and survival of eubacteria and yeast under heat stress are therefore linked to the ability of these cells to reverse protein aggregation.

Besides protection of thermolabile proteins, the removal of heat-denatured proteins by ATP-dependent proteases comprises a second cellular mechanism to cope with misfolded proteins. Lon, ClpAP, ClpXP, and HslUV act synergistically *in vivo* in the degradation of abnormal proteins (181). *E. coli* cells lacking these proteases cannot grow at very high temperatures (45 °C) (182). The fate of a thermolabile protein during heat stress should be determined by its affinities for molecular chaperones or proteases and by its refolding or aggregation kinetics. Interestingly MetE, representing one of the most prominent thermolabile DnaK substrates *in vivo*, aggregated in Δ*dnaK* mutant cells upon temperature-upshift, but was degraded in wild type cells in a Lon-dependent manner (29). This finding indicates that aggregation-prone substrates which are kept or rendered soluble by the DnaK system may become susceptible to proteolysis during stress situations.

3.5 Protection of *E. coli* proteins during oxidative stress

Heat shock treatment of cells is accompanied by oxidative stress and damage, one of the major causes of heat shock-induced cell death in pro- and eukaryotes (183, 184). Severe oxidative stress induces not only the expression of the oxidative stress regulon but also that of the heat shock regulon (185, 186). In a recent study Bardwell and co-workers provided strong evidence that Hsp33, an abundant 33 kDa protein originally defined as HslO (187), mounts an efficient cellular defence mechanism against oxidizing conditions. Hsp33 protects thermally unfolded and oxidatively damaged proteins from irreversible aggregation *in vitro* if present in stoichiometric amounts and thus acts as a chaperone (188). The activity of Hsp33 is regulated by the environmental redox conditions. Inactive, reduced Hsp33 has zinc coordinated by conserved cysteines. Oxidizing conditions cause zinc to be released from the cysteines, thereby allowing disulfide bonds to form and the chaperone function to be activated (188). Hsp33 is thus essentially inactive under reducing conditions but becomes activated upon oxidative stress. This mechanism allows pre-existing Hsp33 to respond rapidly to changes in the cell's redox state. This regulatory principle is reminiscent of data obtained for the prokaryotic transcription factors OxyR and SoxR, which control the oxidative stress regulons of *E. coli*. Both proteins are inactive under reducing conditions but become activated when the redox potential of the environment becomes more oxidizing (189, 190).

In order to determine the *in vivo* role of Hsp33, knockout mutant strains were analysed for growth during heat or oxidative stress. *E. coli* mutants lacking Hsp33 exhibit only a slightly increased sensitivity to high temperatures or H_2O_2 exposure compared to wild type cells. Introduction of the Δ*hslO* mutation in cells containing a mutation in the thioredoxin reductase gene *trxB* caused a more severe phenotype (188). The *trxB* mutation causes more oxidizing conditions in the cytosol, potentially leading to formation of nonproductive disulfide bonds and oxidative damage in cytoplasmic proteins (191). *hslO⁻ trxB⁻* double mutants are more temperature sensitive (47 °C) than the corresponding single knockouts and show a strong increase in H_2O_2 sensitivity (188). These results show that Hsp33 plays a significant role in the cellular protein protection system of *E. coli*, especially under oxidative and severe heat stress.

4. Specialized functions of cytosolic *E. coli* chaperones: protein activity control by the DnaK system

Besides its housekeeping activities, including (re)folding of non-native proteins and prevention/reversion of protein aggregation, the DnaK system is also involved in several specialized processes including replication of bacteriophages and low-copy-number plasmids, and regulation of the heat shock response and signal transduction pathways. In order to fulfill these specialized roles in the cell, DnaK must interact with folded proteins. These proteins have probably evolved a unique feature, namely the exposure of a hydrophobic DnaK-binding site in their native state, thereby ensuring the transient binding of DnaK and in many cases activation of the protein substrate after release.

The *E. coli* – bacteriophage lambda genetic interaction system has been used to uncover the existence of various biological machines. The starting point of all these studies was the isolation and characterization of *E. coli* mutants that blocked λ growth, and the corresponding λ compensatory mutations. λ replication was shown to depend on the DnaK chaperone system (192). During the initiation of λ DNA replication, the host DnaB helicase is complexed with phage λ P protein in order to be properly positioned near the ori initiation complex of phage λ. The viral protein binds with high affinity to DnaB, since it competes with the *E. coli* encoded DnaC, which directs the helicase to the *E. coli* ori, for binding. Tight binding of λ P to DnaB inhibits the activity of the helicase. The DnaK chaperone system destabilizes the λ P–DnaB interaction by binding of λ P protein, thus liberating DnaB's helicase activity, resulting in unwinding of the DNA template (74, 193). This mode of action is genetically supported by mutant λ P proteins with a reduced affinity to DnaB. These mutant proteins fulfill their function in a DnaK-independent manner (194).

The DnaK chaperone system is also required for P1 and mini-F plasmid replication (75, 195). The RepA protein of plasmid P1 binds to the plasmid origin of replication and mediates initiation of replication and its control. The DnaK system renders RepA 100-fold more active for binding to the P1 origin of replication. Activation is achieved by converting inactive RepA dimers into active RepA monomers. Only the monomeric form of RepA is able to bind with high affinity to oriP1 DNA (196). Replication of mini-F plasmid requires the plasmid-encoded RepE initiator protein and several host factors including DnaJ, DnaK, and GrpE. The RepE protein plays a crucial role in replication and exhibits two major functions: initiation of replication from the origin and auto-repression of *repE* transcription. Analysis of RepE mutants revealed that RepE is a structurally and functionally differentiated protein. The monomeric form of RepE has enhanced initiator activity while the dimeric form functions as a repressor (197). Monomerization and therefore activation of RepE is likely mediated by the DnaK system, comparable to the activation of P1 RepA.

In *E. coli* the DnaK system was shown to be a negative modulator of the heat shock response (70). As described in detail in Chapter 10, this heat shock regulon is controlled by σ^{32}, an alternative sigma factor. σ^{32} binds with high affinity to core RNA polymerase, thereby directing the transcription machinery to heat shock genes. Expression of σ^{32}-dependent heat shock genes is mainly regulated by changes in the activity and stability of σ^{32}. σ^{32} is a highly unstable protein caused by degradation mainly through FtsH, but also through Lon, HslUV (ClpYQ), and ClpXP (181, 198–201). The activity control of σ^{32} is mediated by the DnaK system. Binding of σ^{32} to DnaK/DnaJ was demonstrated *in vivo* and *in vitro* (72, 73, 202). This complex formation decreases σ^{32} activity, presumably by preventing association of σ^{32} with core RNA polymerase. Importantly, binding of σ^{32} to core RNA polymerase was shown to stabilize σ^{32} by protecting it from degradation by FtsH (168, 203). Multiple rounds of σ^{32} binding to and release from DnaK should increase the concentration of 'free' σ^{32} in the *E. coli* cytosol, thereby possibly providing FtsH with the opportunity to bind and degrade σ^{32}. Activity and stability control of σ^{32} are therefore linked processes, regulated by the DnaK system. Thus mutant cells lacking the DnaK system exhibit a strong deregulation of the heat shock response: heat shock genes are highly expressed even under non-stress conditions due to the presence of higher amounts of stable and active σ^{32}.

Finally DnaK was recently shown to be involved in the signal transduction pathway of the two-component system RscB/RscC. This system regulates the expression of the *cps* (capsular polysaccharide) operon, encoding the genes required for synthesis of colanic acid mucoid capsules. Activation of the signal transduction pathway is dependent on DnaK, but independent of DnaJ (86). DnaJ function is replaced by DjlA, a membrane-anchored DnaJ-homologue of *E. coli*. DjlA possesses a J-domain at its extreme C-terminus but shares no additional homology with DnaJ (204). Thus DjlA confers substrate specificity to DnaK and targets the Hsp70 to the cytoplasmic membrane for interaction with RcsB/RcsC, a prerequisite for productive signal transduction.

5. Summary and perspectives

In summary, molecular chaperones are highly versatile molecules being involved in all folding events occurring during the life spans of proteins. Chaperones act with broad promiscuity towards their substrates but, in some cases, they also can function in very specific processes at key checkpoints of cellular events. Recent comparative studies have provided fruitful approaches to understand the individual contributions of chaperone systems in the cellular processes. Although we have already some knowledge about the mechanisms of chaperone action and their cellular functions, we still do not understand the functional interplay among the variety of chaperones present in the cell. Furthermore, the distinguishing marks in substrates that determine chaperone functions remain enigmatic. The rapid progress in the chaperone field during the past decade provides hope of answering these questions in the near future.

Acknowledgements

We thank D. Dougan for critically reading the manuscript. This work was supported by grants from the Deutsche Forschungsgemeinschaft (Sonderforschungsbereich 388) and the Landesforschungsschwerpunkt to B.B.

References

1. Anfinsen, C. B. (1973). Principles that govern the folding of protein chains. *Science*, **181**, 223.
2. Zimmerman, S. B. and Trach, S. O. (1991). Estimation of macromolecule concentrations and excluded volume effects for the cytoplasm of *Escherichia coli*. *J. Mol. Biol.*, **222**, 599.
3. Zimmerman, S. B., and Minton, A. P.(1993). Macromolecular crowding: biochemical, biophysical, and physiological consequences. *Annu. Rev. Biophys. Biomol Struct.*, **22**, 27.
4. van den Berg, B., Ellis, R. J., and Dobson, C. M. (1999). Effects of macromolecular crowding on protein folding and aggregation. *EMBO J.*, **18**, 6927.
5. Jaenicke, R. and Seckler, R. (1999). In *Molecular chaperones and folding catalysts. Regulation, cellular function and mechanism* (ed. B. Bukau), pp. 407–36. Harwood Academic, Amsterdam.
6. Minton, K. W., Karmin, P., Hahn, G. M., and Minton, A. P. (1982). Nonspecific stabilization of stress-susceptible proteins by stress-resistant proteins: a model for the biological role of heat shock proteins. *Proc. Natl. Acad. Sci. USA*, **79**, 7107.
7. Bukau, B. and Horwich, A. L. (1998). The Hsp70 and Hsp60 chaperone machines. *Cell*, **92**, 351.
8. Ellis, R. J. (1997). Molecular chaperones: avoiding the crowd. *Curr. Biol.*, **7**, R531.
9. Hartl, F. U. (1996). Molecular chaperones in cellular protein folding. *Nature*, **381**, 571.
10. Wickner, S., Maurizi, M. R., and Gottesman, S. (1999). Posttranslational quality control: folding, refolding, and degrading proteins. *Science*, **286**, 1888.
11. Craig, E. A., Weissman, J. S., and Horwich, A. L. (1994). Heat shock proteins and molecular chaperones: mediators of protein conformation and turnover in the cell. *Cell*, **78**, 365.
12. Gottesman, S., Wickner, S., and Maurizi, M. R. (1997). Protein quality control: triage by chaperones and proteases. *Genes & Dev.*, **11**, 815.
13. Schirmer, E. C., Glover, J. R., Singer, M. A., and Lindquist, S. (1996). HSP100/Clp proteins: a common mechanism explains diverse functions. *Trends Biochem. Sci.*, **21**, 289.
14. Neuwald, A. F., Aravind, L., Spouge, J. L., and Koonin, E. V. (1999). AAA+: a class of chaperone-like ATPases associated with the assembly, operation, and disassembly of protein complexes. *Genome Res.*, **9**, 27.
15. Lupas, A., Flanagan, J. M., Tamura, T., and Baumeister, W. (1997). Self-compartmentalizing proteases. *Trends Biochem. Sci.*, **22**, 399–404.
16. Bochtler, M., Hartmann, C., Song, H. K., Bourenkov, G. P., Bartunik, H. D., and Huber, R. (2000). The structures of HslU and the ATP-dependent protease HslU-HslV. *Nature*, **403**, 800.
17. Gottesman, S., Maurizi, M. R., and Wickner, S. (1997). Regulatory Subunits of Energy-Dependent Proteases. *Cell*, **91**, 435.

18. Turgay, K., Hahn, J., Burghoorn, J., and Dubnau, D. (1998). Competence in *Bacillus subtilis* is controlled by regulated proteolysis of a transcription factor. *EMBO J.*, **17**, 6730.
19. Wickner, S., Gottesman, S., Skowyra, D., Hoskins, J., McKenney, K., and Maurizi, M. R. (1994). A molecular chaperone, ClpA, functions like DnaK and DnaJ. *Proc. Natl. Acad. Sci. USA*, **91**, 12218.
20. Wang, J., Hartling, J. A., and Flanagan, J. M. (1997). The structure of ClpP at 2.3 Å resolution suggests a model for ATP-dependent proteolysis. *Cell*, **91**, 447.
21. Weber-Ban, E. U., Reid, B. G., Miranker, A. D., and Horwich, A. L. (1999).Global unfolding of a substrate protein by the Hsp100 chaperone ClpA. *Nature*, **401**, 90.
22. Levchenko, I., Luo, L., and Baker, T. A. (1995). Disassembly of the Mu transposase tetramer by the ClpX chaperone. *Genes & Dev.*, **9**, 2399.
23. Kruklitis, R., Welty, D. J., and Nakai, H. (1996). ClpX protein of *Escherichia coli* activates bacteriophage Mu transposase in the strand transfer complex for initiation of Mu DNA synthesis. *EMBO J.*, **15**, 935.
24. Mhammedi-Alaoui, A., Pato, M., Gama, M. J., and Toussaint, A. (1994). A new component of bacteriophage Mu replicative transposition machinery: the *Escherichia coli* ClpX protein. *Mol. Microbiol.*, **11**, 1109.
25. Squires, C. L., Pedersen, S., Ross, B. M., and Squires, C. (1991). ClpB is the *Escherichia coli* heat shock protein F84.1. *J Bacteriol.*, **173**, 4254.
26. Sanchez, Y., Taulin, J., Borkovich, K. A., and Lindquist, S. (1992). Hsp104 is required for tolerance to many forms of stress. *EMBO J.*, **11**, 2357.
27. Parsell, D. A., Kowal, A. S., Singer, M. A., and Lindquist, S. (1994). Protein disaggregation mediated by heat-shock protein Hsp104. *Nature*, **372**, 475.
28. Motohashi, K., Watanabe, Y., Yohda, M., and Yoshida, M. (1999). Heat-inactivated proteins are rescued by the DnaK.J-GrpE set and ClpB chaperones. *Proc. Natl. Acad. Sci. USA*, **96**, 7184.
29. Mogk, A., Tomoyasu, T., Goloubinoff, P., Rüdiger, S., Röder, D., Langen, H., and Bukau, B. (1999). Identification of thermolabile *E. coli* proteins: prevention and reversion of aggregation by DnaK and ClpB. *EMBO J.*, **18**, 6934.
30. Glover, J. R. and Lindquist, S. (1998). Hsp104, Hsp70, and Hsp40: a novel chaperone system that rescues previously aggregated proteins. *Cell*, **94**, 73.
31. Goloubinoff, P., Mogk, A., Peres Ben Zvi, A., Tomoyasu, T., and Bukau, B. (1999). Sequential mechanism of solubilization and refolding of stable protein aggregates by a bichaperone network. *Proc. Natl. Acad. Sci. USA*, **96**, 13732.
32. Zolkiewski, M. (1999). ClpB cooperates with DnaK, DnaJ, and GrpE in suppressing protein aggregation. A novel multi-chaperone system from *Escherichia coli*. *J. Biol. Chem.*, **274**, 28083.
33. Koyasu, S., Nishida, E., Kadowaki, T., Matsuzaki, F., Iida, K., Harada, F., *et al.* (1986). Two mammalian heat shock proteins, HSP90 and HSP100, are actin-binding proteins. *Proc. Natl. Acad. Sci. USA*, **83**, 8054.
34. Stebbins, C. E., Russo, A. A., Schneider, C., Rosen, N., Hartl, F. U., and Pavletich, N. P. (1997). Crystal structure of an Hsp90-geldanamycin complex: targeting of a protein chaperone by an antitumor agent. *Cell*, **89**, 239.
35. Prodromou, C., Roe, S. M., O'Brien, R., Ladbury, J. E., Piper, P. W., and Pearl, L. H. (1997). Identification and structural characterization of the ATP/ADP-binding site in the Hsp90 molecular chaperone. *Cell*, **90**, 65.
36. Young, J. C., Schneider, C., and Hartl, F. U. (1997). *In vitro* evidence that hsp90 contains two independent chaperone sites. *FEBS Letters*, **418**, 139.

37. Scheibel, T., Weikl, T., and Buchner, J. (1998). Two chaperone sites in Hsp90 differing in substrate specificity and ATP dependence. *Proc. Natl. Acad. Sci. USA*, **95**, 1495.

38. Jakob, U., Lilie, H., Meyer, I., and Buchner, J. (1995). Transient interaction of Hsp90 with early unfolding intermediates of citrate synthase. Implications for heat shock *in vivo*. *J. Biol. Chem.*, **270**, 7288.

39. Freeman, B. C. and Morimoto, R. I. (1996). The human cytosolic molecular chaperones Hsp90, Hsp70 (Hsc70) and Hdj-1 have distinct roles in recognition of a non-native protein and protein refolding. *EMBO J.*, **15**, 2969.

40. Cutforth, T. and Rubin, G. M. (1994). Mutations in Hsp83 and cdc37 impair signaling by the sevenless receptor tyrosine kinase in Drosophila. *Cell*, **77**, 1027.

41. Borkovich, K. A., Farrelly, F. W., Finkelstein, D. B., Taulien, J., and Lindquist, S. (1989). hsp82 is an essential protein that is required in higher concentrations for growth of cells at higher temperatures. *Mol. Cell. Biol.*, **9**, 3919.

42. Nathan, D. F., Vos, M. H., and Lindquist, S. (1997). *In vivo* functions of the *Saccharomyces cerevisiae* Hsp90 chaperone. *Proc. Natl. Acad. Sci. USA*, **94**, 12949.

43. Toft, D. O. (1999). In *Molecular chaperones and folding catalysts. Regulation, cellular function and mechanism* (ed. B. Bukau), pp. 313–28. Harwood Academic, Amsterdam.

44. Mayer, M. P. and Bukau, B. (1999). Molecular chaperones: the busy life of Hsp90. *Curr. Biol.*, **9**, R322.

45. Bardwell, J. C. A. and Craig, E. A. (1988). Ancient heat shock gene is dispensable. *J. Bacteriol.*, **170**, 2977.

46. Thomas, J. G. and Baneyx, F. (1998). Roles of the *Escherichia coli* small heat shock proteins IbpA and IbpB in thermal stress management: comparison with ClpA, ClpB, and HtpG *in vivo*. *J. Bacteriol.*, **180**, 5165.

47. Derré, I., Rapoport, G., and Msadek, T. (1999). CtsR, a novel regulator of stress and heat shock response, controls clp and molecular chaperone gene expression in Gram-positive bacteria. *Mol. Microbiol.*, **31**, 117.

48. Schulz, A., Schwab, S., Homuth, G., Versteeg, S., and Schumann, W. (1997). The *htpG* gene of *Bacillus subtilis* belongs to class III heat shock genes and is under negative control. *J. Bacteriol.*, **179**, 3103.

49. Rüdiger, S., Germeroth, L., Schneider-Mergener, J., and Bukau, B. (1997). Substrate specificity of the DnaK chaperone determined by screening cellulose-bound peptide libraries. *EMBO J.*, **16**, 1501.

50. Fourie, A. M., Sambrook, J. F., and Gething, M. J. (1994). Common and divergent peptide binding specificities of Hsp70 molecular chaperones. *J. Biol. Chem.*, **269**, 30470.

51. Langer, T., Lu, C., Echols, H., Flanagan, J., Hayer, M. K., and Hartl, F. U. (1992). Successive action of DnaK, DnaJ and GroEL along the pathway of chaperone-mediated protein folding. *Nature*, **356**, 683.

52. Schröder, H., Langer, T., Hartl, F.-U., and Bukau, B. (1993). DnaK, DnaJ, GrpE form a cellular chaperone machinery capable of repairing heat-induced protein damage. *EMBO J.*, **12**, 4137.

53. Gragerov, A., Nudler, E., Komissarova, N., Gaitanaris, G., Gottesman, M., and Nikiforov, V. (1992). Cooperation of GroEL/GroES and DnaK/DnaJ heat shock proteins in preventing protein misfolding in *Escherichia coli*. *Proc. Natl. Acad. Sci. USA*, **89**, 10341.

54. Buchberger, A., Schröder, H., Hesterkamp, T., Schönfeld, H.-J., and Bukau, B. (1996). Substrate shuttling between the DnaK and GroEL systems indicates a chaperone network promoting protein folding. *J. Mol. Biol.*, **261**, 328.

55. Skowyra, D., Georgopoulos, C., and Zylicz, M. (1990). The *E. coli* dnaK gene product, the

Hsp70 homolog, can reactivate heat-inactivated RNA polymerase in an ATP hydrolysis-dependent manner. *Cell*, **62**, 939.

56. Hwang, D. S., Crooke, E., and Kornberg, A. (1990). Aggregated dnaA protein is dissoci-ated and activated for DNA replication by phospholipase or DnaK protein. *J. Biol. Chem.*, **265**, 19244.

57. Ziemienowicz, A., Skowyra, D., Zeilstra-Ryalls, J., Fayet, O., Georgopoulos, C., and Zylicz, M. (1993). Both the *Escherichia coli* chaperone systems GroEL/Gro/ES and DnaK/DnaJ/GrpE, can activate heat treated RNA polymerase. Different mechanisms for the same activity. *J. Biol. Chem.*, **268**, 25425.

58. Kelley, W. L. (1998). The J-domain family and the recruitment of chaperone power. *TiBS*, **23**, 222.

59. Cyr, D. M., Langer, T., and Douglas, M. G. (1994). DnaJ-like proteins: molecular chaper-ones and specific regulators of Hsp70. *TiBS*, **19**, 176.

60. Silver, P. A. and Way, J. C. (1993). Eukaryotic DnaJ homologs and the specificity of Hsp70 activity. *Cell*, **74**, 5.

61. Laufen, T., Zuber, U., Buchberger, A., and Bukau, B. (1998). In *Molecular chaperones in proteins: structure, function, and mode of action* (ed. A. L. Fink and Y. Goto), pp. 241–74. Marcel Dekker, New York.

62. Gamer, J., Multhaup, G., Tomoyasu, T., McCarty, J. S., Rüdiger, S., Schönfeld, H.-J., *et al.* (1996). A cycle of binding and release of the DnaK, DnaJ and GrpE chaperones regulates activity of the *E. coli* heat shock transcription factor σ32. *EMBO J.*, **15**, 607.

63. Laufen, T., Mayer, M. P., Beisel, C., Klostermeier, D., Reinstein, J., and Bukau, B. (1999). Mechanism of regulation of Hsp70 chaperones by DnaJ co-chaperones. *Proc. Natl. Acad. Sci. USA*, **96**, 5452.

64. Teter, S. A., Houry, W. A., Ang, D., Tradler, T., Rockabrand, D., Fischer, G., *et al.* (1999). Polypeptide flux through bacterial Hsp70: DnaK cooperates with TF in chaperoning nascent chains. *Cell*, **97**, 755.

65. Deuerling, E., Schulze-Specking, A., Tomoyasu, T., Mogk, A., and Bukau, B. (1999). TF and DnaK cooperate in folding of newly synthesized proteins. *Nature*, **400**, 693.

66. Hesterkamp, T. and Bukau, B. (1998).Role of the DnaK and HscA homologs of Hsp70 chaperones in protein folding in *E. coli*. *EMBO J.*, **17**, 4818.

67. Wild, J., Altman, E., Yura, T., and Gross, C. A. (1992). DnaK and DnaJ heat shock proteins participate in protein export in *Escherichia coli*. *Genes & Dev.*, **6**, 1165.

68. Keller, J. A. and Simon, L. D. (1988). Divergent effects of a *dnaK* mutation on abnormal protein degradation in *Escherichia coli*. *Mol. Microbiol.*, **2**, 31.

69. Straus, D. B., Walter, W. A., and Gross, C. A. (1988). *Escherichia coli* heat shock gene mutants are defective in proteolysis. *Genes & Dev.*, **2**, 1851.

70. Tilly, K., McKittrick, N., Zylicz, M., and Georgopoulos, C. (1983). The DnaK protein modulates the heat-shock response of *Escherichia coli*. *Cell*, **34**, 641.

71. Straus, D., Walter, W., and Gross, C. A. (1990). DnaK, DnaJ, and GrpE heat shock proteins negatively regulate heat shock gene expression by controlling the synthesis and stability of σ32. *Genes & Dev.*, **4**, 2202.

72. Liberek, K., Galitski, T. P., Zylicz, M., and Georgopoulos, C. (1992). The DnaK chaperone modulates the heat shock response of *Escherichia coli* by binding to the σ32 transcription factor. *Proc. Natl. Acad. Sci. USA*, **89**, 3516.

73. Gamer, J., Bujard, H., and Bukau, B. (1992). Physical interaction between heat shock proteins DnaK, DnaJ, GrpE and the bacterial heat shock transcription factor σ32. *Cell*, **69**, 833.

74. Hoffmann, H. J., Lyman, S. K., Lu, C., Petit, M. A., and Echols, H. (1992). Activity of the Hsp70 chaperone complex-DnaK, DnaJ, and GrpE-in initiating phage λ DNA replication by sequestering and releasing λ P protein. *Proc. Natl. Acad. Sci. USA*, **89**, 12108.

75. Kawasaki, Y., Wada, C., and Yura, T. (1990). Roles of *Escherichia coli* heat shock proteins DnaK, DnaJ and GrpE in mini-F plasmid replication. *Mol. Gen. Genet.*, **220**, 277.

76. Wickner, S., Skowyra, D., Hoskins, J., and McKenney, K. (1992). DnaJ, DnaK, and GrpE heat shock proteins are required in oriP1 DNA replication solely at the RepA monomerization step. *Proc. Natl. Acad. Sci. USA*, **89**, 10345.

77. Georgopoulos, C., Ang, D., Liberek, K., and Zylicz, M. (1990). In *Stress proteins in biology and medicine* (ed. R. Morimoto, A. Tissieres, and C. Georgopoulos), pp. 191–222. Cold Spring Harbor Press, Cold Spring Harbor, NY.

78. Bukau, B. and Walker, G. (1990). Mutations altering heat shock specific subunit of RNA polymerase suppress major cellular defects of *E. coli* mutants lacking the DnaK chaperone. *EMBO J.*, **9**, 4027.

79. Paek, K. H. and Walker, G. C. (1987). *Escherichia coli dnaK* null mutants are inviable at high temperature. *J. Bacteriol.*, **169**, 283.

80. Bukau, B. and Walker, G. C. (1989). Cellular defects caused by deletion of the *Escherichia coli dnaK* gene indicates roles for heat shock protein in normal metabolism. *J. Bacteriol.*, **171**, 2337.

81. Bukau, B. and Walker, G. C. (1989). Δ*dnaK* mutants of *Escherichia coli* have defects in chromosomal segregation and plasmid maintenance at normal growth temperatures. *J. Bacteriol.*, **171**, 6030.

82. Delaney, J. M. (1990). Requirement of the *Escherichia coli dnaK* gene for thermotolerance and protection against H_2O_2. *J. Gen. Microbiol.*, **136**, 2113.

83. Sell, S. M., Eisen, C., Ang, D., Zylicz, M., and Georgopoulos, C. (1990). Isolation and characterization of *dnaJ* null mutants of *Escherichia coli*. *J. Bacteriol.*, **172**, 4827.

84. Ueguchi, C., Kakeda, M., Yamada, H., and Mizuno, T. (1994). An analogue of the DnaJ molecular chaperone in *Escherichia coli*. *Proc. Natl. Acad. Sci. USA*, **91**, 1054.

85. Ueguchi, C., Shiozawa, T., Kakeda, M., Yamada, H., and Mizuno, T. (1995). A study of the double mutation of *dnaJ* and *cbpA*, whose gene products function as molecular chaperones in *Escherichia coli*. *J. Bacteriol.*, **177**, 3894.

86. Kelley, W. L. and Georgopoulos, C. (1997). Positive control of the two-component RcsC/B signal transduction network by DjlA: a member of the DnaJ family of molecular chaperones in *Escherichia coli*. *Mol. Microbiol.*, **25**, 913.

87. Silberg, J. J., Hoff, K. G., and Vickery, L. E. (1998). The Hsc66-Hsc20 chaperone system in *Escherichia coli*: chaperone activity and interactions with the DnaK-DnaJ-GrpE system. *J. Bacteriol.*, **180**, 6617.

88. Zheng, L., Cash, V. L., Flint, D. H., and Dean, D. R. (1998). Assembly of iron-sulfur clusters. *J. Biol. Chem.*, **273**, 13264.

89. Nakamura, M., Saeki, K., and Takahashi, Y. (1999). Hyperproduction of recombinant ferredoxins in *Escherichia coli* by coexpression of the ORF1-ORF2-iscS-iscU-iscA-hscB-hscA-fdx-ORF3 gene cluster. *J. Biochem.*, **126**, 10.

90. Schilke, B., Voisine, C., Beinert, H., and Craig, E. (1999). Evidence for a conserved system for iron metabolism in the mitochondria of *Saccharomyces cerevisiae*. *Proc. Natl. Acad. Sci. USA*, **96**, 10206.

91. Craig, E. A., Voisine, C., and Schilke, B. (1999). Mitochondrial iron metabolism in the yeast *Saccharomyces cerevisiae*. *Biol. Chem.*, **380**, 1167.

92. Itoh, T., Matsuda, H., and Mori, H. (1999). Phylogenetic analysis of the third *hsp70* homolog in *Escherichia coli*; a novel member of the Hsc66 subfamily and its possible co-chaperone. *DNA Res.*, **6**, 299.

93. Fenton, W. A. and Horwich, A. L. (1997). GroEL-mediated protein folding. *Protein Sci.*, **6**, 743.

94. Horwich, A. L., Weber-Ban, E. U., and Finley, D. (1999) .Chaperone rings in protein folding and degradation. *Proc. Natl. Acad. Sci. USA*, **96**, 11033.

95. Xu, Z. and Sigler, P. B. (1998). GorEL/GroES: structure and function of a two-stroke folding machine. *J. Struct. Biol.*, **124**, 129.

96. Coyle, J. E., Texter, F. L., Ashcroft, A. E., Masselos, D., Robinson, C. V., and Radford, S. E. (1999). GroEL accelerates the refolding of hen lysozyme without changing its folding mechanism. *Nature Struct. Biol.*, **6**, 683.

97. Shtilerman, M., Lorimer, G. H., and Englander, S. W. (1999). Chaperonin function: folding by forced unfolding. *Science*, **284**, 822.

98. Xu, Z., Horwich, A. L., and Sigler, P. B. (1997). The crystal structure of the asymmetric GroEL-GroES-(ADP)7 chaperonin complex. *Nature*, **388**, 741.

99. Fayet, O., Ziegelhoffer, T., and Georgopoulos, C. (1989). The *groES* and *groEL* heat shock gene products of *Escherichia coli* are essential for bacterial growth at all temperatures. *J. Bacteriol.*, **171**, 1379.

100. Horwich, A. L., Brooks Low, K., Fenton, W. A., Hirshfield, I. N., and Furtak, K. (1993). Folding *in vivo* of bacterial cytoplasmic proteins: role of GroEL. *Cell*, **74**, 909.

101. Ewalt, K. L., Hendrick, J. P., Houry, W. A., and Hartl, F. U. (1997). *In vivo* observation of polypeptide flux through the bacterial chaperonin system. *Cell*, **90**, 491.

102. Houry, W. A., Frishman, D., Eckerskorn, C., Lottspeich, F., and Hartl, F. U. (1999). Identification of *in vivo* substrates of the chaperonin GroEL. *Nature*, **402**, 147.

103. Kandror, O., Busconi, L., Sherman, M., and Goldberg, A. L. (1994). Rapid degradation of an abnormal protein in *Escherichia coli* involves the chaperones GroEL and GroES. *J. Biol. Chem.*, **269**, 23575.

104. Kusukawa, N., Yura, T., Ueguchi, C., Akiyama, Y., and Ito, K. (1989). Effects of mutations in heat-shock genes *groES* and *groEL* on protein export in *Escherichia coli*. *EMBO J.*, **8**, 3517.

105. Laminet, A. A., Ziegelhoffer, T., Georgopoulos, C., and Pluckthun, A. (1990). The *Escherichia coli* heat shock proteins GroEL and GroES modulate the folding of the beta-lactamase precursor. *EMBO J.*, **9**, 2315.

106. Friedman, D. I., Olson, E. R., Georgopoulos, C., Tilly, K., Herskowitz, I., and Banuett, F. (1984). Interactions of bacteriophage and host macromolecules in the growth of bacteriophage l. *Micro. Reviews*, **48**, 299.

107. Zeilstra-Ryalls, J., Fayet, O., Baird, L., and Georgopoulos, C. (1993). Sequence analysis and phenotypic characterization of *groEL* mutations that block λ and T4 bacteriophage growth. *J. Bacteriol.*, **175**, 1134.

108. Plesofsky-Vig, N., Vig, J., and Brambl, R. (1992). Phylogeny of the alpha-crystallin-related heat-shock proteins. *J. Mol. Evol.*, **35**, 537.

109. Jakob, U. and Buchner, J. (1994). Assisting spontaneity: the role of Hsp90 and small Hsps as molecular chaperones. *TiBS*, **19**, 205.

110. Shearstone, J. R. and Baneyx, F. (1999). Biochemical characterization of the small heat shock protein IbpB from *Escherichia coli*. *J. Biol. Chem.*, **274**, 9937.

111. Buchner, J., Ehrnsperger, M., Gaestel, M., and Walke, S. (1998). Purification and characterization of small heat shock proteins. *Methods Enzymol.*, **290**, 339.

112. Kim, K. K., Kim, R., and Kim, S.-H. (1998). Crystal structure of a small heat shock protein. *Nature*, **394**, 595.

113. Ehrnsperger, M., Gräber, S., Gaestel, M., and Buchner, J. (1997). Binding of non-native protein to Hsp25 during heat shock creates a reservoir of folding intermediates for reactivation. *EMBO J.*, **16**, 221.

114. Veinger, L., Diamant, S., Buchner, J., and Goloubinoff, P. (1998). The small heat shock protein IbpB from *Escherichia coli* stabilizes stress-denatured proteins for subsequent refolding by a multichaperone network. *J. Biol. Chem.*, **273**, 11032.

115. Lee, G. J., Roseman, A. M., Saibil, H. R., and Vierling, E. (1997). A small heat shock protein stably binds heat-denatured model substrates and can maintain a substrate in a folding-competent state. *EMBO J.*, **16**, 659.

116. Jakob, U., Gaestel, M., Engel, K., and Buchner, J. (1993). Small heat shock proteins are molecular chaperones. *J. Biol. Chem.*, **268**, 1517.

117. Haslbeck, M., Walke, S., Stromer, T., Ehrnsperger, M., White, H. E., Chen, S., Saibil, H. R., and Buchner, J. (1999). Hsp26: a temperature-regulated chaperone. *EMBO J.*, **18**, 6744.

118. Allen, S. P., Polazzi, J. O., Gierse, J. K., and Easton, A. M. (1992). Two novel heat shock genes encoding proteins produced in response to heterologous protein expression in *Escherichia coli*. *J. Bacteriol.*, **174**, 6938.

119. Laskowska, E., Wawrzynow, A., and Taylor, A. (1996). IbpA and IbpB, the new heat-shock proteins, bind to endogenous *Escherichia coli* proteins aggregated intracellularly by heat shock. *Biochimie*, **78**, 117.

120. Scholz, C., Stoller, G., Zarnt, T., Fischer, G., and Schmid, F. X. (1997). Cooperation of enzymatic and chaperone functions of TF in the catalysis of protein folding. *EMBO J.*, **16**, 54.

121. Lill, R., Crooke, E., Guthrie, B., and Wickner, W. (1988). The 'TF cycle' includes ribosomes, presecretory proteins and the plasma membrane. *Cell*, **54**, 1013.

122. Hesterkamp, T., Hauser, S., Lütcke, H., and Bukau, B. (1996). *Escherichia coli* TF is a prolyl isomerase that associates with nascent polypeptide chains. *Proc. Natl. Acad. Sci. USA*, **93**, 4437.

123. Valent, Q. A., Kendall, D. A., High, S., Kusters, R., Oudega, B., and Luirink, J. (1995). Early events in preprotein recognition in *E. coli*: interaction of SRP and TF with nascent polypeptides. *EMBO J.*, **14**, 5494.

124. Valent, Q. A., de Gier, J.-W. L., von Heijne, G., Kendall, D. A., ten Hagen-Jongman, C. M., Oudega, B., and Luirink, J. (1997). Nascent membrane and presecretory proteins synthesized in *Escherichia coli* associate with signal recognition particle and TF. *Mol. Microbiol.*, **25**, 53.

125. Fedorov, A. N. and Baldwin, T. O. (1997). Cotranslational Protein Folding. *J. Biol. Chem.*, **272**, 32175.

126. Nicola, A. V., Chen, W., and Helenius, A. (1999). Co-translational folding of an alpha-virus capsid protein in the cytosol of living cells. *Nature Cell Biol.*, **1**, 341.

127. Fedorov, A. N. and Baldwin, T. O. (1995). Contribution of cotranslational folding to the rate of formation of native protein structure. *Proc. Natl. Acad. Sci. USA*, **92**, 1227.

128. Frydman, J., Erdjument-Bromage, H., Tempst, P., and Hartl, F. U. (1999). Co-translational domain folding as the structural basis for the rapid *de novo* folding of firefly luciferase. *Nature Struct. Biol.*, **6**, 697.

129. Netzer, W. J. and Hartl, F. U. (1997). Recombination of protein domains facilitated by co-translational folding in eukaryotes. *Nature*, **388**, 343.

130. Hennecke, J., Sebbel, P., and Glockshuber, R. (1999). Random circular permutation of DsbA reveals segments that are essential for protein folding and stability. *J. Mol. Biol.*, **286**, 1197.

131. McLennan, N. and Masters, M. (1998). GroE is vital for cell-wall synthesis. *Nature*, **392**, 139.

132. Keiler, K. C., Waller, P. R. H., and Sauer, R. T. (1996). Role of a peptide tagging system in degradation of proteins synthesized from damaged messenger RNA. *Science*, **271**, 990.

133. Gottesman, S., Roche, E., Zhou, Y., and Sauer, R. T. (1998). The ClpXP and ClpAP proteases degrade proteins with carboxy-terminal peptide tails added by the SsrA-tagging system. *Genes & Dev.*, **12**, 1338.

134. Sherman, M. and Goldberg, A. L. (1992). Involvement of the chaperon DnaK in the rapid degradation of a mutant protein in *Escherichia coli*. *EMBO J.*, **11**, 71.

135. Kroh, H. E. and Simon, L. D. (1991). Increased ATP-dependent proteolytic activity in lon-deficient *Escherichia coli* strains lacking the DnaK protein. *J. Bacteriol.*, **173**, 2691.

136. Jubete, Y., Maurizi, M. R., and Gottesman, S. (1996). Role of the heat shock protein DnaJ in the lon-dependent degradation of naturally unstable proteins. *J. Biol. Chem.*, **271**, 30798.

137. Driessen, A. J., Fekkes, P., and van der Wolk, J. P. (1998). The Sec system. *Curr. Opin. Microbiol.*, **1**, 216.

138. Ito, K. (1996). The major pathways of protein translocation across membranes. *Genes Cells*, **1**, 337.

139. Wickner, W. (1989). Secretion and membrane assembly. *Trends Biochem. Sci.*, **14**, 280.

140. de Gier, J.-W. L., Scotti, P. A., Sääf, A., Valent, Q. A., Kuhn, A., Luirink, J., and von Heijne, G. (1998). Differential use of the signal recognition particle translocase targeting pathway for inner membrane protein assembly in *Escherichia coli*. *Proc. Natl. Acad. Sci. USA*, **95**, 14646.

141. Beck, K., Wu, L.-F., Brunner, J., and Müller, M. (2000). Discrimination between SRP- and SecA/SecB-dependent substrates involves selective recognition of nascent chains by SRP and TF. *EMBO J.*, **19**, 134.

142. Guthrie, B. and Wickner, W. (1990). TF depletion or overproduction causes defective cell division but does not block protein export. *J. Bacteriol.*, **172**, 5555.

143. Pugsley, A. P. (1993). The complete general secretory pathway in gram-negative bacteria. *Microbiol. Reviews*, **57**, 50.

144. Randall, L. L. and Hardy, S. J. S. (1995). High selectivity with low specificity: how SecB has solved the paradox of chaperone binding. *TiBS*, **20**, 65.

145. Kumamoto, C. A. and Beckwith, J. (1983). Mutations in a new gene, *secB*, cause defective protein localization in *Escherichia coli*. *J. Bacteriol.*, **154**, 253.

146. Kumamoto, C. A. and Francetic, O. (1993). Highly selective binding of nascent polypeptides by an *Escherichia coli* chaperone protein *in vivo*. *J. Bacteriol.*, **175**, 2184.

147. Collier, D. N., Bankaitis, V. A., Weiss, J. B., and Bassford Jr., P. J. (1988). The antifolding activity of SecB promotes the export of the *E. coli* maltose-binding protein. *Cell*, **53**, 273.

148. Randall, L. L., Topping, T. B., and Hardy, S. J. S. (1990). No specific recognition of leader peptide by SecB, a chaperone involved in protein export. *Science*, **248**, 860.

149. Randall, L. L., Hardy, S. J. S., Topping, T. C., Smith, V. F., Bruce, J. E., and Smith, R. D. (1998). The interaction between the chaperone SecB and its ligands: evidence for multiple subsites for binding. *Protein Sci.*, **7**, 2384.

150. Knoblauch, N. T. M., Rüdiger, S., Schönfeld, H.-J., Driessen, A. J. M., Schneider-Mergener, J., and Bukau, B. (1999). Substrate specificity of the SecB chaperone. *J. Biol. Chem.*, **274**, 34219.

151. Hardy, S. J. S. and Randall, L. L. (1991). A kinetic partitioning model of selective binding of non native proteins by the bacterial chaperone SecB. *Science*, **251**, 439.

152. Park, S., Liu, G., Topping, T. B., Cover, W. H., and Randall, L. L. (1988). Modulation of folding pathways of exported proteins by the leader sequence. *Science*, **239**, 1033.

153. Liu, G., Topping, T. B., and Randall, L. L. (1989).Physiological role during export for the retardation of folding by the leader peptide of maltose-binding protein. *Proc. Natl. Acad. Sci. USA*, **86**, 9213.

154. Fekkes, P., den Blaauwen, T., and Driessen, A. J. M. (1995). Diffusion-limited interaction between unfolded polypeptides and the *Escherichia coli* chaperone SecB. *Biochemistry*, **34**, 10078.

155. Stenberg, G. and Fersht, A. R. (1997). Folding of barnase in the presence of the molecular chaperone SecB. *J. Mol. Biol.*, **274**, 268.

156. Smith, V. F., Hardy, S. J. S., and Randall, L. L. (1997). Determination of the binding frame of the chaperone SecB within the physiological ligand oligopeptide-binding protein. *Protein Sci.*, **6**, 1746.

157. Khisty, V. J., Munske, G. R., and Randall, L. L. (1995). Mapping of the binding frame for the chaperone SecB within a natural ligand, galactose-binding protein. *J. Biol. Chem.*, **270**, 25920.

158. Phillips, G. J. and Silhavy, T. J. (1990). Heat-shock proteins DnaK and GroEL facilitate export of LacZ hybrid proteins in *E. coli*. *Nature (London)*, **344**, 882.

159. Wild, J., Rossmeissl, P., Walter, W. A., and Gross, C. A. (1996). Involvement of the DnaK-DnaJ-GrpE chaperone team in protein secretion in *Escherichia coli*. *J. Bacteriol.*, **178**, 3608.

160. Wild, J., Walter, W. A., Gross, C. A., and Altman, E. (1993). Accumulation of secretory protein precursors in *Escherichia coli* induces the heat shock response. *J. Bacteriol.*, **175**, 3992.

161. Müller, J. P. (1996). Influence of impaired chaperone or secretion function on SecB production in *Escherichia coli*. *J. Bacteriol.*, **178**, 6097.

162. Morimoto, R., Tissieres, A., and Georgopoulos, C. (1994). Progress and Perspectives on the Biology of Heat Shock Proteins and Molecular chaperones. In *The Biology of Heat Shock Proteins and Molecular chaperones* (ed. R. Morimoto, A. Tissieres, C. Georgopoulos), pp. 1–30. Cold Spring Harbor Laboratory Press, Cold Spring Harbor, NY.

163. Lindquist, S. and Schirmer, E. C. (1999). In *Molecular chaperones and folding catalysts. regulation, cellular function and mechanism* (ed. B. Bukau), pp. 347–380. Harwood Academic, Amsterdam.

164. Gross, C. A. (1996). In *Escherichia coli and salmonella*, Vol. 1 (ed. F. C. Neidhardt), ASM Press, Washington.

165. Mogk, A., Völker, A., Engelmann, S., Hecker, M., Schumann, W., and Völker, U. (1998). Non-native proteins induce expression of the *Bacillus subtilis* CIRCE regulon. *J. Bacteriol.*, **180**, 2895.

166. Craig, E. A. and Gross, C. A. (1991). Is Hsp70 the cellular thermometer? *TiBS*, **16**, 135.

167. Mogk, A., Homuth, G., Scholz, C., Kim, L., Schmid, F. X., and Schumann, W. (1997). The GroE chaperonin machine is a major modulator of the CIRCE heat shock regulon of *Bacillus subtilis*. *EMBO J.*, **16**, 4579.

168. Tomoyasu, T., Ogura, T., Tatsuta, T., and Bukau, B. (1998). Levels of DnaK and DnaJ provide tight control of heat shock gene expression and protein repair in *E. coli*. *Mol. Microbiol.*, **30**, 567.

169. Tatsuta, T., Tomoyasu, T., Bukau, B., Kitagawa, M., Mori, H., Karata, K., and Ogura, T. (1998). Heat shock regulation in the *ftsH* null mutant of *Escherichia coli*: dissection of stability and activity control mechanisms of sigma32 *in vivo*. *Mol. Microbiol.*, **30**, 583.

170. Mitraki, A. and King, J. (1989). Protein folding intermediates and inclusion body formation. *BioTechnology*, **7**, 690.

171. Seckler, R. and Jaenicke, R. (1992). Protein folding and protein refolding. *FASEB J.*, **6**, 2545.

172. Kucharczyk, K., Laskowska, E., and Taylor, A. (1991). Response of *Escherichia coli* cell membranes to induction of lambda c1857 prophage by heat shock. *Mol. Microbiol.*, **5**, 2935.

173. Laskowska, E., Kuczynska-Wisnik, D., Skórko-Glonek, J., and Taylor, A. (1996). Degradation by proteases Lon, Clp and HtrA, of *Escherichia coli* proteins aggregated *in vivo* by heat shock; HtrA protease action *in vivo* and *in vitro*. *Mol. Microbiol.*, **22**, 555.

174. Kedzierska, S., Staniszewska, M., Wegrzyn, A., and Taylor, A. (1999). The role of DnaK/DnaJ and GroEL/GroES systems in the removal of endogenous proteins aggregated by heat-shock from *Escherichia coli* cells. *FEBS Letters*, **446**, 331.

175. Falah, M. and Gupta, R. S. (1997). Phylogenetic analysis of mycoplasmas based on Hsp70 sequences: cloning of the dnaK (hsp70) gene region of *Mycoplasma capricolum*. *Int. J. Syst. Bacteriol.*, **47**, 38.

176. Osipiuk, J. and Joachimiak, A. (1997). Cloning, sequencing, and expression of *dnaK*-operon proteins from the thermophilic bacterium *Thermus thermophilus*. *Biochim. Biophys. Acta*, 1353, 253.

177. Grandvalet, C., de Crecy-Lagard, V., and Mazodier, P. (1999). The ClpB ATPase of Streptomyces albus G belongs to the HspR heat shock regulon. *Mol. Microbiol.*, **31**, 521.

178. Allan, E., Mullany, P., and Tabaqchali, S. (1998). Construction and characterization of a *Helicobacter pylori clpB* mutant and role of the gene in the stress response. *J. Bacteriol.*, **180**, 426.

179. Lindquist, S. and Kim, G. (1996). Heat-shock protein 104 expression is sufficient for thermotolerance in yeast. *Proc. Natl. Acad. Sci. USA*, **93**, 5301.

180. Sanchez, Y. and Lindquist, S. L. (1990). HSP104 required for induced thermotolerance. *Science*, **248**, 1112.

181. Kanemori, M., Nishihara, K., Yanagi, H., and Yura, T. (1997). Synergistic Roles of HslVU and other ATP-dependent proteases in controlling *in vivo* turnover of σ^{32} and abnormal proteins in *Escherichia coli*. *J. Bacteriol.*, **179**, 7219.

182. Kanemori, M., Yanagi, H., and Yura, T. (1999). The ATP-dependent HslVU/ClpQY protease participates in turnover of cell division inhibitor SulA in *Escherichia coli*. *J. Bacteriol.*, **181**, 3674.

183. Benov, L. and Fridovich, I. (1995). Superoxide dismutase protects against aerobic heat shock *Escherichia coli*. *J. Bacteriol.*, **177**, 3344.

184. Davidson, J. F., Whyte, B., Bissinger, P. H., and Schiestl, R. H. (1996). Oxidative stress is involved in heat-induced cell death in *Saccharomyces cerevisiae*. *Proc. Natl. Acad. Sci. USA*, **93**, 5116.

185. VanBogelen, R. A., Kelley, P. M., and Neidhardt, F. C. (1987). Differential induction of heat shock, SOS, and oxidation stress regulons and accumulation of nucleotides in *Escherichia coli*. *J. Bacteriol.*, **169**, 26.

186. McDuffee, A. T., Senisterra, G., Huntley, S., Lepock, J. R., Sekhar, K. R., Meredith, M. J., *et al.* (1997). Proteins containing non-native disulfide bonds generated by oxidative stress can act as signals for the induction of the heat shock response. *J. Cell Physiol.*, **171**, 143.

187. Chuang, S.-E., and Blattner, F. R. (1993). Characterization of twenty-six new heat shock genes of *Escherichia coli*. *J. Bacteriol.*, **175**, 5242.

188. Jakob, U., Muse, W., Eser, M., and Bardwell, J. C. (1999). Chaperone activity with a redox switch. *Cell*, **96**, 341.

189. Ding, H. and Demple, B. (1997). *In vivo* kinetics of a redox-regulated transcriptional switch. *Proc. Natl. Acad. Sci. USA*, **94**, 8445.

190. Zheng, M., Aslund, F., and Storz, G. (1998). Activation of the OxyR transcription factor by reversible disulfide bond formation. *Science*, **279**, 1718.

191. Derman, A. I., Prinz, W. A., Belin, D., and Beckwith, J. (1993). Mutations that allow disulfide bond formation in the cytoplasm of *Escherichia coli. Science*, **262**, 1744.

192. Georgopoulos, C. P., Lam, B., Lundquist-Heil, A., Rudolph, C. F., Yochem, J., and Feiss, M. (1979). Identification of *E. coli* dnaK (groPC756) Gene Product. *Mol. Gen. Genet.*, **172**, 143.

193. Zylicz, M., Ang, D., Liberek, K., and Georgopoulos, C. (1989).I nitiation of λ DNA replication with purified host- and bacteriophage-encoded proteins: the role of the DnaK, DnaJ and GrpE heat shock proteins. *EMBO J.*, **8**, 1601.

194. Konieczny, I. and Marszalek, J. (1995). The requirement for molecular chaperones in lambda DNA replication is reduced by the mutation pi in lambda P gene, which weakens the interaction between lambda P protein and DnaB helicase. *J. Biol. Chem.*, **270**, 9792.

195. Wickner, S. H. (1990). Three *Escherichia coli* heat shock proteins are required for P1 plasmid replication: formation of an active complex between *E. coli* DnaJ protein and the P1 initiator protein. *Proc. Natl. Acad. Sci. USA*, **87**, 2690.

196. Wickner, S., Hoskins, J., and McKenny, K. (1991). Function of DnaJ and DnaK as chaperones of origin specific DNA binding by RepA. *Nature (London)*, **350**, 165.

197. Ishiai, M., Wada, C., Kawasaki, Y., and Yura, T. (1994). Replication initiator protein RepE of mini-F plasmid: functional differentiation between monomers (initiator) and dimers (autogenous repressor). *Proc. Natl. Acad. Sci. USA*, **91**, 3839.

198. Tomoyasu, T., Gamer, J., Bukau, B., Kanemori, M., Mori, H., Rutman, A. J., *et al.* (1995). *Escherichia coli* FtsH is a membrane-bound, ATP-dependent protease which degrades the heat shock transcription factor σ^{32}. *EMBO J.*, **14**, 2551.

199. Yura, T., Nagai, H., and Mori, H. (1993). Regulation of the heat-shock response in bacteria. *Annu. Rev. Microbiol.*, **47**, 321.

200. Straus, D. B., Walter, W. A., and Gross, C. A. (1987). The heat shock response of *E. coli* is regulated by changes in the concentration of σ^{32}. *Nature*, **329**, 348.

201. Tilly, K., Spence, J., and Georgopoulos, C. (1989). Modulation of stability of the *Escherichia coli* heat shock regulatory factor σ^{32}. *J. Bacteriol.*, **171**, 1585.

202. Liberek, K. and Georgopoulos, C. (1993). Autoregulation of the *Escherichia coli* heat shock response by the DnaK and DnaJ heat shock proteins. *Proc. Natl. Acad. Sci. USA*, **90**, 11019.

203. Blaszczak, A., Georgopoulos, C., and Liberek, K. (1999). On the mechanism of FtsH-dependent degradation of the σ^{32} transcriptional regulator of *Escherichia coli* and the role of the DnaK chaperone machine. *Mol. Microbiol.*, **31**, 157.

204. Clarke, D. J., Jacq, A., and Holland, I. B. (1996). A novel DnaJ-like protein in *Escherichia coli* inserts in the cytoplasmic membrane with a Type III topology. *Mol. Microbiol.*, **20**, 1273.

2 | Chaperones in secretion pathways of *E. coli*

NELLIE HARMS, JOEN LUIRINK, and BAUKE OUDEGA

1. General introduction

Cytoplasmic proteins once synthesized at the ribosome fold into their native conformation, whereas proteins with destinations outside the cytoplasm are targeted to the cytoplasmic membrane and have to be prevented from premature folding and from degradation. For the majority of these latter proteins, folding occurs outside the cytoplasm. The precision and speed of these processes is ensured by different chaperones, targeting factors and folding catalysts. Most of these helper proteins, present in the cytoplasm, the periplasm, and in the membranes have a general role; some, however, are dedicated to special processes. This review focuses on the targeting and folding processes that function in a number of secretion pathways of *E. coli*. Since the field of protein secretion is very broad, not all pathways and their concomitant chaperones will be described.

2. Cytoplasmic chaperones and secreted proteins

2.1 Introduction

In Gram-negative bacteria, proteins that have been synthesized at the ribosome have in principle three destinations. They can either fold in their native conformation and stay in the cytoplasm, they can be incorporated into the inner membrane (IM), or they can be translocated across the IM into the extracytoplasmic environment, that is into the periplasm (see below). The process of decision making and sorting starts at the ribosomal exit site where nascent polypeptides interact with folding catalysts, chaperones, or targeting factors.

Information of a protein's destiny resides in the primary sequence of the polypeptide. Many secretory proteins are equipped with a short amino-terminal extension, the signal sequence. This sequence is recognized by targeting factors that direct the preprotein to a protein conducting channel consisting of at least SecYEG, the core complex of the translocase (1). The *E. coli* signal sequence ranges in length from 18 to about 30 amino acid residues. It is composed of three domains: a positively charged amino-terminus, a non-polar, hydrophobic core region, and a more polar region containing a cleavage site. All these domains have in common that there is little conservation in amino acid sequence, but their physicochemical properties are conserved. Inner membrane proteins (IMPs) are anchored in the IM by hydrophobic membrane spanning α-helices (trans-membrane domains). They are either synthesized with a cleavable signal sequence or with a stop transfer signal that functions as a targeting signal.

In *E. coli* the targeting to the IM involves two major cytosolic pathways converging at the same translocase (2, 3). The hydrophobicity of the signal sequence is thought to play an important role in deciding which targeting pathway will be followed (4). Proteins with relatively hydrophobic targeting signals are particularly dependent on the signal recognition particle (SRP) pathway. Less hydrophobic signals escape binding by the SRP allowing the secretion specific molecular chaperone SecB to bind the proteins in their mature region.

The next sections will deal with the different folding catalysts, targeting factors, and chaperones that interact with secreted and IMPs in the cytoplasm. They will be discussed in the expected order of their interaction with the newly synthesized proteins.

2.2 Trigger factor

One of the first proteins that interacts with nascent proteins of either cytosolic, secreted or IMPs is the peptidyl prolyl isomerase (PPIase) trigger factor (TF) (5, 6). As discussed in the previous chapter, TF can be considered as a general chaperone with PPIase activity.

TF is an abundant cytosolic protein with a high efficiency in catalysing *trans* to *cis* prolyl bond isomerization during the refolding of denatured proteins. TF is associated with the 50S subunit of the *E. coli* ribosome (7). A comparison of the amino acid sequences of TF with those of other PPIase families showed little similarities, but in its activity towards oligopeptide substrates TF resembles the so-called FK506-binding proteins (FKBPs) (8). TF is a modular protein, and consists of at least three domains, which show varying degrees of interactions at the level of function and stability (9). The N-domain of TF binds to the ribosome (10) and the middle or M-domain is a FKBP-type prolyl isomerase (11, 12). The function of the C-domain is still unknown. TF preferentially recognizes unstructured protein chains, which bind with high affinity to a site distinct from the catalytic prolyl isomerase centre in the FKBP domain (13).

The initial interaction with TF has shown to be one of the first events that take place when a nascent polypeptide emerges from the ribosome (5, 6, 14). TF was found to cross-link to nascent PhoE chains as short as 57 amino acids. From the *in vitro* data it has been suggested that TF functions as a cytoplasmic chaperone very early in translation to prevent improper intra- and inter-molecular interactions of the nascent polypeptide emerging on the surface of the large ribosomal subunit (5). Recent *in vivo* studies presented evidence for this hypothesis (15, 16). It was shown that TF and DnaK cooperate in folding of newly synthesized proteins. Under non-stress conditions DnaK transiently associated with a wide variety of nascent and newly synthesized polypeptides, with a preference for chains longer than 30 kDa. Deletion of the non-essential gene encoding TF resulted in doubling of the fraction of nascent polypeptides interacting with DnaK. Nascent chains shorter than 30 kDa were now also found to associate with DnaK. These findings showed *in vivo* activity for TF in general protein folding, and functional cooperation of this protein with DnaK.

It remains to be determined whether TF has a role in protein targeting and protein translocation. TF was originally implicated in protein translocation due to its interaction with purified proOmpA (the precursor to the outer membrane protein OmpA) upon the removal of urea by dialysis or dilution (17). This interaction stimulated the *in vitro* translocation of proOmpA into inverted IM vesicles leading to the hypothesis that TF maintains the translocation competent conformation of proOmpA (17–19). Further *in vivo* studies, however, in which the cellular content of TF was reduced, revealed no secretion defect for proOmpA (20). In the same study it was shown that TF could not overcome the export defects present in a SecB deleted strain. The defect was aggravated, rather than relieved, by the overproduction of trigger factor. Results from our laboratory (5, 14, and unpublished data) showed that TF binds to nascent chains of either secreted or IMPs. It is conceivable that TF, being strategically positioned at the nascent chain exit site of the ribosomal tunnel, just plays a role as a chaperone and protects the nascent chain from aggregation and improper interactions without making a distinction between cytosolic, secreted or IMPs. However, it also remains possible that TF has a regulatory role in targeting.

2.3 Signal recognition particle

Proteins with a targeting signal that emerge from the ribosome are 'scanned' by the signal recognition particle (SRP). Strongly hydrophobic targeting signals bind the SRP with high affinity and are targeted to the IM via the SRP pathway. SRP can be considered as a chaperone dedicated to secretion with a targeting function.

2.3.1 Structural aspects

The SRP pathway in *E. coli* involves a SRP and its receptor FtsY. These compounds resemble the machinery involved in protein targeting to the eukaryotic ER membrane (21, 22). A 'minimal' SRP has been identified that consists of a 4.5S RNA and P48, a 48 kDa GTPase, which are homologous to the eukaryotic 7S RNA (23) and SRP54 (24), respectively. P48 has a modular organization consisting of three domains. The amino-terminal or N-domain is followed by a GTPase module (the G domain). The carboxy terminal or M-domain contains binding sites both for signal peptides and for the 4.5S RNA (25–27; see Fig. 1). The high methionine content of the M-domain is conserved in P48 and SRP54 homologues found in all three kingdoms of life. These methionine residues are predicted to reside on a single face of several discrete amphipathic α-helices. Since methionine is a bulky hydrophobic amino acid with significant con-

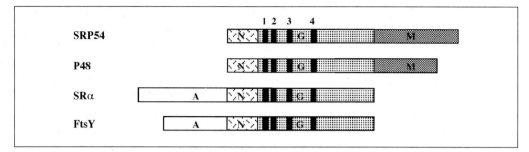

Fig. 1 Structural domains and homology of SRP-type GTPases. Letters indicate the separate structural domains. The four motifs that together constitute the GTP binding site are indicated as black boxes. See the text for discussion.

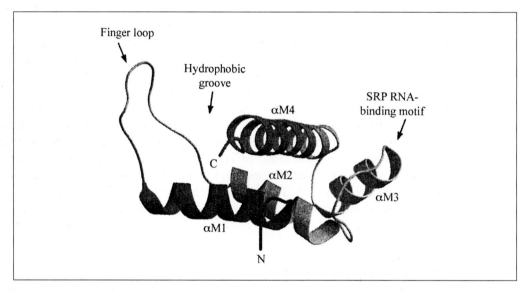

Fig. 2 Structure of the M-domain. Ribbon presentation of the M-domain, shaded dark to light from the N- to the C-terminus. Hydrophobic residues lining the proposed signal sequence binding groove are contributed from helix αM1, the flexible finger loop, helix αM2, and helix αM4. The highly conserved SRP RNA-binding motif is centred around helix αM3. Reprinted with permission from ref. 30.

formational flexibility (28), it has been proposed that the predicted α-helices fold together to produce a hydrophobic signal-sequence binding pocket lined with flexible methionine 'bristles' (29). The recently resolved crystal structure of the M-domain from *Thermus aquaticus* (30) provides support for this bristle hypothesis. Thus, a hydrophobic groove lined with side chains of flexible amino acids is a conserved feature of the M-domain and is likely to contribute to the ability of P48 to bind signal sequences of different length and sequence (See Fig. 2). The structure also reveals a helix–turn–helix motif containing an arginine-rich α-helix that is required for binding to the 4.5S RNA. This motif has been implicated in forming the core of an extended RNA binding surface. The structure of SRP RNA has been divided into four domains

(I–IV) (23), of which domain IV is the only one found in all SRP RNA homologues. RNA-protein contacts mostly involve the internal loop regions in domain IV (31). Recently, the structure of this loop has been resolved by using NMR (32). 4.5S RNA stabilizes the *E. coli* M-domain when it interacts with the signal sequence (33). In addition, it was found that 4.5S RNA stimulates the interaction of P48 with the receptor, FtsY (34), and there is evidence suggesting that the RNA facilitates communication of the M-domain with the N- and G-domains (33). The structural juxtaposition of the proposed signal sequence and SRP RNA-binding sites in the M-domain might suggest that changes in signal sequence occupancy could lead to conformational changes in the M-domain. These latter changes are then communicated to the 4.5S RNA, which in turn could affect interaction with the NG-domain and/or with other SRP ligands, including the ribosome and FtsY. Structural data of a M-domain in complex with a signal sequence or with the 4.5S RNA are needed to present evidence for these hypotheses.

Apart from the M-domain, the N-domain also has a function in signal sequence binding. Mutations in a conserved motif of the N-domain produced significant defects in signal sequence binding that correlate with the severity of the mutation (35). By contrast, mutations in the G-domain had no effect on signal sequence binding, but instead severely impaired protein translocation activity. The NG-domain is as well conserved between prokaryotic P48 and SRP54, as the M-domain. In addition, the NG-domain is also found in the SRP receptors. The crystal structures of the NG-domains of *E. coli* P48 (36) and FtsY (37) have been resolved and indeed exhibit similar conformations. The structure revealed the existence of three subdomains: the α-helical N-domain, the G-domain that is related to the Ras-GTPases but distinguished from them by an extension of the central β-sheet by two additional strands, and two α-helices. This latter subdomain is called the surface exposed insertion box-domain (I-box).

2.3.2 Recognition by SRP: who and where

Consistent with the notion that the SRP-substrate interaction is of hydrophobic nature, the signal sequence hydrophobicity plays an important role in determining SRP binding. In *E. coli* most cleavable targeting signals of secreted proteins possess a hydrophobic core region that is less hydrophobic than that of most IMPs. *In vivo* as well as *in vitro* data indicated that the SRP is involved in targeting of these IMPs to their final destination (14, 38–40). Ulbrandt *et al.* (40) used a genetic approach to identify potential SRP substrates in *E. coli*. They made use of the very limited amount of SRP available in the cell (~50 copies per cell; (41)), which implies that it will be readily titrated out by an excess of substrate. A subset of IMPs was identified, which presumably have a high affinity for SRP. These IMPs conferred synthetic lethality upon moderate over-production at limiting concentrations of P48 (40), suggesting that these IMPs are natural substrates of the SRP. By using an *in vitro* cross-link approach, Valent *et al.* (5, 14) showed that the *E. coli* SRP is most efficiently cross-linked to the targeting signals of IMPs. Eukaryotic SRPs display similar substrate specificity characteristics (42–44). Apparently, hydrophobicity is a conserved property for targeting substrates to SRP.

An additional requirement for binding of the *E. coli* SRP to a hydrophobic targeting signal is the context of the ribosome. Ribosome nascent chain complexes (RNCs) are recognized and bound by the *E. coli* SRP (5, 14). When released from the ribosome by puromycin or EDTA treatment these polypeptides did not remain associated with the SRP (5, 14, 45). In eukaryotes, the translation rate of the bound polypeptide is retarded by the SRP to increase the time window for the RNC to interact productively with translocation sites in the ER membrane (reviewed in (46)). The effect of the translation rate on the targeting reaction in *E. coli* is not clear. The *E. coli* 4.5S RNA lacks the so-called Alu domain of mammalian SRP RNA. Homologues of SRP9 and SRP14, that bind to this region and that are required for translation arrest (47) are also not present in *E. coli*. Consistently, the *E. coli* SRP was not able to retard translation upon targeting signal recognition in a heterologous system (48). It remains to be determined whether the *E. coli* SRP retards translation of its substrates in a homologous system. Possibly no such mechanism is required for the *E. coli* cell, because physical distances and therefore travel times from ribosome to membrane are much shorter.

Thus far, the role of the *E. coli* ribosome during SRP mediated protein targeting and translocation has remained unclear. No data are available on the interaction of the SRP with the ribosome. Is this interaction mediated by the RNA or does P48 directly interact with ribosomal proteins? What is the role of the ribosome in the interaction of the SRP with its receptor? Furthermore, it is not known whether the ribosome has an direct interaction with the translocase during co-translational targeting. Research on these issues has just started.

2.3.3   The targeting pathway

As soon as a hydrophobic targeting signal emerges from the ribosome (the ribosome nascent chain complex, RNC) it is recognized and bound by the SRP (See Fig. 3). In mammalian cells, this RNC–SRP complex is targeted to the α subunit of the SRP receptor (SRα) that is associated with the ER membrane via SRβ (49). In contrast, the SRP-receptor in *E. coli*, FtsY, was found to be located in the cytosol as well as in or at the cytoplasmic membrane (50 and it appeared that membrane association of FtsY is fundamentally different from that of SRα (51, 52).

Cytoplasmic FtsY was shown to interact with the RNC–SRP complex in the absence of membranes (2). A large complex was found that contains the RNC, SRP, and FtsY as identified by immunoprecipitation. The mechanism of the association of this complex with the membrane is not clear, but FtsY presumably mediates the membrane association. FtsY was shown to interact directly with *E. coli* phospholipids and this interaction involved at least two lipid binding sites, one of which is present in the NG-domain (52a). Lipid association induced a conformational change in FtsY and greatly enhanced its GTPase activity. These data suggest that the affinity of FtsY for lipids may contribute to the targeting of nascent chains.

Upon interaction of the large complex with the membrane, the nascent chain was found to be released at the translocation site in a reaction that requires the binding of GTP to at least FtsY (2). Hydrolysis of GTP at both SRP and FtsY probably serves to dissociate and recycle the targeting components.

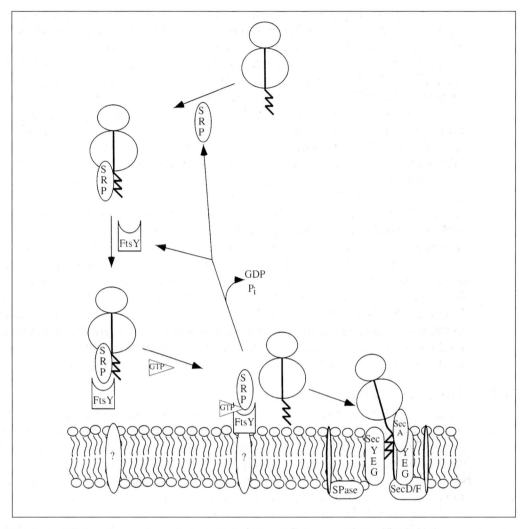

Fig. 3 The SRP targeting pathway. A hydrophobic targeting signal is recognized by the SRP as it emerges from the ribosome. The SRP–RNC complex is recognized by FtsY and subsequently targeted to the IM. The RNC is released from the SRP, an event preceded or accompanied by the binding of GTP to at least FtsY. The released RNC inserts into the SecYEG translocon. GTP hydrolysis serves to dissociate and recycle SRP and FtsY.

2.4 General chaperones, the heat shock proteins

Extracytoplasmic proteins with less hydrophobic signal sequences do not bind the SRP or bind SRP with lower affinity, and fail to be targeted by the SRP. While translation proceeds, the proteins risk aggregation. They have to be prevented from premature folding and have to be targeted to the cytoplasmic membrane. Molecular chaperones are required for these processes. The chaperones involved in protein localization in eukaryotes include members of the Hsp70 and Hsp60 families. In *E. coli*, it was shown that GroEL (the Hsp60 representative in bacteria) is important for the translocation of the SecB-independent precursor of β-lactamase (53, 54). Recently, it was shown that GroEL interacts with SecA, and it was suggested that GroEL might be involved in the release of SecA from the membrane (55). Also DnaK (an Hsp70 isologue), DnaJ, and GrpE were shown to be involved in the export of a number of SecB-independent proteins, such as alkaline phosphatase, β-lactamase, and ribose-binding protein (56, 57).

GroEL and DnaK, as discussed in Chapter 1, clearly function as general chaperones that do not make a distinction between cytosolic and secreted proteins. The latter proteins are kept in an unfolded conformation until they reach the translocation site at the IM. The general chaperones do not appear to have a specialized targeting function.

2.5 Secretion dedicated chaperone, SecB

SecB is a molecular chaperone dedicated to export. It contains a dual function in preprotein translocation. It helps to keep preproteins in a translocation competent state and it functions in targeting to the SecA/ translocon complex(58, 59).

2.5.1 Structural aspects

SecB is a cytoplasmic homotetrameric protein with a molecular mass of 64 kDa. At physiological ionic strength and pH, the SecB tetramer is stable. It can be dissociated directly into monomers, but under some conditions dimeric intermediates were observed (58) indicating that SecB is a dimer of dimers. This model of structural organization was reinforced by the identification of mutations that shift the equilibrium to favour a dimeric form of SecB (59). Amino acids 76–80 are probably involved in forming contacts at one of the two interfaces in the tetramer. However, the available data do not rule out the possibility that mutations in these amino acids cause a conformational change that is propagated to the interface. Interestingly, the residues at position 76, 78, and 80 were also found to be the binding site for ligand preproteins (60). Muren *et al.* (59) presently suggested that the interface of dimers is a likely candidate for the site of ligand binding. The same region, but at alternating positions compared to the residues involved in preprotein binding, also appears to harbour the mutations that interfere with the recognition of SecB by SecA (61).

2.5.2 Preprotein recognition

SecB associates with the nascent chains as they emerge from the ribosome (62, 63). In contrast to SRP, SecB was found to interact with the mature part of the preprotein (64–67). It has been suggested that SecB recognizes preproteins by binding their signal sequence domain (68–70), but other studies clearly showed that signal sequence recognition is not a prerequisite for the binding per se (66, 71–73). *In vivo*, SecB binds selectively a subset of nascent and fully elongated species of protein precursors (62). However, *in vitro*, SecB interacts with all kinds of proteins provided that they are in a non-native conformation (64, 66, 67). SecB prevents premature folding and aggregation and thereby maintains the translocation competence of the preprotein (74, 75).

A fundamental feature common to all chaperones is the remarkable ability to selectively bind non-native polypeptides. It has been proposed that the formation of a complex between SecB and a protein that possesses non-native structure will depend on the rate of folding of the polypeptide relative to its rate of association with SecB (the kinetic partitioning hypothesis) (76). Since SecB has no affinity for native, stably folded proteins, only those proteins that fold slowly are favoured to bind SecB and enter the export pathway. However, as reviewed by Driessen *et al.* (77) there are several observations that are in conflict with this hypothesis and the molecular basis for the specificity of the SecB-preprotein interaction *in vivo* remains unclear.

2.5.3 Interaction with SecA

The preprotein–SecB complex is targeted to SecA, a subunit of the translocase (see Fig. 4). The SecA–SecB interaction appeared weak in solution (78), but when SecA is bound to the IM at SecYEG, it was found to be activated for high affinity recognition of the SecB–preprotein complex. A SecB-binding site was found to be present in the extreme carboxy terminus of SecA (79). Deletion of this domain resulted in a SecA protein that is still active in *in vitro* protein translocation, but that is unable to bind

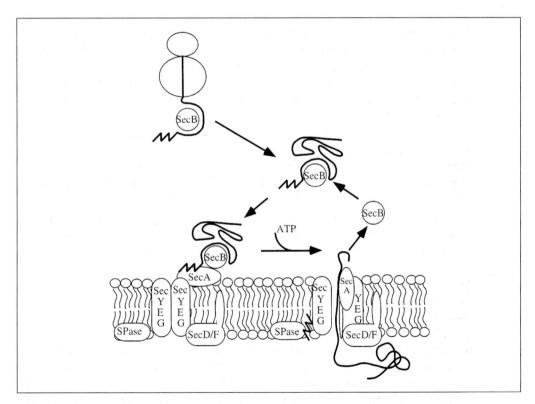

Fig. 4 The SecB targeting pathway. SecB binds to the mature region of proteins destined for secretion and targets them to translocon (SecYEG) bound SecA which has affinity for both SecB and the preprotein. ATP binding by SecA initiates translocation and releases SecB.

SecB. The SecB-binding site was found to be composed of 20 amino acid residues with a high content of lysine and arginine giving rise to a strongly electropositive surface. Since the SecA-binding site on SecB appeared to be composed, at least partially, of negatively charged residues it has been suggested that the binding of the two Sec proteins is electrostatic in nature (61). Once translocation is initiated upon the binding of ATP, SecB appeared to be released into the cytosol (79).

3. Membrane chaperones

Once at the cytoplasmic membrane, IMPs have to insert into the membrane and to obtain their native conformation. Although there is significant knowledge of protein folding for water soluble proteins, there is little information available for the folding of multispanning polytopic integral membrane proteins. Some membrane bound chaperones are discussed below.

FtsH, a member of the AAA family (**A**TPase **a**ssociated with different cellular **a**ctivities), is a membrane anchored metalloprotease (reviewed in (80), and in Chapter 9). Besides being a protease, there is evidence that FtsH also acts as a molecular chaperone (81), since it was found to influence protein assembly into the cytoplasmic membrane.

Oxa1p is a polytopic protein that is located in the inner mitochondrial membrane of *Saccharomyces cerevisiae* (82). Its role in mitochondria is to chaperone and correctly insert nuclear and mitochondrially encoded proteins into the IM. Homologues of Oxa1p have been found in both prokaryotes and eukaryotes (83). Data from our laboratory suggested that the *E. coli* homologue, YidC, is involved in IMP assembly (83a).

Studies on the assembly of *E. coli* lactose permease show a role for phospholipids as molecular chaperones. Phosphatidyl-ethanolamine (PE) was found to correct *in vitro* a LacY folding defect caused by *in vivo* assembly in PE deficient membranes. Once properly folded, PE apparently is no longer required to maintain the native conformation (84). In subsequent experiments it was found that membrane insertion of LacY and early steps in its conformational maturation appear independent of membrane phospholipid composition. However, late steps in the final folding of LacY required a specific phospholipid-assisted process (85).

4. Periplasmic chaperones and secreted proteins

4.1 Introduction

Proteins that have been translocated across the IM enter the periplasm. This subcellular compartment is sandwiched by the IM at the cytoplasmic side and by the outer membrane (OM) at the cell surface. The periplasm contains the highly hydrated polymer peptidoglycan, which is the skeleton of the bacterial cell, and which reduces the diffusional mobility of periplasmic proteins. Besides the peptidoglycan, the periplasm contains a variety of proteins, among which are several folding catalysts, dedicated or specialized chaperones, targeting factors, binding factors and transport proteins.

As in the cytoplasm, secreted proteins being translocated across the IM have in principle three possibilities. They can either fold into their native conformation and stay in this environment, they can be incorporated into the OM, or they can be translocated across the OM and end up at the outer surface of the cell or in the extracellular environment. In all of these cases, periplasmic chaperones that prevent folding, misfolding, and aggregate formation, as well as folding catalysts that assist in proper folding, play important roles. In a number of instances chaperone function and folding catalyst function appear to be combined in one protein/enzyme. The chaperones are often specific for their protein substrate and are referred to as dedicated chaperones. Folding catalysts accelerate specific rate-limiting steps in the folding reaction (86) of secreted periplasmic proteins and they can contribute to the folding of outer membrane proteins (OMPs) and extracellular proteins when these proteins pass through the periplasm on their way into the OM or on their way out.

OMPs have an exceptional structure. The amino acid composition of OMPs is often more polar than that of water soluble proteins. They contain predominantly β-structure, in contrast to the α-helical structure of IMPs. The porins, which are found in the OM, are homotrimers of intimately associated monomers. OmpA is a monomeric OMP.

Once the precursors of OMPs have been transported across the IM and processed, the proteins have to be folded, targeted, and inserted to the OM. The details of this sorting process remain elusive. Two different pathways have been proposed for the transport of OMPs from the IM to the OM. The transport might proceed via contact or adhesion sites between the IM and the OM, the so-called Bayer's junctions, whereby the OMPs remain in a membranous environment (87, 88), or via a free periplasmic intermediate, the so-called periplasmic intermediate pathway. Surprisingly it is still a matter of debate which pathway is used by OMP (89).

Currently, four assembly intermediates have been described for the trimeric OMPs: (i) the mature, processed unfolded monomer; (ii) the folded monomer which has a higher mobility on SDS-PAGE than the denatured form; (iii) the dimer; and (iv) the metastable trimer, which is less resistant to heat and SDS than the native trimer. It is now becoming clear that the metastable trimer is formed before the process of OM insertion. For monomeric OMPs, one assembly intermediate has been observed, namely the immature processed intermediate, which is the unfolded monomer (89).

Another class of membrane proteins are the lipoproteins, of which the major OM protein Lpp has been studied most extensively. Lipoproteins were found to be translocated across the IM in a Sec-dependent manner (90). Lipid modification and processing to yield a mature lipoprotein takes place at the IM, and then the localization of the mature protein to either the inner or the outer membrane follows. The amino acid residue next to the lipid-modified Cys at the N-terminus of a mature lipoprotein is thought to function as a sorting signal (91). A periplasmic chaperone, LolA, is involved in targeting of lipoproteins to the OM.

For the transport of proteins across the OM at least seven different pathways have been described. With the exception of one beautiful example, these mechanisms and the chaperones functioning in their pathways will not be described here due to a lack of space. At the cell surface, structures are present that enable the bacterial cells to attach to host tissue, cells, or inanimate surfaces. The so-called chaperone/usher pathway is used for assembling architecturally diverse surface structures, like pili, fimbria, or non-pilus adhesins (92). A dedicated periplasmic chaperone plays an essential role in this pathway.

The next sections will deal with the different chaperones and folding catalysts that function in the folding of periplasmic and OM proteins. They will be discussed in random order, since not much is known about the order of their interactions. The periplasmic chaperone that plays an essential role in the chaperone/usher pathway will be described in detail.

4.2   LPS and phospholipids

Lipopolysaccharide and phospholipids play a role in the folding and assembly process of OMPs. An assembly-competent folded monomer of PhoE was found to be formed *in vitro* in the presence of low amounts of Triton X-100, LPS, and divalent cations (93). Especially the negative charges in the inner core region of LPS and a non-lamellar structure of lipidA were implicated (94). Consistently, LPS of deep rough mutants was much less efficient in folding of PhoE (95). Similarly, the unfolded, processed form of OmpA which accumulated in cells overproducing this protein, could be converted into a heat-modifiable form after addition of LPS (96). The role of LPS in the folding of PhoE *in vitro* is consistent with *in vivo* data, which showed a severe defect in the biogenesis of porins in deep-rough mutants (97, 98). Furthermore, the biogenesis of the porins, but not of OmpA, was strongly affected *in vivo* by the drug cerulenin, which inhibits fatty acid synthesis, consistent with an important role of LPS and/or phospholipids in porin biogenesis (99–101). Finally, the observation that an assembly defect of an OmpF mutant protein could be suppressed *in vivo* by mutations that affect the LPS/phospholipid ratio (89) suggested an important role for LPS and/or phospholipids in OMP biogenesis.

Trimerization and OM insertion of the folded monomers of PhoE generated *in vitro* could be achieved by the addition of OMs and increasing the Triton X-100 concentration to 0.08% (95). The role of Triton X-100 in this *in vitro* folding and assembly process remains elusive, but it might mimic the role of a periplasmic chaperone, since Triton X-100 affected the assembly-competent state of PhoE (93). *In vitro* synthesized OmpF could be converted into trimers in the presence of purified LPS even in the absence of Triton X-100 (102).

4.3   Skp

The periplasmic cationic protein Skp (**s**eventeen **kDa** **p**rotein) appeared to be involved in the folding of OMPs during their passage through the periplasm. It has been observed that purified Skp binds *in vitro* to denatured OMPs (103, 104) and to

LPS (105). An *skp* mutant strain was found to contain reduced concentrations of OMPs (104). Missiakas *et al.* (106) proposed a role of Skp late in the assembly process, based on the observation that Skp over-expression, unlike that of SurA, did not suppress the phenotypic defects conferred by the expression of altered LPS. Skp may help to remove or exchange the original LPS molecule associated with folded mono-mers or pro-trimers of OMPs, thus facilitating its assembly. An opposing view was presented by de Cock *et al.* (103) who suggested an early role of Skp in the biogenesis of OMPs, since Skp interacts specifically with non-native OMPs and has a high affinity for phospholipids. Schäfer *et al.* (107) showed that OmpA interacts with Skp in close vicinity of the IM and that Skp is involved in release of OMP from the IM.

4.4 Disulfide oxidoreductases (Dsbs)

Disulfide bonds can play different roles in protein structure and activity. For many proteins, disulfide bonds are permanent features of the final folded structure, which enhance the stability of the protein. So far, six so-called Dsb proteins, DsbA, B, C, D, E, and G, involved in periplasmic disulfide bond formation, reduction and isomerization, have been identified and characterized. These Dsb proteins lack overall sequence homology, but they have two properties in common. They contain the active site Cys-X-X-Cys and, with the exception of DsbB, they have the thioredoxin fold (108). The activities of the some of the major Dsb proteins are summarised in Fig. 5.

The periplasmic DsbA protein, which was the first discovered Dsb protein (109), catalyses disulfide bond formation in unfolded substrates (110, 111). The catalysing capacity of the protein is caused by the destabilizing effect of the active site disulfide (112) and by the low pK_a of Cys30 (113), which is the essential residue of the Cys-X-X-Cys motif of DsbA (114). After catalysis, DsbA is reduced and reoxidation is caused by the integral membrane protein DsbB (115, 116). DsbB has two pairs of essential cysteines (117). The disulfide bond between Cys104 and Cys130 in the carboxy-termi-nal periplasmic loop acts directly to oxidize DsbA (118). The second pair of cysteines, Cys41 and Cys44, which are located in the N-terminal loop, reoxidizes the first pair. Oxidation of the second pair requires the presence of a functional respiratory chain (119).

When a protein contains only two cysteines, DsbA will inevitably form the correct disulfide bond. However, DsbA may promote incorrect disulfide bonds in substrate proteins with multiple cysteines. Incorrect pairing has to be corrected for the protein to assume its native conformation. This isomerization of disulfide bonds is catalysed by the periplasmic DsbC protein (120–122). The active site cysteines of the dimeric DsbC are normally in the reduced state (123) by the action of the IM protein DsbD (124–126). The reduction of DsbD depends on the cytoplasmic proteins thioredoxin and thioredoxin reductase (127).

The DsbE protein is required to maintain the heme-binding site of apo-cytochrome c reduced and, therefore, the active site cysteines are in the reduced state (128). DsbE has been localized in the periplasm (129) and in the IM with its active site exposed to the periplasm (128).

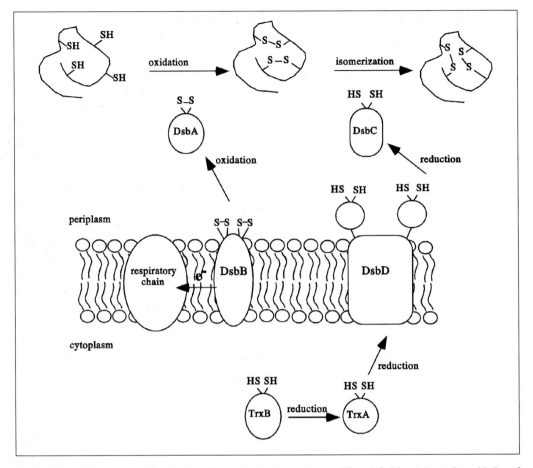

Fig. 5 Schematic overview of disulfide bond formation in the periplasm of *E. coli*. DsbA catalyses the oxidation of free thiols to disulfide bridges and DbsC catalyses the isomerization of these disulfides. DsbB reoxidises reduced DsbA, with electron flow being coupled to the respiratory chain in the inner membrane. DsbD maintains DsbC in a reduced state. DsbD depends on thioredoxin and thioredoxin reductase in the cytoplasm to maintain its reduced state.

Finally, DsbG has been identified recently, but it remains under debate whether it functions as an oxidant or displays isomerase activity (130, 131).

Studies on the involvement of the Dsb proteins in the assembly of OMPs have been limited to DsbA. DsbA catalysed the formation of a disulfide bond in OmpA (109). The expression of OmpF was drastically reduced in a *dsbA* mutant at the transcriptional level, possibly as a consequence of feedback inhibition (90). However, this protein is devoid of disulfide bonds. Possibly, DsbA has additional functions and might act as a chaperone in the biogenesis of OmpF. Alternatively, it might catalyse disulfide bond formation in an as yet unidentified chaperone required for the OmpF assembly or in an as yet unidentified envelope protein involved in the regulation of *ompF* expression.

4.5 Peptidyl prolyl *cis/trans* isomerases (PPIases)

PPIases catalyse *cis/trans* isomerization of peptidylprolyl bonds (132). *In vitro* experiments have shown that the (re-)folding of proteins can be accelerated by PPIases from various sources. However, the exact role of these PPIases in protein folding *in vivo* remains to be established. PPIases have been identified both in the cytoplasm and in the periplasm, and they have been divided into three subfamilies on basis of their binding capacity to different drugs (132): (i) the cyclophilins with RotA as a periplasmic representative; (ii) FK506-binding proteins with FkpA as periplasmic member; and (iii) the parvulins with SurA in the periplasm and PpiD in the IM with its catalytic domain in the periplasm.

Of the PPIases, SurA and PpiD have been implicated in OMP folding. SurA was initially identified as a protein essential for stationary phase survival (133). By using a *surA* mutant strain, a role of SurA in the assembly of OMPs but not of periplasmic proteins was demonstrated in pulse-chase experiments (Rouvière and Gross, 1996; (134). In the absence of SurA, the periplasmic folding of LamB appeared to be affected. The IMP PpiD was isolated as a multicopy suppressor of a *surA* mutant (135). Its role in folding was demonstrated by the fact that the levels of OMPs were reduced in a *ppiD* mutant, although less severe than in a *surA* null mutant. A *ppiD/surA* double mutant proved to be lethal, suggesting an overlapping function. PpiD showed significantly higher PPIase activity *in vitro* than SurA, in spite of the less pronounced phenotype of the *ppiD* mutant. Hence, SurA and PpiD might have another role in addition to their PPIase activity in folding.

4.6 LolA

Lipoproteins are localized in the outer or inner membrane of *E. coli*, depending on the type of amino acid located next to the N-terminal fatty-acylated Cys. An Asp at this position (+2) causes localization in the IM, whereas any other amino acid residue changes the final location from the inner to the OM. The major OM lipoprotein (Lpp) expressed in spheroplasts was, however, retained in the IM as a mature form. A periplasmic protein, LolA, was found to mediate the release of Lpp (136) or other lipoproteins (137) from the IM by forming a soluble lipoprotein-LolA complex. No release could be found when an Asp was present on the N-terminal second position of Lpp (136), indicating that the LolA dependent release is dependent on the OM-specific sorting signal. In addition, the LolA dependent release was found to require energy since it is dependent on NTP hydrolysis (138). The soluble lipoprotein–LolA complex is targeted to the OM. The OM receptor LolB mediates the outer membrane incorporation of the lipoprotein via the transient formation of a LolB–lipoprotein complex (137, 139).

4.7 The PapD chaperone family

By far the best studied dedicated chaperone in the periplasm that functions in the secretion of proteins across the OM is the PapD chaperone that functions in the chaperone/usher pathway for Pap or P pili biosynthesis. These pili are composite fibres at the cell surface consisting of a rod, 7 nm thick, joined to a thinner tip fibrillum. The rod consists of identical subunits arranged in a right-handed helical cylinder. The tip fibrillum primarily consists of repeating subunits arranged in a open helical configuration At the distal end of the fibrillum the so-called adhesin is bound via an adapter subunit. This adhesin is able to bind to receptors of human kidney cells (140).

PapD is the prototype of the periplasmic chaperone in a family that includes more than 30 members. The proteins binds and caps interactive surfaces of pilus subunits and prevents them for having non-productive interactions as they enter the periplasm during translocation across the IM. Furthermore, the chaperones might assist in folding of the subunits and delivering them at the OM molecular usher for further steps in the assembly pathway. The chaperone is a boomerang-shaped two domain molecule with each domain having an immunoglobulin-like fold (141). The two β-barrel domains form a cleft which is highly conserved among the family and which contains highly invariant residues that protrude into this cleft. Most of the other conserved amino acid residues are concentrated in the β-strands in the cleft region and they maintain the overall immunoglobulin-like structures of the domains. The first indications for the functioning of these chaperones and especially for the role of the cleft region came from site-directed mutagenesis studies, binding tests of peptides and from the co-crystallization of the chaperone with a 19 amino acid carboxy-terminal peptide of a pilus subunit PapG (142). The structure of the carboxy-terminal region of practically all pili subunits is highly conserved. The PapG carboxy-terminal peptide binds along the edge of the G1 β-strand of the chaperone's first β-barrel domain by a backbone of hydrogen bonds. They form a β-sheet structure (β-zipper) between the chaperone and the peptide. However, the binding of the peptide to the chaperone does not rely solely on the backbone hydrogen bonds. Part of the binding specificity of a chaperone with one of its pili or fimbrial subunits appeared to come from the alternating pattern of hydrophobic residues in the carboxy-terminus of the subunits that fit with the conserved hydrophobic G1 β-strand of the chaperone. The role of the cleft of the chaperone and of the G1 β-strand of the chaperone was recently confirmed by the analysis of the crystal structure of the complex of the PapD chaperone with one of its subunits (PapK) and that of the FimC–FimH chaperone–adhesin complex (see Fig. 6) (143, 144). These studies revealed that the chaperones function by donating their G1 β-strand to complete the immunoglobulin-like fold of the subunits, via a mechanism termed donor strand complementation. The pili subunits studied also have an immunoglobulin-like fold, except that the seventh strand is missing, leaving a part of the hydrophobic core exposed. The chaperones donates their G1 β-strand to complete the fold of the subunits. In the final pilus structure the subunits interact with each other in a similar way, a mechanism termed donor strand exchange.

The K88 chaperone FaeE and the PapD chaperone have been shown to form dimers (145, 146). Recent studies on the PapD dimer (146) revealed that chaperone dimerization involves the same surface that is otherwise used to bind subunits.

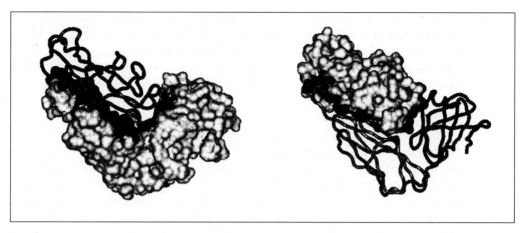

Fig. 6 Structure of the PapD–PapK complex. On the left, PapD is shown as a space-filling model and PapK as a ribbon. On the right, PapK is shown as a space-filling model and PapD as a ribbon. Reprinted with permission from ref. 143.

5. Summary and conclusions

In the *E. coli* cytoplasm two main targeting pathways exist. The co-translational SRP pathway targets proteins with strongly hydrophobic targeting signals, mainly IMPs, to the cytoplasmic membrane. The post-translational pathway that uses general chaperones as DnaK (an Hsp70 isologue), GroEL (the Hsp 60 representative in bacteria), or SecB, targets proteins that escape the SRP pathway. Both targeting pathways converge at the same translocase. Studies on the SRP-mediated targeting pathway have benefited in recent years from the development of *in vitro* targeting assays and an *in vitro* reconstituted *E. coli* translocase. In addition, the solved crystal structure of domains of both P48 and the SRP receptor FtsY enlarged our knowledge of the pathway. Crystal structures of the complete SRP and of complexes of SRP with a nascent chain and/or with FtsY are likely to follow. In addition the role of the ribosome during SRP-mediated protein targeting and translocation will be subject for future research.

Once targeted, IMPs have to insert into the membrane and to obtain their native conformation. Research on these processes has recently led to the discovery of YidC. This IMP was proposed to be involved in the insertion of hydrophobic sequences into the lipid bilayer after initial recognition by the translocase. Apart from the two main targeting pathways, a new secretion pathway (the Tat pathway) has been identified in recent years, reviewed by others (147). Little is known about chaperones and targeting factors that are involved in this pathway, but it is in the centre of interest in the field.

In the periplasm, chaperones and folding catalysts assist proteins in obtaining their native conformation and/or their final destination. Six proteins were identified that play a role in the disulfide bond formation (Dsb). DsbA and B are mainly involved in the formation of disulfide bonds and reoxidation of these Dsb proteins is coupled to a functional electron transfer chain. The others are involved in isomerization of disulfide bonds and in cytochrome c biogenesis. Genetic analysis revealed that the reduction of these Dsbs depends on the cytoplasmic proteins thioredoxin and thioredoxin reductase. Genetic screens also revealed the presence of several PPIases in the periplasm. In addition to their PPIase activity these proteins might have an additional chaperone function. Future biochemical studies and structural analysis are necessary to increase our knowledge on protein folding in general in the periplasm. The best studied dedicated chaperone in the periplasm, PapD, functions in the secretion of proteins across the OM in the chaperone/usher pathway for Pap pili. The recent studies on the crystal structure of the complex of the PapD chaperone with one of its subunits is an example of how structural analysis can contribute to the understanding of folding and targeting pathways.

References

1. Rapoport, T. A., Jungnickel, B., and Kutay, U. (1996). Protein transport across the eukaryotic endoplasmic reticulum and bacterial inner membranes. *Annu. Rev. Biochem.*, **65**, 271.
2. Valent, Q. A., Scotti, P. A., High, S., de Gier, J.-W. L., von Heijne, G., Lentzen, G., *et al.* (1998). The *Escherichia coli* SRP and SecB targeting pathways converge at the translocon. *EMBO J.*, **17**, 2504.
3. Koch, H.-G., Hengelage, T., Neumann-Haefelin, C., MacFarlane, J., Hoffschulte, H. K., Schimz, K.-L., *et al.* (1999). In vitro studies with purified components reveal signal recognition particle (SRP) and SecA/B as constituents of two independent protein-targeting pathways of Escherichia coli. *Mol. Biol. Cell*, **10**, 2163.
4. DeGier, J. W. L., Valent, Q. A., von Heijne, G., and Luirink, J. (1997). The E-coli SRP: preferences of a targeting factor. *FEBS Lett.*, **408**, 1.
5. Valent, Q. A., Kendall, D. A., High, S., Kusters, R., Oudega, B., and Luirink, J. (1995). Early events in preprotein recognition in *E. coli*: interaction of SRP and trigger factor with nascent polypeptides. *EMBO J.*, **14**, 5494.
6. Hesterkamp, T., Hauser, S., Lütcke, H., and Bukau, B. (1996). Escherichia coli trigger factor is a prolyl isomerase that associates with nascent polypeptide chains. *Proc. Natl. Acad. Sci. USA*, **93**, 4437.
7. Stoller, G., Rucknagel, K. P., Nierhaus, K. H., Schmid, F. X., Fischer, G., and Rahfeld, J. U. (1995). A ribosome-associated peptidyl-prolyl cis/trans isomerase identified as the trigger factor. *EMBO J.*, **14**, 4939.
8. Callebaut, I., and Mornon, J. P. (1995). Trigger factor, one of the Escherichia coli chaperone proteins, is an original member of the FKBP family. *FEBS Lett.*, **374**, 211.
9. Zarnt, T., Tradler, T., Stoller, G., Scholz, C., Schmid, F. X., and Fischer, G. (1997). Modular structure of the trigger factor required for high activity in protein folding. *J. Mol. Biol.*, **271**, 827.
10. Hesterkamp, T., Deuerling, E., and Bukau, B. (1997). The aminoterminal 118 amino acids of *E. coli* trigger factor constitute a domain that is necessary and sufficient for binding to ribosomes. *J. Biol. Chem.*, **272**, 21865.
11. Stoller, G., Tradler, T., Rücknagel, J.-U., and Fisher, G. (1996). An 11.8KDa proteolytic fragment of the *E. coli* trigger factor represents the domain carrying the peptidyl-prolyl cis/trans isomerase activity. *FEBS Lett.*, **384**, 117.

12. Hesterkamp, T. and Bukau, B. (1996). Identification of the prolyl isomerase domain of Escherichia coli trigger factor. *FEBS Lett.*, **385**, 67.

13. Scholz, C., Mücke, M., Rape, M., Pecht, A., Pahl, A., Bang, H., and Schmid, F. X. (1998). Recognition of protein substrates by the prolyl isomerase trigger factor is independent of proline residues. *J. Mol. Biol.*, **277**, 723.

14. Valent, Q. A., deGier, J. W. L., von Heijne, G., Kendall, D. A., ten Hagen-Jongman, C. M., Oudega, B., and Luirink, J. (1997). Nascent membrane and presecretory proteins synthesized in Escherichia coli associate with signal recognition particle and trigger factor. *Mol. Micobiol.*, **25**, 53.

15. Deuerling, E., Schulze-Specking, A., Tomoyasu, T., Mogk, A., and Bukau, B. (1999). Trigger factor and DnaK cooperate in folding of newly synthesized proteins. *Nature*, **400**, 693.

16. Teter, S. A., Houry, W. A., Ang, D., Tradler, T., Rockabrand, D., Fischer, G., *et al.* (1999). Polypeptide flux through bacterial Hsp70: DnaK cooperates with trigger factor in chaperoning nascent chains. *Cell*, **97**, 755.

17. Crooke, E. and Wickner, W. (1987). Trigger factor: a soluble protein that folds pro-OmpA into a membrane-assembly competent form. *Proc. Natl. Acad. Sci. USA*, **84**, 5216.

18. Crooke, E., Guthrie, B., Lecker, S., Lill, R., and Wickner, W. (1988). ProOmpA is stabilized for membrane translocation by either purified *E. coli* trigger factor or canine signal recognition particle. *Cell*, **54**, 1003.

19. Crooke, E., Brundage, L., Rice, M., and Wickner, W. (1988). ProOmpA spontaneously folds in a membrane assembly competent state which trigger factor stabilizes. *EMBO J.*, **7**, 1831.

20. Guthrie, B. and Wickner, W. (1990). Trigger factor depletion or overproduction causes defective cell division but does not block protein export. *J. Bacteriol.*, **172**, 5555.

21. Luirink, J. and Dobberstein, B. (1994). Mammalian and Escherichia Coli Signal Recognition Particles. *Mol. Micobiol.*, **11**, 9.

22. Lutcke, H. (1995). Signal recognition particle (SRP), a ubiquitous initiator of protein translocation. *Eur. J. Biochem.*, **228**, 531.

23. Poritz, M. A., Strub, K., and Walter, P. (1988). Human SRP RNA and E. coli 4.5S RNA contain a highly homologous structural domain. *Cell*, **55**, 4.

24. Römisch, K., Webb, J., Herz, J., Prehn, S., Frank, R., Vungron, M., and Dobberstein, B. (1989). Homology of 54K protein of signal-recognition particle, docking protein and two *E. coli* proteins with putative GTP-binding domains. *Nature*, **340**, 478.

25. Römisch, K., Webb, J., Lingelbach, K., Gausepohl, H., and Dobberstein, B. (1990). The 54-kD protein of signal recognition particle contains a methionine-rich RNA binding domain. *J. Cell Biol.*, **111**, 1793.

26. Zopf, D., Bernstein, H. D., Johnson, A. E., and Walter, P. (1990). The methionine-rich domain of the 54 kd protein subunit of the signal recognition particle contains an RNA binding site and can be crosslinked to a signal sequence. *EMBO J.*, **9**, 4511.

27. Lütcke, H., High, S., Römisch, K., Ashford, A. J., and Dobberstein, B. (1992). The methionine-rich domain of the 54 kDa subunit of signal recognition particle is sufficient for the interaction with signal sequences. *EMBO J.*, **11**, 1543.

28. Gellman, S. (1991). On the role of methionine residues in the sequence independent recognition of nonpolar protein surfaces. *Biochem.*, **30**, 6633.

29. Bernstein, H. D., Poritz, M. A., Strub, K., Hoben, P. J., Brenner, S., and Walter, P. (1989). Model for signal sequence recognition from amino-acid sequence of 54K subunit of signal recognition particle. *Nature*, **340**, 482.

30. Keenan, R. J., Freymann, D. M., Walter, P., and Stroud, R. M. (1998). Crystal structure of the signals sequence binding subunit of the signal recognition particle. *Cell*, **94**, 181.

31. Schmitz, U., Freymann, D., James, T., Keenan, R., Vinayak, R., and Walter, P. (1996). NMR studies of the most conserved RNA domain of the mammalia signal recognition particle (SRP). *RNA*, **2**, 1213.

32. Schmitz, U., James, T. L., Lukavsky, P., and Walter, P. (1999). Structure of the most conserved internal loop in SRP RNA. *Nature Struct. Biol.*, **6**, 634.

33. Zheng, N. and Gierasch, L. M. (1997). Domain interactions in E. coli SRP: stabilization of M-domain by RNA is required for effective signal sequence modulation of NG domain. *Mol. Cell*, **1**, 1.

34. Miller, J. D., Bernstein, H. D., and Walter, P. (1994). Interaction of E-Coli Ffh/4.5S ribonucleoprotein and Ftsy mimics that of mammalian signal recognition particle and its receptor. *Nature*, **367**, 657.

35. Newitt, J. A. and Bernstein, H. D. (1997). The N-domain of the signal recognition article 54-kDa subunit promotes efficient signal sequence binding. *Eur. J. Biochem.*, **245**, 720.

36. Freymann, D. M., Keenan, R. J., Stroud, R. M., and Walter, P. (1997). Structure of the conserved GTPase domain of the signal recognition particle. *Nature*, **385**, 361.

37. Montoya, G., Svensson, C., Luirink, J., and Sinning, I. (1997). Crystal structure of the NG domain from the signal-recognition particle receptor FtsY. *Nature*, **385**, 365.

38. de Gier, J. W. L., Mansournia, P., Valent, Q. A., Phillips, G. J., Luirink, J., and von Heijne, G. (1996). Assembly of a cytoplasmic membrane protein in *Escherichia coli* is dependent on the signal recognition particle. *FEBS Lett.*, **399**, 307.

39. Mac Farlane, J. and Muller, M. (1995). Functional integration of a polytopic membrane protein of E-coli requires the bacterial signal recognition particle. *Eur. J. Biochem.*, **223**, 766.

40. Ulbrandt, N. D., Newitt, J. A., and Bernstein, H. D. (1997). The E. coli signal recognition particle is required for the insertion of a subset of inner membrane proteins. *Cell*, **88**, 187.

41. Wikström, P. M. and Björk, G. R. (1988). Noncoordinate translation-level regulation of ribosomal and nonribosomal protein genes in the Escherichia coli trmD operon. *J. Bacteriol.*, **170**, 3025.

42. Hatsuzawa, K., Tagaya, M., and Mizushima, S. (1997). Hydrophobic region of signal peptides is a determinant for SRP recognition and protein translocation across the ER membrane. *J. Biochem.*, **121**, 270.

43. High, S., Henry, R., Mould, R., Valent, Q. A., Meacock, S., Cline, K., *et al.* (1997). Chloroplast SRP54 interacts specifically with a subset of thylakoid precursor proteins. *J. Biol. Chem.*, **272**, 11622.

44. Ng, D. T. W., Brown, J. D., and Walter, P. (1996). Signal sequences specify the targeting route to the endoplasmic reticulum membrane. *J. Cell Biol.*, **134**, 269.

45. Luirink, J., High, S., Wood, H., Giner, A., Tollervey, D., and Dobberstein, B. (1992). Signal sequence recognition by an *Escherichia coli* ribonucleoprotein complex. *Nature*, **359**, 741.

46. Brodsky, J. L., Bauerle, M., Horst, M., and McClellan, A. J. (1998). Mitochondrial Hsp70 cannot replace BiP in driving protein translocation into the yeast endoplasmic reticulum. *FEBS Lett.*, **435**, 183.

47. Siegel, V. and Walter, P. (1988). Functional dissection of the signal recognition particle. *TIBS*, **13**, 314.

48. Powers, T. and Walter, P. (1997). Co-translational protein targeting catalyzed by the Escherichia coli signal recognition particle and its receptor. *EMBO J.*, **16**, 4880.

49. Young, J. C., Ursini, J., Legate, K. R., Miller, J. D., Walter, P., and Andrews, D. W. (1995). An amino-terminal domain containing hydrophobic and hydrophilic sequences binds the signal recognition particle receptor alpha subunit to the beta subunit on the endoplasmic reticulum membrane. *J. Biol. Chem.*, **270**, 15650.

50. Luirink, J., ten Hagen-Jongman, C. M., Van der Weijden, C. C., Oudega, B., High, S., Dobberstein, B., and Kusters, R. (1994). An alternative protein targeting pathway in Escherichia coli: Studies on the role of FtsY. *EMBO J.*, **13**, 2289.

51. de Leeuw, E., Poland, D., Mol, O., Sinning, I., ten Hagen-Jongman, C. M., Oudega, B., and Luirink, J. (1997). Membrane association of FtsY, the E-coli SRP receptor. *FEBS Lett.*, **416**, 225.

52. Millman, J. S. and Andrews, D. W. (1999). A site-specific, membrane-dependent cleavage event defines the membrane binding domain of FtsY. *J. Biol. Chem.*, **274**, 33227.

52a de Leeuw, E., te Kaat, K., Moser, C., Menestrina, G., Demel, R., de Kruijff, B., *et al.* (2000). Anionic phospholipids are involved in membrane association of FtsY and stimulate its GTPase activity. *EMBO J.*, **19**, 531.

53. Bochkareva, E. S., Lissin, N. M., and Girshovich, A. S. (1988). Transient association of newly synthesized unfolded proteins with the heat-shock GroEL protein. *Nature*, **336**, 254.

54. Kusukawa, N., Yura, T., Ueguchi, C., Akiyama, Y., and Ito, K. (1989). Effects of mutations in heat-shock genes groES and groEL on protein export in Escherichia coli. *EMBO J.*, **8**, 3517.

55. Bochkareva, E. S., Solovieva, M. E., and Girshovich, A. S. (1998). Targeting of GroEL to secA on the cytoplasmic membrane of *Escherichia coli*. *Proc. Natl. Acad. Sci. USA*, **95**, 478.

56. Wild, J., Rossmeissl, P., Walter, W. A., and Gross, C. A. (1996). Involvement of the DnaK-DnaJ-GrpE chaperone team in protein secretion in *Escherichia coli*. *J. Bacteriol.*, **178**, 360.

57. Wild, J., Altman, E., Yura, T., and Gross, C. A. (1992). DnaK and DnaJ heat shock proteins participate in protein export in *Escherichia coli*. *Genes Dev.*, **6**, 1165.

58. Smith, V. F., Schwartz, B. L., Randall, L. L., and Smith, R. D. (1996). Electrospray mass spectrometric investigation of the chaperone secB. *Protein Science*, **5**, 488.

59. Muren, E. M., Suciu, D., Topping, T., Kumamoto, C. A., and Randall, L. L. (1999). Mutational alterations in the homotetrameric chaperone SecB that implicate the structure as dimer of dimers. *J. Biol. Chem.*, **274**, 19397.

60. Kimsey, H. H., Dagarag, M. D., and Kumamoto, C. A. (1995). Diverse effects of mutation on the activity of the Escherichia coli export chaperone SecB. *J. Biol. Chem.*, **270**, 22831.

61. Fekkes, P., de Wit, J. G., van der Wolk, J. P., Kimsey, H., Kumamoto, C. A., and Driessen, A. J. M. (1998). Preprotein transfer to the *Escherichia coli* translocase requires the cooperative binding of secB and the signal sequence to SecA. *Mol. Microbiol.*, **29**, 1179.

62. Kumamoto, C. A. and Francetic, O. (1993). Highly selective binding of nascent polypeptides by an *Escherichia coli* chaperone protein in vivo. *J. Bacteriol.*, **175**, 2184.

63. Randall, L. L., Topping, T. B., Hardy, S. J. S., Pavlov, M. Y., Freistroffer, D. V., and Ehrenberg, M. (1997). Binding of SecB to ribosome-bound polypeptides has the same characteristics as binding to full-length, denatured proteins. *Proc. Natl. Acad. Sci. USA*, **94**, 802.

64. Collier, D. N., Bankaitis, V. A., Weiss, J. B., and P. J. Bassford, Jr. (1988). The antifolding activity of SecB promotes the export of the *E. coli* maltose-binding protein. *Cell*, **53**, 273.

65. Gannon, P. M., Li, P., and Kumamoto, C. A. (1989). The mature part of *Escherichia coli* maltose-binding protein (MBP) determines the dependence of MBP on SecB for export. *J. Bacteriol.*, **171**, 813.

66. Randall, L. L., Topping, T. B., and Hardy, S. J. S. (1990). No specific recognition of leader peptide by SecB, a chaperone involved in protein export. *Science*, **248**, 860.

67. de Cock, H. and Tommassen, J. (1992). SecB-binding does not maintain the translocation-competent state of prePhoE. *Mol. Microbiol.*, **6**, 599.

68. Watanabe, M. and Blobel, G. (1989). Cytosolic factor purified from *Escherichia coli* is necessary and sufficient for the export of a preprotein and is a heterotramer of SecB. *Proc. Natl. Acad. Sci. USA*, **86**, 2728.

69. Watanabe, M., and Blobel, G. (1995). High-affinity binding of Escherichia coli SecB to the signal sequence region of a presecretory protein. *Proc. Natl. Acad. Sci. USA*, **92**, 10133.

70. Altman, E., Emr, S. D., and Kumamoto, C. A. (1990). The presence of both the signal sequence and a region of mature LamB is required for the interaction of LamB with export factor SecB. *J. Biol. Chem.*, **265**, 18154.

71. Lecker, S., Lill, R., Ziegelhoffer, T., Georgopoulos, C., Bassford, P. J., Kumamoto, C. A., and Wickner, W. (1989). Three pure chaperone proteins of *Escherichia coli*—SecB, trigger factor and GroEL—form soluble complexes with precursor proteins in vitro. *EMBO J.*, **8**, 2703.

72. Liu, G., Topping, T. B., and Randall, L. L. (1989). Physiological role during export for the retardation of folding by the leader peptide of maltose-binding protein. *Proc. Natl. Acad. Sci. USA*, **86**, 9213.

73. Hardy, S. J. S. and Randall, L. L. (1991). A kinetic partitioning model of selective binding of nonnative proteins by the bacterial chaperone SecB. *Science*, **251**, 439.

74. Kusters, R., de Vrije, T., Breukink, E., and de Kruijff, B. (1989). SecB stabilizes a translocation-competent state of purified prePhoE protein. *J. Biol. Chem.*, **264**, 20827.

75. Lecker, S. H., Driessen, A. J. M., and Wickner, W. (1990). ProOmpA contains secondary and tertiary structure and is shielded from aggregation by association with SecB. *EMBO J.*, **9**, 2309.

76. Randall, L. L. and Hardy, S. J. S. (1995). High selectivity with low specificity: how SecB has solved the paradox of chaperone binding. *Trends in Biochemical Science*, **29**, 65.

77. Driessen, A. J., Fekkes, P., and van der Wolk, J. P. (1998). The Sec system. *Curr. Opin. Microbiol.*, **1**, 216.

78. Hoffschulte, H. K., Drees, B., and Müller, M. (1994). identification of a soluble SecA/secB complex by means of a subfractionated cell-free export system. *J. Biol. Chem.*, **269**, 12833.

79. Fekkes, P., van der Does, C., and Driessen, A. J. M. (1997). The molecular chaperone SecB is released from the carboxy-terminus of SecA during initiation of precursor protein translocation. *EMBO J.*, **16**, 6105.

80. Schumann, W. (1999). FtsH- a single-chain charonin? *FEMS Micriob. Rev.*, **23**, 1.

81. Akiyama, Y., Ogura, T., and Ito, K. (1994). Involvement of Ftsh in protein assembly into and through the membrane. 1. Mutations that reduce retention efficiency of a cytoplasmic reporter. *J. Biol. Chem.*, **269**, 5218.

82. Bauer, M., Behrens, M., Esser, K., Michaelis, G., and Pratje, E. (1994). PET1402, a nuclear gene required for proteolytic processing of cytochrome oxidase biogenesis. *Mol. Gen. Genet.*, **245**, 272.

83. Bonnefoy, N., Chalvet, F., Hamel, P., Slonimski, P., and Dujardin, G. (1994). OXA1, a *Saccharomyces cerevisiae* nuclear gene whose sequence is conserved from prokaryotes to eukaryotes controls cytochrome oxidase biogenesis. *J. Mol. Biol.*, **239**, 201.

83a Scotti, P. A., Urbanus, M. L., Brunner, J., de Gier, J-W., von Heijne, G., van der Does, C., *et al.* (2000). YidC, the *Escherichia coli* homologue of mitochondrial Oxa1p, is a component of the Sec translocase. *EMBO J.*, **19**, 542.

84. Bogdanov, M., Sun, J., Kaback, H. R., and Dowhan, W. (1996). A phospholipid acts as a chaperone in assembly of a membrane transport protein. *J. Biol. Chem.*, **271**, 11615.

85. Bogdanov, M. and Dowhan, W. (1998). Phospholipid-assisted protein folding: phosphatidylethanolamine is required at a late step of the conformational maturation of the polytopic membrane protein lactose permease. *EMBO J.*, **17**, 5255.

86. Ellis, R. J. and Van der Vies, S. M. (1991). Molecular chaperones. *Annu. Rev. Biochem.*, **60**, 321.

87. Bayer, M. E. (1968). Areas of adhesion between wall and membrane of *Escherichia coli. J. Gen. Microbiol.*, **53**, 395.

88. Bayer, M. E. (1994). Periplasm. In *Bacterial cell wall* (ed. J. Ghuysen and R. Hakenbeck), pp. 464–474. Elsevier, Amsterdam.

89. Danese, P. N. and Silhavy, T. J. (1998). Targeting and assembly of periplasmic and outer-membrane proteins in Escherichia coli. *Annu. Rev. Genet.*, **32**, 59.

90. Pugsley, A. P. (1993). The complete general secretory pathway in Gram-negative bacteria. *Microbiol. Rev.*, **57**, 50.

91. Yamaguchi, K., Yu, F., and Inouye, M. (1988). A single amino acid determinant of the membrane localization of lipoproteins in E. coli. *Cell*, **53**, 423.

92. Hultgren, S. J., Abraham, S. N., Caparon, M. G., Falk, P., St Geme, J. W., and Normark, S. (1993). Pilus and non-pilus bacterial adhesins: assembly and function in cell recognition. *Cell*, **73**, 887.

93. De Cock, H., Van Blokland, S., and Tommassen, J. (1996). In vitro insertion and assembly of outer membrane protein PhoE of *Escherichia coli* K-12 into the outer membrane. *J. Biol. Chem.*, **271**, 12885.

94. De Cock, H., Brandenburg, K., Wiese, A., Holst, O., and Seydel, U. (1999). Non-lamellar structure and negative charges of lipopolysaccharides required for efficient folding of outer membrane protein PhoE of *Escherichia coli. J. Biol. Chem.*, **274**, 5114.

95. De Cock, H. and Tommassen, J. (1996). Lipopolysaccharide and divalent cations are involved in the formation of an assembly-competent intermediate of outer membrane protein PhoE of *E. coli. EMBO J.*, **15**, 5567.

96. Freudl, R., Schwarz, H., Stierhof, Y.-D. G. K., Hindennach, I., and Henning, U. (1986). An outer membrane protein (OmpA) of *Escherichia coli* K-12 undergoes a conformational change during export. *J. Biol. Chem.*, **261**, 11355.

97. Tommassen, J. and Lugtenberg, B. (1981). localization of *phoE*, the structural gene for outer membrane protein E in *Escherichia coli. J. Bacteriol.*, **147**, 118.

98. Ried, G., Hindennach, I., and Henning, U. (1990). Role of lipopolysaccharide in assembly of *Escherichia coli* outer membrane proteins OmpA, OmpC, and OmpF. *J. Bacteriol.*, **172**, 6048.

99. Bocquet-Pages, C., Lazdunski, C., and Lazdunski, A. (1981). Lipid-synthesis-dependent biosynthesis (or assembly) of major outer membrane proteins of *Escherichia coli. Eur. J. Biochem.*, **118**, 105.

100. Pages, C., Lazdunsk, C., and Lazdunski, A. (1982). The receptor of bacteriophage lambda: evidence for a biosynthesis dependent on lipid synthesis. *Eur. J. Biochem.*, **122**, 381.

101. Bolla, J.-M., Lazdunski, C., and Pages, J.-M. (1988). The assembly of the major outer membrane protein OmpF of *Escherichia coli* depends on lipid synthesis. *EMBO J.*, **7**, 3595.

102. Sen, K. and Nikaido, H. (1991). Lipopolysaccharide structure required for in vitro trimerization of *Escherichia coli* OmpF porin. *J. Bacteriol.*, **173**, 926.

103. De Cock, H., Schafer, U., Potgeter, M., Demel, R., Muller, M., and Tommassen, J. (1999). Affinity of the periplasmic chaperone Skp of Escherichia coli for phospholipids, lipopolysaccharides and non-native outer membrane proteins—role of Skp in the biogenesis of outer membrane protein. *Eur. J. Biochem.*, **259**, 96.

104. Chen, R. and Henning, U. (1996). A periplasmic protein (Skp) of Escherichia coli selectively binds a class of outer membrane proteins. *Mol. Microbiol.*, **19**, 1287.

105. Geyer, R., Galanos, C., Westphal, O., and Golecki, J. R. (1979). A lipopolysaccharide-binding cell-surface protein from *Salmonella minnesota. Eur. J. Biochem.*, **98**, 27.

106. Missiakas, D., Betton, J., and Raina, S. (1996). new components of protein folding in extracytoplasmic compartments of *Escherichia coli*: SurA, FkpA, Skp/OmpH. *Mol. Microbiol.*, **21**, 871.
107. Schäfer, U., Beck, K., and Müller, M. (1999). Skp, a molecular chaperone of Gram-negative bacteria, is required for the formation of soluble periplasmic intermediates of outer membrane proteins. *J. Biol. Chem.*, **274**, 24567.
108. Martin, J. L. (1995). Thioredoxin – a fold for all reasons. *Struct.*, **3**, 245.
109. Bardwell, J. C. A., McGovern, K., and Beckwith, J. (1991). Identification of a protein required for disulfide bond formation in vivo. *Cell*, **65**, 581.
110. Joly, J. C. and Schwartz, J. R. (1994). Protein folding activities of *Escherichia coli* protein disulfide isomerase. *Biochem.*, **33**, 4231.
111. Frech, C., Wunderlich, M., Glockshuber, R., and Schmid, F. X. (1996). Preferential binding of an unfolded protein to DsbA. *EMBO J.*, **15**, 392.
112. Zapun, A., Missiakas, D., Raina, S., and Creighton, T. E. (1993). The reactive and de-stabilizing disulfide bond of DsbA, a protein required for protein disulfide bond formation *in vivo*. *Biochem.*, **32**, 5983.
113. Guddat, L. W., Bardwell, J. C. A., and Martin, J. L. (1998). Crystal structures of reduced and oxidized DsbA: investigation of domain motion and thiolate stabilization. *Struct.*, **6**, 757.
114. Wunderlich, M., Otto, A., Maskos, K., Mucke, M., Seckler, R., and Glockshuber, R. (1995). Efficient catalysis of disulfide formation during protein folding with a single active-site cysteine. *J. Mol. Biol.*, **247**, 28.
115. Missiakas, D., Georgopoulos, C., and Raina, S. (1993). Identification and characterization of the *Escherichia coli dsbB*, whose product is involved in the formation of disulfide bonds *in vivo*. *Proc. Natl. Acad. Sci. USA*, **90**, 7084.
116. Kishigami, S., Akiyama, Y., and Ito, K. (1995). Redox states of DsbA in the periplasm of *Escherichia coli*. *FEBS Lett.*, **364**, 55.
117. Jander, G., Martin, N. L., and Beckwith, J. (1994). Two cysteines in each periplasmic domain of the membrane protein DsbB are required for its function in protein disulfide bond formation. *EMBO J.*, **13**, 5121.
118. Guilhot, R., Jander, G., Martin, N. L., and Beckwith, J. (1995). Evidence that the pathway of disulfide bond formation in *Escherichia coli* involves interactions between cysteines of DsbB and DsbA. *Proc. Natl. Acad. Sci. USA*, **92**, 9895.
119. Kobayashi, T. and Ito, K. (1999). Respiratory chain strongly oxidizes the CXXC motif of DsbB in the *Escherichia coli* disulfide bond formation pathway. *EMBO J.*, **18**, 1192.
120. Missiakas, D., Georgopoulos, C., and Raina, S. (1994). The *Escherichia coli dsbC(xprA)* gene encodes a periplasmic protein involved in disulfide bond formation. *EMBO J.*, **13**, 2013.
121. Rietsch, A., Belin, D., Martin, N., and Beckwith, J. (1996). An *in vivo* pathway for disulfide bond isomerization in *Escherichia coli*. *Proc. Natl. Acad. Sci. USA*, **93**, 13048.
122. Sone, M., Akiyama, Y., and Ito, K. (1998). Differential in vivo roles played by DsbA and DsbC in the formation of protein disulfide bonds. (Vol. **272**, p. 10349, 1997). *J. Biol. Chem.*, **273**, 27756.
123. Zapun, A., Missiakas, D., Raina, S., and Creighton, T. E. (1995). Structural and functional; characterization of DsbC, a protein involved in disulfide bond formation in *Escherichia coli*. *Biochem.*, **34**, 5075.
124. Sambongi, Y. and Ferguson, S. J. (1994). Specific thiol components complement deficiency in c-type cytochrome biogenesis in *Escherichia coli* carrying a mutation in membrane bound disulphide isomerase-like protein. *FEBS Lett.*, **353**, 235.

125. Missiakas, D., Schwager, F., and Raina, S. (1995). Identification and characterization of a new disulfide isomerase-like protein (DsbD) in *Escherichia coli*. *EMBO J.*, **14**, 3415.

126. Crooke, H. and Cole, J. (1995). The biogenesis of c-type cytochrome in *Escherichia coli* requires a membrane-bound protein, DipZ, with a protein disulphide isomerase-like domain. *Mol. Microbiol.*, **15**, 1139.

127. Rietsch, A., Bessette, P., Georgiou, G., and Beckwith, J. (1997). Reduction of the periplasmic disulfide bond isomerase, DsbC, occurs by passage of electrons from cytoplasmic thioredoxin. *J. Bacteriol.*, **179**, 6602.

128. Fabianek, R. A., Hennecke, H., and Thony-Meyer, L. (1998). The active-site cysteines of the periplasmic thioredoxin-like protein CcmG of *Escherichia coli* are important but not essential for cytochrome c maturation *in vivo*. *J. Bacteriol.*, **180**, 1947.

129. Raina, S. and Missiakas, D. (1997). Making and breaking disulfide bonds. *Annu. Rev. Microbiol.*, **51**, 179.

130. Andersen, C. L., Matthey-Dupraz, A., Missiakas, D., and Raina, S. (1997). A new *Escherichia coli* gene, *dsbG*, encodes a periplasmic protein involved in disulphide bond formation, required for recycling DsbA / DsbB and DsbC redox proteins. *Mol. Microbiol.*, **26**, 121.

131. Besette, P. H., Cotto, J. J., Gilbert, H. F., and Georgiou, G. (1999). *In vivo* and *in vitro* function of the *Escherichia coli* periplasmic cysteine oxidoreductase DsbG. *J. Biol. Chem.*, **274**, 7774.

132. Göthel, S. F. and Marahiel, M. A. (1999). Peptidyl-prolyl cis-trans isomerases, a superfamily of ubiquitous folding catalysts. *CMLS Cell. and Mol. Life Sci.*, **55**, 423.

133. Tormo, A., Almiron, M., and Kolter, R. (1990). *surA*, an *Escherichia coli* gene essential for survival in stationary phase. *J. Bacteriol.*, **172**, 4339.

134. Lazar, S. W. and Kolter, R. (1996). SurA assists the folding of Escherichia coli outer membrane proteins. *J. Bacteriol.*, **178**, 1770.

135. Dartigalongue, C. and Raina, S. (1998). A new heat-shock gene, ppiD, encodes a peptidyl prolyl isomerase required for folding of outer membrane proteins in *Escherichia coli*. *EMBO J.*, **17**, 3968.

136. Matsuyama, S., Tajima, T., and Tokuda, H. (1995). A novel periplasmic carrier protein involved in the sorting and transport of Escherichia coli lipoproteins destined for the outer membrane. *EMBO J.*, **14**, 3365.

137. Yokota, N., Kuroda, T., Matsuyama, A., and Tokuda, H. (1999). Characterization of the LolA-lolB system as the general lipoprotein localization mechanism in *Escherichia coli*. *J. Biol. Chem.*, **274**, 30995.

138. Yakushi, T., Yokota, N., Matsuyama, S., and Tokuda, H. (1998). LolA-dependent release of a lipid-modified protein from the inner membrane of Escherichia coli requires nucleoside triphosphate. *J. Biol. Chem.*, **273**, 32576.

139. Matsuyama, S., Yokota, N., and Tokuda, H. (1997). A novel outer membrane lipoprotein, LolB (HemM), involved in the LolA (p20)-dependent localization of lipoproteins to the outer membrane of *Escherichia coli*. *EMBO J.*, **16**, 6947.

140. Soto, G. E. and Hultgren, S. J. (1999). Bacterial adhesins: common themes and variations in architecture and assembly. *J. Bacteriol.*, **181**, 1059.

141. Holmgren, A. and Bränden, C.-I. (1989). Crystal structure of chaperone protein PapD reveals an immunoglobulin fold. *Nature (London)*, **342**, 248.

142. Soto, G. E., Dodson, K. W., Ogg, D., Liu, C., Heuser, J., Knight, S., *et al.* (1998). Periplasmic chaperone recognition motif of subunits mediates quaternary interactions in the pilus. *EMBO J.*, **17**, 6155.

143. Sauer, F. G., Fütterer, K., Pinkner, J. S., Dodson, K. W., Hultgren, S. J., and Waksman, G. (1999). Structural basis of chaperone function and pilu biogenesis. *Science*, **285**, 1058.
144. Choudhury, D., Thompson, A., Stojanoff, V., Langermann, S., Pinkner, J., Hultgren, S. J., and Knight, S. D. (1999). X-ray structure of the FimC-FimH chaperone-adhesin complex from uropathogenic *Escherichia coli*. *Science*, **285**, 1061.
145. Mol, O., Visschers, R. W., Degraaf, F. K., and Oudega, B. (1994). Escherichia coli periplasmic chaperone FaeE is a homodimer and the chaperone-K88 subunit complex is a heterotrimer. *Mol. Microbiol.*, **11**, 391.
146. Hung, D. L., Pinkner, J. S., Knight, S. D., and Hultgren, S. J. (1999). Structural basis of chaperone self-capping in P Pilus biogenesis. *Proc. Natl. Acad. Sci. USA*, **96**, 8178.
147. Berks, B. C., Sargent, F., and Palmer, T. (2000). The Tat protein export pathway. *Mol. Microbiol.*, **35**, 260.

3 | The role of chaperone proteins in the import and assembly of proteins in mitochondria and chloroplasts

WOLFGANG VOOS and NIKOLAUS PFANNER

1. Introduction

Mitochondria and chloroplasts are the two major organelles of eukaryotic cells that have been generated by the endosymbiotic uptake of prokaryotic cells during the early stages of eukaryote evolution (1). Both perform crucial functions mainly in the energy metabolism of the host cells. During the million of years of mutual inter-dependence with the host cell, both chloroplasts and mitochondria have lost their ability to synthesize most of their endogenous proteins. Also during this process, most of their genetic information was transferred to the nucleus. Concomitantly, mechanisms had to evolve that allow a specific transport of proteins from the cytosol to their organellar destination. Due to the endosymbiotic origin, mitochondria and chloroplasts show a special arrangement of several membranes, resulting in multiple subcompartments. Apart from the general problem of how large polypeptides are transferred across biomembranes per se, the particular organellar structure exacer-bates the problem of specific preprotein targeting.

It is assumed that mitochondria and chloroplasts have been generated by two independent endosymbiotic uptake events. Despite the various similarities, their machineries responsible for uptake of proteins turned out to be remarkably different. Most proteins involved in key steps of recognition and membrane translocation show no homology between mitochondria and chloroplasts. However, a recurring motif of all protein membrane translocation systems, including the endoplasmic reticulum, is the crucial function of heat shock proteins (Hsps). Typical eukaryotic cells contain numerous types of Hsps. The families involved in organellar biogenesis are mainly

Hsp70, Hsp60, and Hsp100/Clp. Additionally, there are several types of co-chaperones, stimulating and/or regulating Hsp activity. Co-chaperones for Hsp70 belong to the GrpE and DnaJ families, named after the well studied homologues in *Escherichia coli*. Hsp60 activity is regulated by the co-chaperone Hsp10. Apart from the different types of Hsps, for each family several members can exist in a single cell, including specific co-chaperones for each member. For example, in *Saccharomyces cerevisiae*, 14 members of the Hsp70 and 21 members of the DnaJ co-chaperone families have been identified. One reason for this variety is the compartmentalization of a eukaryotic cell. Each compartment has its own set of Hsps. Mitochondria contain three Hsp70s, one Hsp60, and at least two members of the Hsp100/Clp family. Chloroplasts have three or four types of Hsp70, two members of the Hsp60 and at least one of the Hsp100/Clp family. Since polypeptide transport across membranes requires multiple folding and unfolding processes, Hsps are mostly found as soluble components of the translocation machineries. Recent results even indicate that in addition to promoting protein folding, Hsps also have crucial functions in various cellular protein translocation processes.

2. Mitochondrial biogenesis

2.1 Protein membrane translocation

The vast majority of mitochondrial proteins are synthesized as precursor proteins at cytosolic ribosomes. The genetic information for only a handful of mainly very hydrophobic proteins of the inner membrane has been retained inside mitochondria. It is generally assumed that preproteins destined for mitochondria are imported after their synthesis is completed. Having a double membrane system, at least four compartments can be distinguished in mitochondria, each containing its specific set of protein components. For each submitochondrial compartment specific targeting or sorting pathways are required. A model for the general pathway of matrix targeted preproteins is shown in Fig. 1.

2.1.1 Preproteins: Targeting signals and import competence

Specific targeting of mitochondrial preproteins is in many cases accomplished by an amino-terminal sequence extension, the presequence (2). This targeting sequence typically forms an amphipathic α-helix with a positively charged and a hydrophobic surface. In most cases, the presence of the targeting sequence inhibits the folding of the precursor to the native and enzymatically active conformation. After entering the mitochondrial matrix, the amino-terminal signal sequence is removed by a specific metallo-protease. Other mitochondrial preproteins, mainly from the conserved class of metabolite carrier proteins, lack an amino-terminal signal sequence but contain internal targeting information. Proteins of this family are highly hydrophobic membrane proteins integrated into the inner mitochondrial membrane and follow a different import pathway from matrix targeted preproteins.

Which cytosolic components stabilize the import-competent conformation of pre-

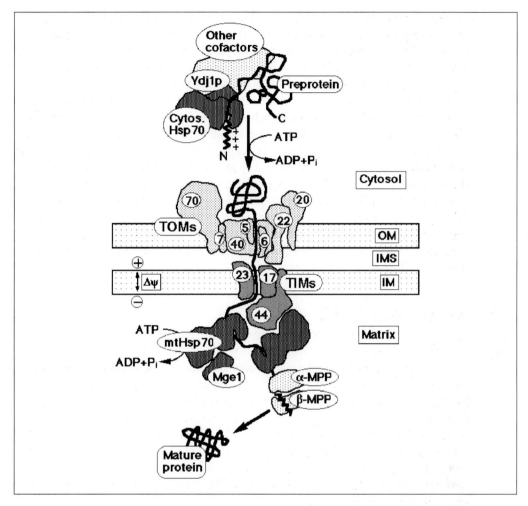

Fig. 1 Model of the mitochondrial preprotein translocation machineries. After synthesis in the cytosol, a mitochondrial precursor protein (black line) is targeted to mitochondria. The amino terminal presequence (zig-zag line) is recognized by receptors and translocated across the outer membrane (OM). Receptors and membrane channel components form the outer membrane translocase (TOM, yellow). After crossing the intermembrane space (IMS), the presequence inserts into the inner membrane (IM) driven by the mitochondrial membrane potential (Δv). The bulk polypeptide chain is transported across the inner membrane with the help of the inner membrane translocase (TIM, green). The driving force is supplied by the mtHsp70 system in the matrix. Chaperone proteins and their co-factors are coloured red. After removal of the presequence by a specific processing peptidase (blue), the preprotein can fold to the mature conformation. See colour plate section?

proteins and assist targeting to mitochondria is not yet established completely. Depending on the properties of the precursor, several cytosolic factors may play a role. As early as during synthesis, ribosome associated proteins like the nascent polypeptide-associated complex (NAC) may provide a environment for proper folding and presentation of the targeting information (3, 4). Cytosolic chaperones, mainly Hsps of the 70 kDa-family (Hsp70), have been implicated in maintaining import competence by stabilizing the partially folded or unfolded conformation of precursors (5–9). In that respect, cytosolic Hsp70s perform a rather general function, which is not specific for mitochondrial protein import but is also involved in import into the endoplasmic reticulum (10) and probably also the chloroplast (see below).

The functional connection of cytosolic Hsp70s with protein membrane transloca-tion processes was indirectly confirmed by the identification of a co-chaperone of the DnaJ-type, Hsp40 or Ydj1p in yeast, which is involved in protein import into the ER and mitochondria. Ydj1p is localized in the cytosol, but is also found associated with intracellular membranes due to addition of a farnesyl group (11, 12). As was ex-tensively studied in prokaryotes (13), co-chaperones of the DnaJ-type stimulate the ATPase activity of their corresponding Hsp70 and also stimulate interaction with unfolded substrate proteins (14). Indeed, conditional mutants of *YDJ1* were defective for the import of several mitochondrial and ER preproteins (12). In mammalian cells, the mitochondrial import stimulation factor (MSF) has been identified (15, 16), which may have a chaperone-like activity. It is mainly implicated in the import of hydro-phobic proteins via the receptor Tom70 (see below).

The main role of chaperones in preprotein targeting seems to be to prevent irregular reactions like misfolding and/or aggregation directly after synthesis by direct interaction with the polypeptide chain. Removal of bound chaperones before insertion into the outer membrane is assumed to be the major cause for the ATP-requirement of the import reaction on the outside of the mitochondria. However, a possible role of the cytosolic cofactors in specific targeting to mitochondria is still unclear since chemically pure preproteins can be specifically imported into mito-chondria in the absence of any soluble factor (17).

2.1.2 Translocase complexes in outer and inner membrane

In the outer mitochondrial membrane a dedicated machinery has been identified that is responsible for preprotein recognition and insertion into the outer membrane (18). This machinery consists of receptor proteins and a hetero-oligomeric protein complex of 400 kDa that forms a protein conducting channel through the outer membrane (TOM; translocase of the **o**uter **m**embrane). There are two types of receptor proteins, both anchored in the outer membrane but exposing their major portion to the cytosol. Tom20 mainly recognizes precursors with amino-terminal presequences and Tom70 those with internal targeting information. The receptor Tom70 associates primarily with preproteins requiring cytosolic chaperones for import and may even have a chaperone function by itself. Subsequent to the initial recognition, the preproteins are inserted into the membrane channel. The core component of the TOM complex is Tom40, which forms the protein conducting channel (19, 20). Further components are Tom22, Tom5, Tom6, and Tom7. Tom22 and Tom5 have also receptor function and Tom6 and Tom7 regulate the activity of the receptor complex. Tom40 and Tom22 additionally expose sequences to the intermembrane space that have been implicated in formation of *trans* binding sites on the inner face of the outer membrane. It has been proposed that the driving force for outer membrane polypeptide translocation is generated by a sequential binding of the positively charged presequence with rising affinity to components of the TOM complex and the outer face of the inner membrane ('Acid-chain' hypothesis) (21, 22).

After the outer membrane translocase has mediated the translocation of the amino-terminal moiety of the preprotein, the presequence interacts with translocase components in the inner membrane, thereby functionally coupling the outer and the inner membrane translocase. The core components of the inner membrane translocase are Tim23 and Tim17, both integral membrane proteins with several membrane-spanning α-helices (23). Tim23 and Tim17 are the major components of a translocase complex of 90 kDa (TIM23 complex). Although direct evidence is missing, the biochemical characteristics of Tim23 and Tim17 indicate that the two (and maybe other) proteins form a translocation channel across the inner membrane.

Recent results have established a separate import machinery in the inner membrane dedicated to the import of proteins belonging to the metabolite carrier family (24). After crossing the outer membrane, metabolite carrier proteins and other hydrophobic proteins interact with several small soluble proteins, Tim8, Tim9, Tim10, and Tim13 in the intermembrane space (IMS). After this step, they are handed over to a specific inner membrane translocase, the TIM22 complex, which comprises at least two integral membrane proteins, Tim22 and Tim54, and the soluble IMS protein Tim12. The TIM22 complex then catalyses insertion of the metabolite carrier proteins into the inner membrane.

2.1.3 Membrane potential ($\Delta\psi$)

A functional membrane potential was established early as a prerequisite for successful preprotein translocation for both matrix and inner membrane proteins (25). The membrane potential, positive outside, negative inside, is thought to drive the movement of positively charged sequence patches, like mitochondrial presequences, across the inner membrane by an electrophoretic mechanism (26). In addition to this direct action, it was proposed that the membrane potential might modulate the activity of the inner membrane channel complex via Tim23, thereby facilitating preprotein insertion into the inner membrane (27).

2.2 Function of mtHsp70 in driving membrane translocation

After the amino-terminal part of a preprotein has reached the inner face of the inner membrane driven by the membrane potential, the subsequent translocation of the polypeptide chain is driven by the mitochondrial Hsp 70 (mtHsp70). The absolute requirement for mtHsp70 function during preprotein translocation was already indicated by the essential nature of the mtHsp70 gene (*SSC1*) in yeast (28). Analysis of conditional mutants then demonstrated that the presequence of preproteins reaches the matrix under nonpermissive conditions but further translocation is blocked and the preproteins become stuck as membrane spanning intermediates (29, 30). Various techniques showed that mtHsp70 binds directly to the incoming polypeptide chain (30, 31). This binding reaction is promoted by the unfolded state of the precursor polypeptide chain during translocation (32). If the binding to the preprotein in transit is disturbed, as in mtHsp70 mutant mitochondria, polypeptide translocation will be aborted completely, except for the membrane potential dependent step of pre-sequence transport (30, 33).

2.2.1   Tim44 as membrane anchor

Binding to the preprotein, however, is not sufficient to explain mtHsp70 action completely. By a genetic selection searching for components of the mitochondrial import apparatus, another essential gene was identified that encodes a protein associated with the inner membrane, called Tim44 (34). At the same time, Tim44 was identified by a biochemical approach screening an antibody library against IMPs for influence on import reaction (35, 36). Tim44 is in the direct vicinity of the preprotein in transit, as was shown by cross-linking experiments. Surprisingly, it turned out that over-expression of mtHsp70 can rescue *tim44* mutants in yeast. This genetic inter-action was confirmed by biochemical studies showing a direct nucleotide-dependent interaction of mtHsp70 with Tim44 (37, 38). The model of Tim44/mtHsp70 function was based on three results: the binding of mtHsp70 to the preprotein is required for translocation; Tim44 is in the direct neighbourhood of the preprotein in transit; and mtHsp70 transiently interacts with Tim44. It was concluded that Tim44 acts as a membrane anchor directing the soluble mtHsp70 to the inner membrane import site and assists its binding to the incoming polypeptide chain.

In addition to the interaction with Tim44, mtHsp70 was also found to interact with the core components of the inner membrane translocase itself (39). mtHsp70 over-expression can suppress the growth defect of conditional mutants of Tim17, and co-precipitation experiments revealed a binding of mtHsp70 to the complex formed by Tim23 and Tim17. The interaction of mtHsp70 with the core translocase is independ-ent of the presence of Tim44. The functional implication of this interaction is not yet clear, but it is speculated that mtHsp70 might have an additional role in regulating the activity of the import channel itself.

2.2.2   Driving force for polypeptide translocation

The discovery that mtHsp70 is interacting with a protein closely associated with the inner membrane translocase had profound implications for the mechanism providing the driving forces for polypeptide membrane translocation. mtHsp70 function is critical for providing the driving force since perturbation of mtHsp70 function results in accumulation of membrane spanning translocation intermediates. Before having identified mtHsp70 as a component of the translocation machinery, it was already established that matrix ATP is required for protein import. ATP depletion and mutations in mtHsp70 that abolish ATPase activity lead to a similar transloca-tion defect (30, 33). Indeed, mtHsp70 is the only ATPase identified in mitochondrial import and therefore the major source of translocation energy.

How does mtHsp70 accomplish the movement of the polypeptide chain in transit? One of the earlier models stated that the movement of the polypeptide in the channel is generated by random Brownian motion. Binding of mtHsp70 inside the mito-chondria to the precursor protein would make the movement vectorial by preventing the backward movement. Several cycles of binding, random movement, and release result in a net transport, a model for mtHsp70 function that was summarized by the expressions 'brownian ratchet' model or 'trapping' model (38, 40). This model was then extended by the Tim44–mtHsp70 interaction, stating that by binding to Tim44, more mtHsp70 becomes available directly at the import site, increasing the efficiency of preprotein binding. Additionally, Tim44 might regulate the ATPase activity so that mtHsp70 at the import site is in its high-affinity state for substrate interaction.

However, several observations subsequently required an extension of this model. Based firstly on theoretical considerations, a more active role of mtHsp70 in translocation was proposed (41). The combination of three properties of mtHsp70, binding of unfolded preproteins (29, 30), ATP-dependent conformational change (42), and anchoring at the inner membrane (37, 38, 43, 44), pointed to a mechanism that results in a force generation on the preprotein in the channel. In this model, mtHsp70 interacts with Tim44, then binds the incoming preprotein and undergoes an ATP-regulated conformational change, pulling some residues of the preprotein into the matrix. The subsequent release of mtHsp70 from Tim44 and the preprotein, and a repetition of the whole cycle would result in polypeptide transport and unfolding of the polypeptide on the outside. This model for the mechanism of mtHsp70 was named the 'active motor' model or 'pulling' model. This 'pulling' model is supported by an extensive analysis of a set of conditional mutants in mtHsp70 (44, 45). To summarize these data, mutants that retain preprotein binding or even show increased binding to preproteins in transit are not capable of importing precursors efficiently that have a conformational restriction, like a folded carboxy-terminal domain. These mutants also do not show the generation of an inward directed force, which is required for unfolding of preproteins during translocation. Hence, the 'trapping' mechanism alone is not sufficient to describe the full mtHsp70 function. Since these mutants are also defective *in vivo*, it can be inferred that mtHsp70 also has this active translocation function in a living cell. Additionally, even without severe conformational restrictions, the interaction of the unfolded polypeptide chain with the components of the translocation channel might be too strong to allow movement without the generation of a force on the preprotein (46). Indeed it could be shown that a protein complex consisting of a polypeptide arrested in transit, the outer membrane translocase (TOM) and the inner membrane translocase (TIM23 complex) is stable enough even in the absence of mtHsp70 to survive native electrophoresis (47). In a unified model, both mechanisms contribute to polypeptide movement. In case of conformational restrictions of the precursor, which may be the case for most precursors *in vivo*, an active role of mtHsp70 is required (44, 45).

2.2.3 Mge1p as nucleotide exchange factor for mtHsp70

The matrix protein Mge1p is the yeast homologue of the Hsp70 co-chaperone GrpE of *E. coli*. It was initially identified by its highly conserved sequence homology to its bacterial relative. A first analysis showed that a null mutant is not viable. Again, this indicates that Mge1p is involved in a key step of mitochondrial biogenesis (48, 49). The function of Mge1p in protein translocation was confirmed by experiments using a conditional mutant of Mge1p, which accumulates uncleaved precursor proteins in the cytosol *in vivo* (49). In addition, it was shown that Mge1p interacts via mtHsp70 with translocation intermediates spanning the inner membrane (50). This interaction is specific for the translocation process, since no interaction with Mge1p could be detected with fully imported proteins. Other experiments using conditional mutants of Mge1p showed decreased binding of mtHsp70 to incoming precursor proteins and reduced ATP-dependent dissociation of mtHsp70 from Tim44 (51, 52). This lead to the conclusion that Mge1p function is necessary at an intermediate import stage for the activity of mtHsp70, resulting in the transfer of the polypeptide chain across the inner membrane.

It was shown that Mge1p specifically binds to mtHsp70 (50, 53, 54). The complex is very stable but dissociated by ATP. Non-hydrolyzable nucleotides are not effective in releasing Mge1p from mtHsp70. Mge1p is also found as a component of the translocase motor complex at the inner face of the inner membrane which is formed by mtHsp70 bound to Tim44 (42). A protruding loop of the ATPase domain of mtHsp70 is involved in the interaction site between Mge1p and mtHsp70 (55, 56). Based on experiments with purified Mge1p and mtHsp70 it was concluded that the biochemical function of Mge1p is as a nucleotide exchange factor for mtHsp70. Mge1p specifically increases the rate of ADP release from mtHsp70, thereby enhancing the intrinsic ATPase activity (55, 57) similar to its prokaryotic counterpart GrpE. Stimulating release of ADP (and thereby binding of ATP) may be especially important during polypeptide translocation reactions by increasing the amount of functional mtHsp70 at the translocation site.

The details of the reaction cycle of mtHsp70 and the role of ATP binding/hydrolysis at the import site still have to be established. Generally, the ATP state of Hsp70 has high affinity to substrates, but also a high off rate, whereas the ADP state has a higher binding stability (58). These features of substrate binding, ATP-binding, and regulation by Mge1p (GrpE) seem to be conserved between mtHsp70 and the closely related DnaK protein from *E. coli*. The importance of other specific binding partners like Tim44 indicate possible differences to the reaction cycle. Although it was shown that the interaction of mtHsp70 with Tim44 is ATP-dependent, the actual function of Tim44 in the reaction cycle of mtHsp70 is not yet known. The stable complex of Tim44 and mtHsp70 contains mainly ADP (42), and addition of ATP or non-hydrolyzable analogues dissociates the complex (42, 59). In the presence of ADP, mtHsp70 can bind to Tim44 and to a peptide substrate simultaneously, although the binding properties to substrate and Tim44 are clearly distinct (44). One can conclude that the role of mtHsp70 in protein import is clearly distinct from the role of DnaK in *E. coli*. Hence, it can be expected that some biochemical properties of mtHsp70 are also unique. One indication already is that Ssc1p (mtHsp70 from yeast) is not able to functionally complement a DnaK mutant of *E. coli* (60).

In summary, this system, which is made up of the three proteins Tim44, mtHsp70, and Mge1p, is the central ATP-consuming motor complex responsible for translocation of the bulk polypeptide chain through the import channels and for unfolding of the carboxy-terminal part of the precursor on the cytosolic side of the mitochondria.

2.3 Preprotein folding in the mitochondrial matrix

2.3.1 mtHsp70: coupling between translocation and folding process

Apart from its essential role in preprotein translocation, mtHsp70 also has a more traditional chaperone function. Similar to its prokaryotic counterpart DnaK, it is involved in protein folding processes in the mitochondrial matrix. After the crucial interaction with mtHsp70 during import preproteins are released mostly as unfolded proteins (61). Only then can subsequent folding and assembly reaction take place. Depending on their biochemical properties, the newly imported proteins can interact with different types of mitochondrial chaperones, most prominently Hsp60 or again mtHsp70. Since the mtHsp70 molecules that are involved in translocation per se are only a small minority, unfolded proteins are likely to interact with soluble mtHsp70 in the matrix. Indeed two independent protein complexes can be identified in mitochondria. One at the inner membrane, and a second, soluble complex (62). Both are also distinct in the type of co-chaperones involved in mtHsp70 function. The soluble complex, but not the membrane-bound complex, contains Mdj1p, the genuine homologue of the *E. coli* co-chaperone DnaJ. The ATPase activity of mtHsp70 is regulated by Mge1p in both cases.

2.3.2 Mdj1p/Mdj2p: co-chaperones of mtHsp70

Deletion of Mdj1p in yeast results in a petite phenotype (growth defect on non-fermentable carbon sources) due to loss of mitochondrial DNA and growth inhibition at elevated temperature (37°C). This clearly demonstrates a severely compromised mitochondrial metabolism. However, the phenotype of *mdj1* mutations is not lethal as compared with null mutants of *SSC1* or *MGE1*, indicating that Mdj1p might not be involved in a crucial step of protein translocation. Indeed the mutant shows normal protein translocation efficiency but folding defects of the newly imported preproteins (63). Analysis of the aggregation behaviour of the reporter protein firefly luciferase *in vitro* and *in organello* lead to the hypothesis that Mdj1p plays a major role in prevention of thermal damage to mitochondria (64, 65). Similar to its prokaryotic homologue DnaJ, Mdj1p is also able to stimulate the ATPase activity of mtHsp70 in cooperation with Mge1p to a high extent. Mdj1p and mtHsp70 were found associated with nascent polypeptide chains of mitochondrial ribosomes and the *mdj1* mutant mitochondria show increased aggregation of a mitochondrially encoded protein named var1. It was concluded that Mdj1p is involved in the folding of the proteins synthesized by mitochondria themselves (66). Taken together, these experiments demonstrate Mdj1p as a co-chaperone for mtHsp70. It is involved in folding of newly synthesized or imported mitochondrial proteins and in protecting mitochondrial preproteins from thermal denaturation.

Recently, a second mitochondrial homologue of DnaJ was identified. This protein, named Mdj2p, shares 30% identity in its J-homologue domain, but in contrast to Mdj1p it has an amino-terminal transmembrane anchor that directs this protein to the inner membrane. Since a deletion mutant of Mdj2p does not show any growth phenotype, not much is known about its molecular function, except that Mdj2p might be involved in protection against thermal damage together with Mdj1p (67).

2.3.3 Other mitochondrial Hsp70s

After completion of the yeast genome sequencing project, two further Hsp70 genes with a potential mitochondrial presequence and hence mitochondrial localization were identified. Sequence homology of these new members of the yeast Hsp70 family to Ssc1p is quite pronounced. No data are available in the literature about the closest homologue of Ssc1p, encoded by the open reading frame YEL030w. Except for the extreme amino-terminal part, its amino acid sequence is virtually identical to mtHsp70. The other homologue, eventually named Ssq1p, is 47% identical to mtHsp70 and also localized in the mitochondrial matrix. Null mutations result in a cold-sensitive growth defect but show normal rates of preprotein import (68). It was found that Ssq1p plays a critical role in maintaining mitochondrial iron homeostasis (69). *SSQ1* mutants accumulate iron inside the mitochondria. Processing of yeast frataxin, a protein implicated in mitochondrial iron metabolism, is also impeded. The actual cellular function of Ssq1p and its relation to the major Hsp70 remains to be established.

2.3.4 Hsp60/Hsp10: high molecular weight 'folding machinery'

The oligomeric chaperone Hsp60 performs a prominent role in mitochondrial biogenesis and is involved in folding and assembly of imported proteins in yeast (70). Newly imported proteins can be found in association with the Hsp60 complex (71). The bound proteins are released in an ATP-dependent reaction regulated by the co-chaperone Hsp10 (72, 73). With the high similarity of Hsp60 to its prokaryotic homologue, GroEL, it is generally assumed that mitochondrial Hsp60 works by a principally similar mechanism (74). In short, unfolded proteins can be bound inside the central cavity of the large double-ring shaped Hsp60 14mer. Binding of ATP and hydrolysis induces a large conformational shift, changing the properties of the binding sites and opening up the cavity. The substrate will be eventually released in the folded conformation, possibly after several cycles of binding and release. The main function of Hsp60 is to provide a protected environment for proteins during the folding process in order to avoid unproductive reactions with other proteins. Recent work established that in contrast to GroEL from *E. coli* mitochondrial Hsp60 forms a single-ring structure active in protein folding, thereby utilizing a somewhat simpler mechanism (75, 76). In mitochondria there is a clear directionality of the folding reaction. First mtHsp70 interacts with preproteins that are transferred as unfolded substrate proteins to Hsp60 (77). However, protein folding in the matrix is a very complex process (Fig. 2). Not all imported proteins require the chaperone function of Hsp60, and even concerning preproteins associated with the chaperonin complex, the degree of folding catalysis or the dependence on Hsp10 varies substantially (78, 79). Additionally, other folding catalysts, like the prolyl-isomerase cyclophilin play a major role in mitochondria (80). Cyclophilin and Hsp60 have been shown to act sequentially during the folding process and complement each other to obtain maximum folding efficiency (81).

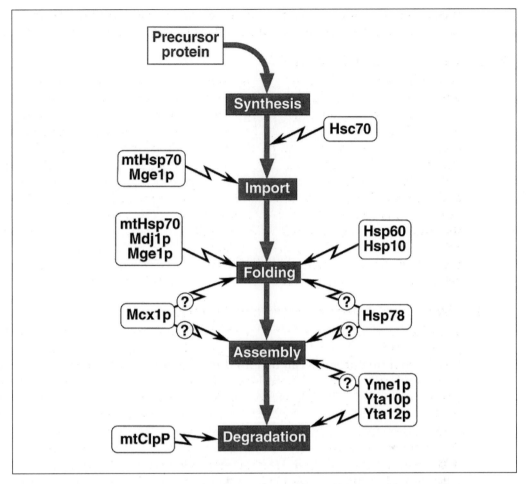

Fig. 2 Chaperone proteins function in several key steps of mitochondrial biogenesis. Preproteins are synthesized, imported into mitochondria, folded, assembled into protein complexes, and eventually degraded (grey). Chaperone proteins of different structural classes are required at each step (black arrows).

2.4 Protein complex assembly and degradative processes

Recent results have established that the different chaperone proteins form an extensive functional network, reaching from the biosynthesis of a protein at the ribosome via intracellular transport and assembly processes to its eventual proteolytic degradation (Fig. 2). Depending on the metabolic state of the cell and their biochemical properties, organellar proteins will fold, unfold, aggregate, and/or be degraded. Chaperone proteins of all major classes are involved in these processes, positively cooperating which each other (58). The interplay between chaperone action in folding and unfolding of proteins and degradative processes also becomes apparent in mitochondria (82).

2.4.1 Hsp78/Mcx1p: homologues of bacterial Clp family of chaperones

A family of closely related proteins has been identified in *E. coli* as regulatory components of the ClpP serine protease (**c**aseino**l**ytic **p**rotease). These Clp-ATPases (A, B, X, and Y) are not only involved in protein degradation but have also chaperone function independent of the protease activity (83). In the meantime, this family has become very extended, with homologues in all classes of organisms (84). Clp-ATPases contain two conserved nucleotide-binding motifs and also a so called PDZ-domain implicated in protein–protein interactions (85). In *E. coli* at least ClpA and ClpX have been shown to form a high molecular weight complex with the proteolytically active component ClpP. ClpA/X and ATP-hydrolysis are required for proteolysis of larger proteins by ClpP (84). In contrast to eubacteria, where the function of Clp-like proteins is closely coupled to the function of the protease ClpP for protein degradation, no Clp-type protease was discovered in chloroplasts or mitochondria of lower eukaryotes. This is indicative of a separate function of the organellar Clp proteins as genuine chaperones in their own right. Indeed, recent *in vitro* experiments demonstrate a possible function of Clp proteins in suppressing protein aggregation (86, 87).

ClpA, ClpB, and ClpX have all been shown to be protein-activated ATPases. So far two members of the Clp-ATPase protein family have been identified in mitochondria. Hsp78 was identified as a soluble matrix protein with a high sequence homology to the ClpB proteins (88). Its expression is tightly regulated in response to heat stress, but surprisingly, null mutants grow quite normally under all conditions. It could be shown, that under circumstances where mtHsp70 function is ineffective, Hsp78 might have a chaperone function on its own (89). However, Hsp78 cannot substitute for mtHsp70 function but only alleviate some defects of conditional mtHsp70 mutants. Based on these data, it was proposed that its function becomes necessary for preprotein translocation only under circumstances where the function of mtHsp70 is compromised (89–91). Concerning its cellular function, it is rather likely that Hsp78 has an important role for maintenance of mitochondrial function by protecting thermo-sensitive proteins from inactivation (90). Additional evidence comes from the behaviour of the cytosolic ClpB homologue of yeast, Hsp104, which has been shown to dissociate insoluble proteins (92). Hsp78 is able to substitute for Hsp104 action in the cytosol, suggesting a conserved activity between both proteins (91).

Based on sequence homology, mitochondrial homologues of the ClpX family were recently identified in yeast (Mcx1p) and mouse (93, 94). Apart from having ATPase activity no functional data about the ClpX homologues are available.

Although no homologue for the bacterial protease ClpP could be identified in yeast, higher eukaryotes apparently contain a ClpP protein in mitochondria (95). In humans, the mitochondrial localization of the ClpP homologue was clearly demonstrated *in vivo* and *in vitro* (96). A recent biochemical characterization showed that the mitochondrial ClpP forms a double ring system with a sevenfold symmetry similar to *E. coli* ClpP (97). The function of ClpP in mitochondria remains to be established.

2.4.2 Chaperone functions involved in protein assembly and degradation

Three proteins have been identified in yeast mitochondria that are involved in proteolytic processes but also exhibit chaperone activities. They all belong to the AAA protein family (**A**TPases **a**ssociated with diverse cellular **a**ctivities). Yme1p, Yta10p, and Yta12p are members of large protein complexes in the mitochondrial inner membrane and seem to participate in the degradation of abnormal or unassembled membrane anchored proteins in the inner membrane (98–101). Biogenesis of inner membrane respiratory complexes is dependent on the activity of these proteins (102). Interestingly, proteolytically inactive mutants of Yta10p and Yta12p still show activity in the assembly of respiratory protein complexes (103–105). This distinct chaperone activity seems to be correlated with their ability to specifically bind non-native proteins.

Consistent with the role of mtHsp70 in protein folding, it was demonstrated that it is also involved in folding and assembly of at least part of the small set of proteins synthesized by mitochondria themselves (106). Mitochondrial translation products are mainly very hydrophobic subunits of the respiratory chain and due to their biochemical properties prone to misfolding or aggregation. Their biochemical properties make them ideal substrates for mtHsp70 chaperone activity.

3. Chloroplast biogenesis

As was described in the previous section, Hsps are prominently involved in several key steps of mitochondrial biogenesis. The investigation of mitochondrial protein import thereby revealed many of the essential cellular functions of Hsps or chaperones under normal growth conditions. The comparison of Hsp function in the other eukaryotic organelle of endosymbiotic origin, the chloroplast, will make it possible to draw general conclusions about the molecular mechanisms of polypeptide membrane translocation and the role of chaperones.

3.1 Translocation across the envelope membranes

Due to their endosymbiotic origin, chloroplasts are surrounded by a double membrane system, the outer and inner envelope membrane. Additionally, they contain a third membrane system, the thylakoid, which contains the photosynthetic machinery. Chloroplasts have retained the ability to synthesize much more proteins by themselves, but the majority is still encoded by the nucleus and has to be imported. Similar to the situation in mitochondria, the chloroplast envelope membranes contain large oligomeric protein complexes with translocase function (Fig. 3). The outer membrane components are named Toc (**t**ranslocase in the **o**uter envelope of **c**hloroplasts), the inner envelope components Tic (**t**ranslocase in the **i**nner envelope of **c**hloroplasts).

Due to the thylakoid membrane system, two completely independent membrane translocation steps can occur in chloroplasts, one across the envelope membranes and a second across the thylakoid membrane. The thylakoid transport seems to be

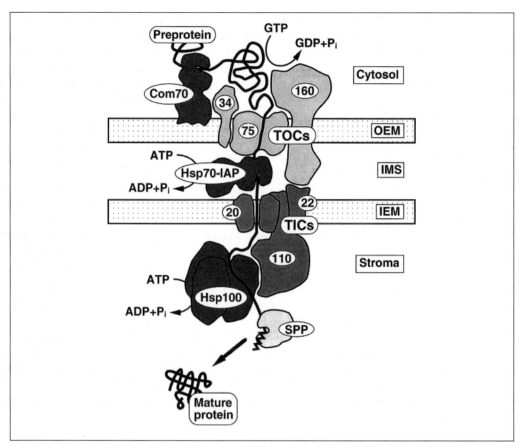

Fig. 3 Model of the chloroplast preprotein translocation machinery. Precursor proteins (black line) are recognized by putative receptors in the outer envelope membrane (OEM) and inserted through an aqueous channel, formed by the chloroplast outer envelope translocase (TOC, yellow). Heat shock proteins (Hsp, red) on the outer face of the outer envelope membrane and in the intermembrane space (IMS) are involved in translocation. Transport into the stroma across the inner envelope membrane (IEM) involves the chloroplast inner envelope translocase (TIC) and is ATP-dependent. In the stroma the precursor proteins interact with a member of the Hsp100 class (red) and the presequence is removed by the stromal processing peptidase (SPP). See colour plate section?

very similar to the ancient bacterial export pathway, both in regard to the involved proteins and the mechanism of translocation (107, 108). Several independent targeting pathways have been described recently. So far there is only limited information available about Hsps in thylakoid biogenesis. Based on immunological detection methods, Hsps of the Hsp70 and of the Hsp60/Hsp10 families have been identified (109). In the following text, only the translocation of proteins across the envelope membranes will be discussed .

3.1.1 Targeting signals and import competence

Chloroplast targeting sequences, also called transit sequences, are generally amino-terminal extensions characterized by an overall basic charge, a high proportion of hydroxylated amino acids, and the absence of acidic amino acids. In aqueous solution the transit sequences do not form any regular secondary structure (110). However, interaction with the chloroplast envelope lipid bilayer induces an α-helical structure. This structural feature probably supports the selective partitioning into the lipid environment of the outer envelope membrane (OEM). This interaction is especially pronounced with membranes containing chloroplast envelope lipid composition, thereby promoting targeting and membrane translocation (111). Chloroplast transit sequences can be specifically phosphorylated by a specific cytosolic kinase, assisting targeting efficiency. After binding to the chloroplast surface, the phosphate group has to be removed for membrane transfer (112). Apart from the cytosolic kinase, no other proteins of the cytosol have been identified to be required for chloroplast protein targeting. However, *in vitro* experiments showed that cytosolic Hsp70s stimulates import efficiency at least of some chloroplast precursor proteins (113).

Surprisingly, preproteins that are chemically cross-linked to a folded, albeit small, protein domain can be imported across the outer envelope and also the thylakoidal membranes, demonstrating that there is no absolute requirement for preprotein unfolding during chloroplast membrane translocation (114). This is in clear contrast to mitochondrial preprotein translocation, where any folded domains strongly decrease translocation efficiency (115, 116). This might be an indication that stromal Hsp70 function does not need to provide an unfolding force on the polypeptide during membrane transport as is the case in mitochondria (30, 44). However, sufficiently bulky and stably folded domains cannot be transported across the chloroplast envelope membranes (117, 118). In accordance with the data obtained about the folding state of precursors, the chloroplast import channel may be rather wide (119).

3.1.2 Docking and outer envelope translocation

The initial interaction of preproteins with the OEM appears to be of low affinity and independent of nucleotides. Tight binding of the preprotein (docking) is then accompanied by the hydrolysis of low levels of ATP/GTP (120). Most proteins in the OEM have been identified by their close association with preproteins trapped as translocation intermediates in the presence of low levels of ATP (121). Two components of the outer envelope translocase, Toc34 (122) and Toc86 (123), contain GTPase motifs. Since no direct interaction with preproteins could be demonstrated so far, it is speculated that Toc34 has a gating function of the OEM translocation channel (124). As demonstrated by cross-linking experiments, Toc86 is probably the major surface receptor for precursor proteins. Due to its GTPase activity it might have an additional regulatory role. Recently, it was established that Toc86 is a proteolytic fragment of a larger protein named Toc160 (125). Toc34 and Toc160 are the core components of the OEM translocase, together with a third protein that forms the actual translocation channel, Toc75 (126). Similar to the mitochondrial channel protein Tom40, Toc75 belongs to the family of β-barrel pore proteins, and indeed forms an ion channel after reconstitution into lipid vesicles (127).

Further insertion into the envelope membrane system seems also to be dependent on ATP hydrolysis on the inner side of the chloroplast double membrane. Both in the intermembrane space (128) and in the stroma (129), nucleotides are required for membrane translocation. This requirement for ATP hydrolysis might be attributed to the function of the chaperones in the intermembrane space and in the stroma (see below). Several results also indicate that outer and inner envelope translocation are functionally and physically coupled. Outer and inner envelope translocation complexes seem to interact directly even in the absence of preproteins (130).

3.1.3   Inner envelope translocation

The translocase of the inner envelope membrane (IEM) is not as well characterized as the OEM translocase. Based on cross-linking data, the peripheral membrane protein Tic22 is probably the first IEM protein making contact with the preprotein after insertion into the OEM (124). As it is bound to the outer face of the inner envelope, it might also function in establishing contact to the outer envelope machinery (131). The core components of the IEM translocase are Tic20 and Tic110. In contrast to Tic22, Tic20 is an integral membrane protein with limited contact to the surrounding solution. From its topology and by cross-linking to preproteins in later stages of import (131), it was concluded that Tic20 is involved in formation of the protein conducting channel in the inner envelope. Tic110 is another integral protein of the IEM with rather interesting properties. The topology of Tic110 in the IEM is still controversial. Experimental evidence indicates that its major part (> 90 kDa) is exposed to the stroma (132) or to the IMS (133). Since Tic110 can be co-immunoprecipitated with two stromal chaperones, Hsp100/ClpC (130, 134) and Cpn60 (135), it was suggested that it acts as a docking site for stromal chaperones. However, the interaction with Cpn60, the stromal Hsp60 homologue, is assumed to be potentially indirect due to preproteins that are still in transit, but which have already started to fold in the stromal compartment. Depending on its actual orientation, Tic110 might also serve in mediating contact between the translocases of outer and inner envelope membrane.

Another protein, Tic55, has been identified by its close association with the inner envelope channel components (136). Surprisingly, it contains an iron–sulphur cluster and a mononuclear iron-binding site. The functional implications of Tic55 involvement are still unclear.

3.2   Chaperones involved in chloroplast polypeptide transport and folding

3.2.1   Com70 at the outer surface

Although quite distinct in its biochemical properties, chloroplast protein import employs a homologue of the cytosolic Hsp70, named Com70 (chloroplast **o**uter envelope **m**embrane protein) (137, 138). In contrast to cytosolic Hsp70, Com70 is stably bound to the outer envelope membrane despite its overall hydrophilic character. Com70 is accessible from the cytosol and is in close proximity to the import machinery. Cross-linking experiments revealed an interaction of Com70 with external precursor proteins in an early state of translocation but also when arrested in an membrane spanning fashion (139). It may support binding of precursors to the envelope membrane dependent on the presence of the specific targeting signals. A membrane associated Hsp70 exposed to the cytosol is neither found in mitochondria or the ER. The function of this intriguing member of the Hsp70 family remains to be established. In contrast to the pulling/relay mechanism established for the mito-chondrial Hsp70, Com70 may prevent precursors from diffusing back into the cytosol or by facilitating precursor unfolding during entry into the import channel. In mitochondria, a membrane association could be demonstrated for the cytosolic DnaJ-homologue, Ydj1p, which indeed could indirectly result in an orientation of the cytosolic Hsp70 to membranes. It may be possible that Com70 is also be involved in a similar manner in maintaining import competence of chloroplast targeted preproteins.

3.2.2 Hsp70-IAP in the intermembrane space

Studies of chloroplast protein import revealed a second unique member of the Hsp70 family. Purification of import intermediate associated proteins (IAPs) by accumula-tion of import intermediates of a chimeric precursor protein resulted in the identi-fication of a membrane-bound Hsp70 (121). This Hsp70-IAP is very tightly anchored to the OEM, although the precise nature of the membrane anchor is not known. Its resistance to external proteases shows that the bulk of the polypeptide is located in the intermembrane space between outer and inner envelope. It was proposed that it functions as chaperone for preproteins as they emerge from the outer envelope translocation channel en route to the inner envelope membrane. It would prevent misfolding into an import-incompetent conformation and could also drive vectorial transport by stabilization of import intermediates after insertion into the outer envelope.

3.2.3 Hsp100/ClpC in the stroma

Precursor proteins can be arrested in the docking stage bound to the chloroplast membranes under low ATP conditions. This stable complex contains both outer and inner envelope membrane components and can be purified by immunoprecipitation. Surprisingly, one of the major proteins components of this complex belongs to the family of Hsp100 proteins (134). This protein, also called ClpC due to its homology to the Clp-ATPases from *E. coli*, is localized to the stroma but strongly interacts with the inner envelope translocation machinery. Initially, it was identified in a screen for chloroplast envelope associated proteins (140). Hsp100 directly interacts with pre-proteins in transit, as could be demonstrated by co-immunoprecipitation with radio-labelled precursors. This interaction is early during import since only unprocessed precursors could be detected bound to Hsp100. Since it interacts with preproteins destined for several internal chloroplast subcompartments, Hsp100 has a general function in inner envelope translocation. Similar interactions with precursor proteins during import were also detected with stromal Hsp70 (stHsp70) but they seem to be less stable. Hsp100 also seems to be a part of the IEM translocase complex in the absence of preproteins, shown by chemical cross-linking experiments (130) and by blue-native gel electrophoresis (136). The interaction of Hsp100 with precursor proteins is disrupted by high levels of ATP as was expected from the behaviour of the related Clp-ATPases (83). It is assumed, that ATP disrupts the interaction with substrate proteins but leaves the association of Hsp100 with the translocation machinery intact. An additional function of Hsp100 in protein degradation is likely since it is found in association with the chloroplast ClpP protease and at least *in vitro* enhances the ATP-dependent degradation of substrate proteins (141).

3.2.4 Stromal Hsp70

Initially, three members of the Hsp70 family have been identified in chloroplasts by immunological methods (142). One was found associated with the envelope membranes (Hsp70-IAP, see above) and two other as soluble components in the stroma. However, further biochemical experiments so far only give evidence about the function of one stromal Hsp70, also called S78. Upon translocation into the stromal compartment, maturation and folding of newly imported preproteins is assisted by chaperones of the Hsp70 and Hsp60 families. Their function is similar to the mechanism proposed for mitochondria, with the substrate protein first interacting with stromal Hsp70 (stHsp70) and then with stromal Hsp60 (Cpn60) (143). Stromal Hsp70 was also found in association with precursors in transit across the outer and inner envelope (134). Since it could not be excluded that this interaction might be due to aggregated or unfolded substrate proteins, the role of stHsp70 during the translocation itself remains to be clarified. Similar to prokaryotic systems, there is evidence for DnaJ- and GrpE-like proteins in the chloroplast stroma (144). Lacking functional data, it can be assumed that both function in regulating stHsp70 activity during folding reactions in the stromal compartment.

Because import into the thylakoid membrane system is a multiple step translocation process, it is essential to keep the preproteins in an import-competent conformation after reaching the stromal compartment. It was reported that further transit to the thylakoid requires soluble stromal components to maintain import competence (145). stHsp70 has been implicated in stabilizing thylakoid precursors, but its function might not be sufficient for all types of preproteins. It could be shown that stHsp70 is required for the efficient insertion of integral membrane proteins into thylakoid membranes *in vitro* (146). It is likely that hydrophobic proteins particularly require the chaperone action of stHsp70 for import competence.

3.2.5 Chaperonin 60 system

Compared to prokaryotes and mitochondria, the chaperonin system in the chloroplast stroma displays some unique features. There are two isoforms of stromal Cpn60 (147, 148) having distinct amino acid sequences but both related to GroEL. Despite the existence of two isoforms, the biochemical and structural properties of stromal Cpn60 seem to be very similar to those of bacterial GroEL (149). It forms an oligomer of 14 subunits in two stacked rings surrounding a central cavity. Also the co-chaperonin related to GroES is unusual in having two Hsp10-like domains that are fused together, therefore being called Cpn21 (150). Despite this unusual topology, *in vitro* experiments demonstrated that both domains are functional as single proteins and also the complete 'double'-Cpn10 protein can interact with the *E. coli* chaperonin GroEL, similar to its homologue GroES (151). As mentioned above, Cpn60 has been found associated with preproteins during translocation (135) and also with completely imported proteins, assisting in the folding process after they are released from Hsp70 (143).

4. Comparison of chloroplast and mitochondrial membrane translocation

Both, chloroplasts and mitochondria need to address very similar problems in targeting their respective preproteins and keeping them in an import-competent conformation after synthesis. Here, the role of Hsps seems to be comparable. Receptor recognition and membrane insertion follow at least similar principles, albeit with different types of regulation. The major difference between import across the double-membrane system of mitochondria and chloroplasts is the absence of a proton motive force in chloroplasts. This has major implications on the driving forces for polypeptide membrane translocation. Since there is no electrophoretic effect on the presequence in chloroplasts, preprotein transfer from outer to inner membrane might need the support by the intermembrane space Hsp70-IAP, a function that is completely absent in mitochondria.

On the inner face of the envelope membrane, the involvement of Hsps in chloroplasts seems to be quite different from those in mitochondria. Whereas Hsp70 is the major Hsp interacting with the incoming precursor proteins in mitochondria, the dominant interaction partner of translocation intermediates in chloroplasts is Hsp100 (ClpC). Although a Hsp100 homologue, Hsp78, exists in mitochondria, it is unlikely that Hsp78 can directly substitute for mtHsp70 function. One interesting similarity is the presence of an IMP serving as membrane anchor for chaperones in both organelles. This is Tic110 in chloroplasts (130, 134) and Tim44 in mitochondria (37). This might indicate a general requirement for an immediate chaperone function at the import site at the inner face of the membrane, even if different types of chaperones fulfil this function. In mitochondria there is general agreement that mtHsp70 function preprotein binding enzyme and ATPase drives the actual polypeptide chain translocation, even if details of the mechanism are still controversial. The driving forces in chloroplast envelope translocation into the stroma are still unclear. It is possible that, due to its ATPase activity, stromal Hsp100 can take over this function, similar to mtHsp70. As an alternative possibility, the presence of the Hsp70-IAP in the chloroplast intermembrane space could provide a driving force for translocation. The unfolding of the preprotein for translocation across the OEM, especially, might be caused by the interaction of Hsp70-IAP with the inserted precursor.

Both similarities and differences between mitochondrial and chloroplast preprotein membrane translocation are mirrored in the involvement of the respective chaperones or Hsps. These proteins, as far as they have been identified, are listed in Table 1

Table 1 (Putative) chaperones involved in mitochondrial biogenesis

Local.	Name	Putative function	Reference(s)
Cytosol	Hsc70/Ssa solubility	Maintenance of import competence, stabilization of preprotein	(5, 6)
	MSF	Maintenance of import competence, targeting factor	(152, 153)
Matrix	mtHsp70/Ssc1p	Inner membrane translocation motor, protein folding in the matrix	(28, 29)
	Mge1p	Co-chaperone of mtHsp70	(48, 49, 54)
	Mdj1p	Co-chaperone of mtHsp70	(63, 64)
	Mdj2p	Membrane-bound co-chaperone	(67)
	Ssq1p	Involved in mitochondrial iron metabolism	(68, 69)
	Hsp60	Preprotein folding in the matrix	(70, 78)
	Hsp10	Co-chaperone of Hsp60	(73)
	Hsp78	Maintenance of mitochondrial function under stress conditions	(88–90)
	Mcx1p	unknown	(93)
IM	Yme1p	Protein degradation, assembly of membrane complexes	(103, 154)
	Yta10p	Protein degradation, assembly of membrane complexes	(98, 103, 155)
	Yta12p	Protein degradation, assembly of membrane complexes	(103–105)

Table 2 Chaperones and co-chaperones involved in chloroplast biogenesis

Local.	Name	Putative function	Ref.
OEM	Com70	Maintaining import competence at OEM surface	(138)
IMS	Hsp70-IAP	Maintaining import competence in the IMS, possibly generation of translocation driving force	(121)
Stroma	Hsp100/ClpC	Component of IEM translocation complex, ATPase	(134)
	stHsp70	Protein folding	(143, 146)
	stDnaJ	Regulation of stHsp70 activity	(144)
	stGrpE	Regulation of stHsp70 activity	(144)
	α-, β-Cpn60	Protein folding	(147, 148)
	Cpn10	Co-chaperone for Cpn60	(150)

for mitochondria and Table 2 for chloroplasts. The chaperones perform crucial functions in import competence, driving membrane translocation and subsequent folding and assembly reactions. Since these functions are absolutely required for the survival of the cell even under normal growth conditions, it can be speculated that the involvement in intracellular transport represents the original role of Hsps.

Acknowledgements

We thank S. Schmidt and A. Strub for critically reading the manuscript. W. Voos and N. Pfanner were supported by grants from the Deutsche Forschungsgemeinschaft (SFB 388) and the Fonds der Chemischen Industrie.

References

1. Gray, M. W., Burger, G., and Lang, B. F. (1999). Mitochondrial evolution. *Science,* **283**, 1476.
2. von Heijne, G. (1990). Protein targeting signals. *Curr. Opin. Cell Biol.,* **2**, 604.
3. George, R., Beddoe, T., Landl, K., and Lithgow, T. (1998). The yeast nascent polypeptide-associated complex initiates protein targeting to mitochondria *in vivo. Proc. Natl. Acad. Sci. USA,* **95**, 2296.
4. Fünfschilling, U. and Rospert, S. (1999). Nascent polypeptide-associated complex stimulates protein import into yeast mitochondria. *Mol. Biol. Cell.,* **10**, 3289.
5. Murakami, H., Pain, D., and Blobel, G. (1988). 70-kD heat shock-related protein is one of at least two distinct cytosolic factors stimulating protein import into mitochondria. *J. Cell. Biol.,* **107**, 2051.
6. Deshaies, R., Koch, B., Werner-Washburne, M., Craig, E. A., and Schekman, R. (1988). A subfamily of stress proteins facilitates translocation of secretory and mitochondrial precursor polypeptides. *Nature,* **332**, 800.
7. Sheffield, W. P., Shore, G. C., and Randall, S. K. (1990). Mitochondrial precursor protein. Effects of 70-kD heat shock protein on polypeptide folding, aggregation, and import competence. *J. Biol. Chem.,* **265**, 11069.
8. Endo, T., Mitsui, S., Nakai, M., and Roise, D. (1996). Binding of mitochondrial presequences to yeast cytosolic heat shock protein 70 depends on the amphiphilicity of the presequence. *J. Biol. Chem.,* **271**, 4161.
9. Roise, D. (1997). Recognition and binding of mitochondrial presequences during the import of proteins into mitochondria. *J. Bioenerg. Biomembr.,* **29**, 19.
10. Zimmermann, R. (1998). The role of molecular chaperones in protein transport into the mammalian endoplasmic reticulum. *Biol. Chem.,* **379**, 275.
11. Caplan, A. J. and Douglas, M. G. (1991). Characterisation of YDJ1: a yeast homologue of the bacterial dnaJ protein. *J. Cell. Biol.,* **114**, 609.
12. Caplan, A., Cyr, D., and Douglas, M. (1992). Ydj1p facilitates polypeptide translocation across different intracellular membranes by a conserved mechanism. *Cell,* **71**, 1143.
13. Hartl, F. U. (1996). Molecular chaperones in cellular protein folding. *Nature,* **381**, 571.
14. Cyr, D., Lu, X., and Douglas, M. (1992). Regulation of Hsp70 function by a eukaryotic DnaJ homolog. *J. Biol. Chem.,* **267**, 20927.
15. Hachiya, N., Komiya, T., Alam, R., Iwahashi, J., Sakaguchi, M., Omura, T., *et al.* (1994). MSF, a novel cytoplasmic chaperone which functions in precursor targeting to mitochondria. *EMBO J.,* **13**, 5146.
16. Komiya, T., Sakaguchi, M., and Mihara, K. (1996). Cytoplasmic chaperones determine the targeting pathway of precursor proteins to mitochondria. *EMBO J.,* **15**, 399.
17. Becker, K., Guiard, B., Rassow, J., Söllner, T., and Pfanner, N. (1992). Targeting of a chemically pure preprotein to mitochondria does not require the addition of a cytosolic signal recognition factor. *J. Biol. Chem.,* **267**, 5637.
18. Ryan, M. T. and Pfanner, N. (1998). The preprotein translocase of the mitochondrial outer membrane. *Biol. Chem.,* **379**, 289.
19. Hill, K., Model, K., Ryan, M. T., Dietmeier, K., Martin, F., Wagner, R., *et al.* (1998). Tom40 forms the hydrophilic channel of the mitochondrial import pore for preproteins. *Nature,* **395**, 516.
20. Künkele, K. P., Heins, S., Dembowski, M., Nargang, F. E., Benz, R., Thieffry, M., *et al.* (1998). The preprotein translocation channel of the outer membrane of mitochondria. *Cell,* **93**, 1009.

21. Hönlinger, A., Kübrich, M., Moczko, M., Gärtner, F., Mallet, L., Bussereau, F., *et al.* (1995). The mitochondrial receptor complex: Mom22 is essential for cell viability and directly interacts with preproteins. *Mol. Cell. Biol.*, **15**, 3382.

22. Schatz, G. (1997). Just follow the acid chain. *Nature*, **388**, 121.

23. Rassow, J., Dekker, P. J. T., van Wilpe, S., Meijer, M., and Soll, J. (1999). The preprotein translocase of the mitochondrial inner membrane: function and evolution. *J. Mol. Biol.*, **286**, 105.

24. Truscott, K. N. and Pfanner, N. (1999). Import of carrier proteins into mitochondria. *Biol. Chem.*, **380**, 1151.

25. Schleyer, M., Schmidt, B., and Neupert, W. (1982). Requirement of a membrane potential for the postranslational transfer of proteins into mitochondria. *Eur. J. Biochem.*, **125**, 109.

26. Martin, J., Mahlke, K., and Pfanner, N. (1991). Role of an energized inner membrane in mitochondrial protein import. *J. Biol. Chem.*, **266**, 18051.

27. Bauer, M. F., Sirrenberg, C., Neupert, W., and Brunner, M. (1996). Role of Tim23 as voltage sensor and presequence receptor in protein import into mitochondria. *Cell*, **87**, 33.

28. Craig, E. A., Kramer, J., and Kosic-Smithers, J. (1987). *SSC1*, a member of the 70-kDa heat shock protein multigene family of *Saccharomyces cerevisiae*, is essential for growth. *Proc. Natl. Acad. Sci. USA*, **84**, 4156.

29. Kang, P. J., Ostermann, J., Shilling, J., Neupert, W., Craig, E. A., and Pfanner, N. (1990). Requirement for hsp70 in the mitochondrial matrix for translocation and folding of precursor proteins. *Nature*, **348**, 137.

30. Gambill, B. D., Voos, W., Kang, P. J., Miao, B., Langer, T., Craig, E. A., *et al.* (1993). A dual role for mitochondrial heat shock protein 70 in membrane translocation of preproteins. *J. Cell Biol.*, **123**, 109.

31. Scherer, P. E., Krieg, U. C., Hwang, S. T., Vestweber, D., and Schatz, G. (1990). A precursor protein partly translocated into yeast mitochondria is bound to a 70 kD mitochondrial stress protein. *EMBO J.*, **9**, 4315.

32. Rassow, J., Hartl, F. U., Guiard, B., Pfanner, N., and Neupert, W. (1990). Polypeptides traverse the mitochondrial envelope in an extended state. *FEBS Lett.*, **275**, 190.

33. Voos, W., Gambill, B. D., Guiard, B., Pfanner, N., and Craig, E. A. (1993). Presequence and mature part of preproteins strongly influence the dependence of mitochondrial protein import on heat shock protein 70 in the matrix. *J. Cell Biol.*, **123**, 119.

34. Maarse, A. C., Blom, J., Grivell, L. A., and Meijer, M. (1992). *MPI1*, an essential gene encoding a mitochondrial membrane protein, is possibly involved in protein import into yeast mitochondria. *EMBO J.*, **11**, 3619.

35. Scherer, P. E., Manning-Krieg, U. C., Jenö, P., Schatz, G., and Horst, M. (1992). Identification of a 45-kDa protein import site of the yeast mitochondrial inner membrane. *Proc. Natl. Acad. Sci. USA*, **89**, 11930.

36. Horst, M., Jenö, P., Kronidou, N. G., Bolliger, L., Oppliger, W., Scherer, P., *et al.* (1993). Protein import into yeast mitochondria: the inner membrane import site protein ISP45 is the *MPI1* gene product. *EMBO J.*, **12**, 3035.

37. Rassow, J., Maarse, A. C., Krainer, E., Kübrich, M., Müller, H., Meijer, M., *et al.* (1994). Mitochondrial protein import: biochemical and genetic evidence for interaction of matrix hsp70 and the inner membrane protein MIM44. *J. Cell Biol.*, **127**, 1547.

38. Schneider, H.-C., Berthold, J., Bauer, M. F., Dietmeier, K., Guiard, B., Brunner, M., *et al.* (1994). Mitochondrial Hsp70/MIM44 complex facilitates protein import. *Nature*, **371**, 768.

39. Bömer, U., Meijer, M., Maarse, A. C., Hönlinger, A., Dekker, P. J., Pfanner, N., *et al.* (1997).

Multiple interactions of components mediating preprotein translocation across the inner mitochondrial membrane. *EMBO J.*, **16**, 2205.

40. Ungermann, C., Guiard, B., Neupert, W., and Cyr, D. M. (1996). The delta psi- and Hsp70/MIM44-dependent reaction cycle driving early steps of protein import into mitochondria. *EMBO J.*, **15**, 735.
41. Glick, B. S. (1995). Can Hsp70 proteins act as force-generating motors? *Cell*, **80**, 11.
42. von Ahsen, O., Voos, W., Henninger, H., and Pfanner, N. (1995). The mitochondrial protein import machinery. Role of ATP in dissociation of the Hsp70.Mim44 complex. *J. Biol. Chem.*, **270**, 29848.
43. Kronidou, N. G., Oppliger, W., Bolliger, L., Hannavy, K., Glick, B. S., Schatz, G., *et al.* (1994). Dynamic interaction between Isp45 and mitochondrial hsp70 in the protein import system of the yeast mitochondrial inner membrane. *Proc. Natl. Acad. Sci. USA*, **91**, 12818.
44. Voos, W., von Ahsen, O., Müller, H., Guiard, B., Rassow, J., and Pfanner, N. (1996). Differential requirement for the mitochondrial Hsp70-Tim44 complex in unfolding and translocation of preproteins. *EMBO J.*, **15**, 2668.
45. Voisine, C., Craig, E. A., Zufall, N., von Ahsen, O., Pfanner, N., and Voos, W. (1999). The protein import motor of mitochondria: unfolding and trapping of preproteins are distinct and separable functions of matrix Hsp70. *Cell*, **97**, 1.
46. Chauwin, J. F., Oster, G., and Glick, B. S. (1998). Strong precursor-pore interactions constrain models for mitochondrial protein import. *Biophys J.*, **74**, 1732.
47. Dekker, P. J. T., Martin, F., Maarse, A. C., Bömer, U., Müller, H., Guiard, B., *et al.* (1997). The Tim core complex defines the number of mitochondrial translocation contact sites and can hold arrested preproteins in the absence of matrix Hsp70-Tim44. *EMBO J.*, **16**, 5408.
48. Ikeda, E., Yoshida, S., Mitsuzawa, H., Uno, I., and Toh-e, A. (1994). YGE1 is a yeast homologue of *Escherichia coli* grpE and is required for maintenance of mitochondrial functions. *FEBS Lett.*, **339**, 265.
49. Laloraya, S., Gambill, B. D., and Craig, E. A. (1994). A role for a eukaryotic GrpE-related protein Mge1p, in protein translocation. *Proc. Natl. Sci. Acad. USA*, **91**, 6481.
50. Voos, W., Gambill, B. D., Laloraya, S., Ang, D., Craig, E. A., and Pfanner, N. (1994). Mitochondrial GrpE is present in a complex with hsp70 and preproteins in transit across membranes. *Mol. Cell. Biol.*, **14**, 6627.
51. Laloraya, S., Dekker, P. J., Voos, W., Craig, E. A., and Pfanner, N. (1995). Mitochondrial GrpE modulates the function of matrix Hsp70 in translocation and maturation of preproteins. *Mol. Cell. Biol.*, **15**, 7098.
52. Westermann, B., Prip-Buus, C., Neupert, W., and Schwarz, E. (1995). The role of the GrpE homologue, Mge1p, in mediating protein import and protein folding in mitochondria. *EMBO J.*, **14**, 3452.
53. Nakai, M., Kato, Y., Toh-e, A., and Endo, T. (1994). Yge1p, a eukaryotic GrpE-homolog, is localized in the mitochondrial matrix and interacts with mitochondrial hsp70. *Biochem. Biophys. Res. Commun.*, **200**, 435.
54. Bolliger, L., Deloche, O., Glick, B. S., Georgopoulos, C., Jenö, P., Kronidou, N., *et al.* (1994). A mitochondrial homolog of bacterial GrpE interacts with mitochondrial hsp70 and is essential for viability. *EMBO J.*, **13**, 1998.
55. Miao, B., Davis, J. E., and Craig, E. A. (1997). Mge1 functions as a nucleotide release factor for Ssc1, a mitochondrial Hsp70 of *Saccharomyces cerevisiae*. *J. Mol. Biol.*, **265**, 541.
56. Sakuragi, S., Liu, Q., and Craig, E. (1999). Interaction between the nucleotide exchange factor Mge1 and the mitochondrial Hsp70 Ssc1. *J. Biol. Chem.*, **274**, 11275.

57. Dekker, P. J. and Pfanner, N. (1997). Role of mitochondrial GrpE and phosphate in the ATPase cycle of matrix Hsp70. *J. Mol. Biol.*, **270**, 321.

58. Beissinger, M. and Buchner, J. (1998). How chaperones fold proteins. *Biol. Chem.*, **379**, 245.

59. Horst, M., Oppliger, W., Feifel, B., Schatz, G., and Glick, B. S. (1996). The mitochondrial protein import motor: dissociation of mitochondrial hsp70 from its membrane anchor requires ATP binding rather than ATP hydrolysis. *Protein Sci.*, **5**, 759.

60. Deloche, O., Kelley, W. L., and Georgopoulos, C. (1997). Structure-function analyses of the Ssc1p, Mdj1p, and Mge1p *Saccharomyces cerevisiae* mitochondrial proteins in *Escherichia coli*. *J. Bacteriol.*, **179**, 6066.

61. Manning-Krieg, U. C., Scherer, P. E., and Schatz, G. (1991). Sequential action of mitochondrial chaperones in protein import into mitochondria. *EMBO J.*, **10**, 3273.

62. Horst, M., Oppliger, W., Rospert, S., Schonfeld, H. J., Schatz, G., and Azem, A. (1997). Sequential action of two hsp70 complexes during protein import into mitochondria. *EMBO J.*, **16**, 1842.

63. Rowley, N., Prip-Buus, C., Westermann, B., Brown, C., Schwarz, E., Barrell, B., *et al.* (1994). Mdj1p, a novel chaperone of the DnaJ family, is involved in mitochondrial biogenesis and protein folding. *Cell*, **77**, 249.

64. Prip-Buus, C., Westerman, B., Schmitt, M., Langer, T., Neupert, W., and Schwarz, E. (1996). Role of the mitochondrial DnaJ homologue, Mdj1p, in the prevention of heat-induced protein aggregation. *FEBS Lett.*, **380**, 142.

65. Kubo, Y., Tsunehiro, T., Nishikawa, S., Nakai, M., Ikeda, E., Toh-e, A., *et al.* (1999). Two distinct mechanisms operate in the reactivation of heat-denatured proteins by the mitochondrial Hsp70/Mdj1p/Yge1p chaperone system. *J. Mol. Biol.*, **286**, 447.

66. Westermann, B., Gaume, B., Herrmann, J. M., Neupert, W., and Schwarz, E. (1996). Role of the mitochondrial DnaJ homolog Mdj1p as a chaperone for mitochondrially synthesized and imported proteins. *Mol. Cell. Biol.*, **16**, 7063.

67. Westermann, B. and Neupert, W. (1997). Mdj2p, a novel DnaJ homolog in the mitochondrial inner membrane of the yeast *Saccharomyces cerevisiae*. *J. Mol. Biol.*, **272**, 477.

68. Schilke, B., Forster, J., Davis, J., James, P., Walter, W., Laloraya, S., *et al.* (1996). The cold sensitivity of a mutant of *Saccharomyces cerevisiae* lacking a mitochondrial heat shock protein 70 is suppressed by loss of mitochondrial DNA. *J. Cell Biol.*, **134**, 603.

69. Knight, S. A., Sepuri, N. B., Pain, D., and Dancis, A. (1998). Mt-Hsp70 homolog, Ssc2p, required for maturation of yeast frataxin and mitochondrial iron homeostasis. *J. Biol. Chem.*, **273**, 18389.

70. Cheng, M. Y., Hartl, F. U., Martin, J., Pollock, R. A., Kalusek, F., Neupert, W., *et al.* (1989). Mitochondrial heat-shock protein hsp60 is essential for assembly of proteins imported into yeast mitochondria. *Nature*, **337**, 620.

71. Ostermann, J., Horwich, A. L., Neupert, W., and Hartl, F. U. (1989). Protein folding in mitochondria requires complex formation with hsp60 and ATP hydrolysis. *Nature*, **341**, 125.

72. Lubben, T., Gatenby, A., Donaldson, G., Lorimer, G., and Viitanen, P. (1990). Identification of a groES-like chaperonin in mitochondria that facilitates protein folding. *Proc. Natl. Acad. Sci. USA*, **87**, 7683.

73. Rospert, S., Junne, T., Glick, B. S.. and Schatz, G. (1993). Cloning and disruption of the gene encoding yeast mitochondrial chaperonin 10, the homolog of *E. coli* groES. *FEBS Lett.*, **335**, 358.

74. Martin, J. and Hartl, F. U. (1997). Chaperone-assisted protein folding. *Curr. Opin. Struct. Biol.*, **7**, 41.

75. Nielsen, K. L. and Cowan, N. J. (1998). A single ring is sufficient for productive chaperonin-mediated folding *in vivo*. *Mol. Cell*, **2**, 93.

76. Nielsen, K. L., McLennan, N., Masters, M., and Cowan, N. J. (1999). A single-ring mito-chondrial chaperonin (Hsp60-Hsp10) can substitute for GroEL-GroES in vivo. *J. Bacteriol.*, **181**, 5871.

77. Heyrovska, N., Frydman, J., Höhfeld, J., and Hartl, F. U. (1998). Directionality of poly-peptide transfer in the mitochondrial pathway of chaperone-mediated protein folding. *Biol. Chem.*, **379**, 301.

78. Rospert, S., Looser, R., Dubaquié, Y., Matouschek, A., Glick, B. S., and Schatz, G. (1996). Hsp60-independent protein folding in the matrix of yeast mitochondria. *EMBO J.*, **15**, 764.

79. Dubaquie, Y., Looser, R., Fünfschilling, U., Jenö, P., and Rospert, S. (1998). Identification of *in vivo* substrates of the yeast mitochondrial chaperonins reveals overlapping but non-identical requirement for hsp60 and hsp10. *EMBO J.*, **17**, 5868.

80. Rassow, J., Mohrs, K., Koidl, S., Barthelmess, I. B., Pfanner, N., and Tropschug, M. (1995). Cyclophilin 20 is involved in mitochondrial protein folding in cooperation with molecular chaperones Hsp70 and Hsp60. *Mol. Cell. Biol.*, **15**, 2654.

81. von Ahsen, O., Tropschug, M., Pfanner, N., and Rassow, J. (1997). The chaperonin cycle cannot substitute for prolyl isomerase activity, but GroEL alone promotes productive folding of a cyclophilin-sensitive substrate to a cyclophilin-resistant form. *EMBO J.*, **16**, 4568.

82. Suzuki, C. K., Rep, M., Maarten van Dijl, J., Suda, K., Grivell, L. A., and Schatz, G. (1997). ATP-dependent proteases that also chaperone protein biogenesis. *Trends Biochem. Sci.*, **22**, 118.

83. Wawrzynow, A., Banecki, B., and Zylicz, M. (1996). The Clp ATPases define a novel class of molecular chaperones. *Mol. Microbiol.*, **21**, 895.

84. Schirmer, E. C., Glover, J. R., Singer, M. A., and Lindquist, S. (1996). HSP100/Clp proteins: a common mechanism explains diverse functions. *Trends Biochem. Sci.*, **21**, 289.

85. Feng, H. P. and Gierasch, L. M. (1998). Molecular chaperones: clamps for the Clps? *Curr. Biol.*, **8**, R464.

86. Weber-Ban, E. U., Reid, B. G., Miranker, A. D., and Horwich, A. L. (1999). Global unfolding of a substrate protein by the Hsp100 chaperone ClpA. *Nature*, **401**, 90.

87. Zolkiewski, M. (1999). ClpB cooperates with DnaK, DnaJ, and GrpE in suppressing pro-tein aggregation. A novel multi-chaperone system from *Escherichia coli. J. Biol. Chem.*, **274**, 28083.

88. Leonhardt, S. A., Fearon, K., Danese, P. N., and Mason, T. L. (1993). *HSP78* encodes a yeast mitochondrial heat shock protein in the Clp family of ATP-dependent proteases. *Mol. Cell. Biol.*, **13**, 6304.

89. Schmitt, M., Neupert, W., and Langer, T. (1995). Hsp78, a Clp homologue within mito-chondria, can substitute for chaperone functions of mt-hsp70. *EMBO J.*, **14**, 3434.

90. Moczko, M., Schönfisch, B., Voos, W., Pfanner, N., and Rassow, J. (1995). The mitochondrial ClpB homolog Hsp78 cooperates with matrix Hsp70 in maintenance of mitochondrial function. *J. Mol. Biol.*, **254**, 538.

91. Schmitt, M., Neupert, W., and Langer, T. (1996). The molecular chaperone Hsp78 confers compartment-specific thermotolerance to mitochondria. *J. Cell Biol.*, **134**, 1375.

92. Parsell, D. A., Kowal, A. S., Singer, M. A., and Lindquist, S. (1994). Protein disaggregation mediated by heat-shock protein Hsp104. *Nature*, **372**, 475.

93. van Dyck, L., Dembowski, M., Neupert, W., and Langer, T. (1998). Mcx1p, a ClpX homologue in mitochondria of *Saccharomyces cerevisiae*. *FEBS Lett.*, **438**, 250.

94. Santagata, S., Bhattacharyya, D., Wang, F. H., Singha, N., Hodtsev, A., and Spanopoulou, E. (1999). Molecular cloning and characterization of a mouse homolog of bacterial ClpX, a novel mammalian class II member of the Hsp100/Clp chaperone family. *J. Biol. Chem.*, **274**, 16311.

95. Bross, P., Andresen, B. S., Knudsen, I., Kruse, T. A., and Gregersen, N. (1995). Human ClpP protease: cDNA sequence, tissue-specific expression and chromosomal assignment of the gene. *FEBS Lett.*, **377**, 249.

96. Corydon, T. J., Bross, P., Holst, H. U., Neve, S., Kristiansen, K., Gregersen, N., *et al.* (1998). A human homologue of Escherichia coli ClpP caseinolytic protease: recombinant expression, intracellular processing and subcellular localization. *Biochem. J.*, **331**, 309.

97. de Sagarra, M. R., Mayo, I., Marco, S., Rodriguez-Vilarino, S., Oliva, J., Carrascosa, J. L., *et al.* (1999). Mitochondrial localization and oligomeric structure of HClpP, the human homologue of E. coli clpP. *J. Mol. Biol.*, **292**, 819.

98. Pajic, A., Tauer, R., Feldmann, H., Neupert, W., and Langer, T. (1994). Yta10p is required for the ATP-dependent degradation of polypeptides in the inner membrane of mitochondria. *FEBS Lett.*, **353**, 201.

99. Guélin, E., Rep, M., and Grivell, L. A. (1996). Afg3p, a mitochondrial ATP-dependent metalloprotease, is involved in degradation of mitochondrially-encoded Cox1, Cox3, Cob, Su6, Su8 and Su9 subunits of the inner membrane complexes III, IV and V. *FEBS Lett.*, **381**, 42.

100. Weber, E. R., Hanekamp, T., and Thorsness, P. E. (1996). Biochemical and functional analysis of the YME1 gene product, an ATP and zinc-dependent mitochondrial protease from S. cerevisiae. *Mol. Biol. Cell*, **7**, 307.

101. Leonhard, K., Herrmann, J. M., Stuart, R. A., Mannhaupt, G., Neupert, W., and Langer, T. (1996). AAA proteases with catalytic sites on opposite membrane surfaces comprise a proteolytic system for the ATP-dependent degradation of inner membrane proteins in mitochondria. *EMBO J.*, **15**, 4218.

102. Arlt, H., Steglich, G., Perryman, R., Guiard, B., Neupert, W., and Langer, T. (1998). The formation of respiratory chain complexes in mitochondria is under the proteolytic control of the m-AAA protease. *EMBO J.*, **17**, 4837.

103. Tzagoloff, A., Yue, J., Jang, J., and Paul, M. F. (1994). A new member of a family of ATPases is essential for assembly of mitochondrial respiratory chain and ATP synthetase complexes in Saccharomyces cerevisiae. *J. Biol. Chem.*, **269**, 26144.

104. Paul, M. F. and Tzagoloff, A. (1995). Mutations in RCA1 and AFG3 inhibit F1-ATPase assembly in Saccharomyces cerevisiae. *FEBS Lett.*, **373**, 66.

105. Arlt, H., Tauer, R., Feldmann, H., Neupert, W., and Langer, T. (1996). The YTA10–12 complex, an AAA protease with chaperone-like activity in the inner membrane of mitochondria. *Cell*, **85**, 875.

106. Herrmann, J. M., Stuart, R. A., Craig, E. A., and Neupert, W. (1994). Mitochondrial heat shock protein 70, a molecular chaperone for proteins encoded by mitochondrial DNA. *J. Cell Biol.*, **127**, 893.

107. Robinson, C., Hynds, P. J., Robinson, D., and Mant, A. (1998). Multiple pathways for the targeting of thylakoid proteins in chloroplasts. *Plant Mol. Biol.*, **38**, 209.

108. Dalbey, R. E. and Robinson, C. (1999). Protein translocation into and across the bacterial plasma membrane and the plant thylakoid membrane. *Trends Biochem. Sci.*, **24**, 17.

109. Schlicher, T. and Soll, J. (1996). Molecular chaperones are present in the thylakoid lumen of pea chloroplasts. *FEBS Lett.*, **379**, 302.

110. Chen, X. and Schnell, D. J. (1999). Protein import into chloroplasts. *Trends Cell Biol.*, **9**, 222.

111. Bruce, B. D. (1998). The role of lipids in plastid protein transport. *Plant Mol. Biol.*, **38**, 223.

112. Waegemann, K. and Soll, J. (1996). Phosphorylation of the transit sequence of chloroplast precursor proteins. *J. Biol. Chem.*, **271**, 6545.

113. Waegemann, K., Paulsen, H., and Soll, J. (1990). Translocation of proteins into isolated chloroplasts requires cytosolic factors to obtain import competence. *FEBS Lett.*, **261**, 89.

114. Clark, S. A. and Theg, S. M. (1997). A folded protein can be transported across the chloroplast envelope and thylakoid membranes. *Mol. Biol. Cell*, **8**, 923.

115. Eilers, M. and Schatz, G. (1986). Binding of a specific ligand inhibits import of a purified precursor protein into mitochondria. *Nature*, **322**, 228.

116. Jascur, T., Goldenberg, D. P., Vestweber, D., and Schatz, G. (1992). Sequential translocation of an artificial precursor protein across the two mitochondrial membranes. *J. Biol. Chem.*, **267**, 13636.

117. Schnell, D. J. and Blobel, G. (1993). Identification of intermediates in the pathway of protein import into chloroplasts and their localization to envelope contact sites. *J. Cell Biol.*, **120**, 103.

118. Wu, C. and Ko, K. (1993). Identification of an uncleavable targeting signal in the 70-kilodalton spinach chloroplast outer envelope membrane protein. *J. Biol. Chem.*, **268**, 19384.

119. van den Wijngaard, P. W. and Vredenberg, W. J. (1997). A 50-picosiemens anion channel of the chloroplast envelope is involved in chloroplast protein import. *J. Biol. Chem.*, **272**, 29430.

120. Olsen, L. J., Theg, S. M., Selman, B. R., and Keegstra, K. (1989). ATP is required for the binding of precursor proteins to chloroplasts. *J. Biol. Chem.*, **264**, 6724.

121. Schnell, D. J., Kessler, F., and Blobel, G. (1994). Isolation of components of the chloroplast protein import machinery. *Science*, **266**, 1007.

122. Seedorf, M., Waegemann, K., and Soll, J. (1995). A constituent of the chloroplast import complex represents a new type of GTP-binding protein. *Plant J.*, **7**, 401.

123. Hirsch, S., Muckel, E., Heemeyer, F., von Heijne, G., and Soll, J. (1994). A receptor component of the chloroplast protein translocation machinery. *Science*, **266**, 1989.

124. Kouranov, A. and Schnell, D. J. (1997). Analysis of the interactions of preproteins with the import machinery over the course of protein import into chloroplasts. *J. Cell Biol.*, **139**, 1677.

125. Bölter, B., May, T., and Soll, J. (1998). A protein import receptor in pea chloroplasts, Toc86, is only a proteolytic fragment of a larger polypeptide. *FEBS Lett.*, **441**, 59.

126. Perry, S. E. and Keegstra, K. (1994). Envelope membrane proteins that interact with chloroplastic precursor proteins. *Plant Cell*, **6**, 93.

127. Hinnah, S. C., Hill, K., Wagner, R., Schlicher, T., and Soll, J. (1997). Reconstitution of a chloroplast protein import channel. *EMBO J.*, **16**, 7351.

128. Olsen, L. J. and Keegstra, K. (1992). The binding of precursor proteins to chloroplasts requires nucleoside triphosphates in the intermembrane space. *J. Biol. Chem.*, **267**, 433.

129. Theg, S. M., Bauerle, C., Olsen, L. J., Selman, B. R., and Keegstra, K. (1989). Internal ATP is the only energy requirement for the translocation of precursor proteins across chloroplastic membranes. *J. Biol. Chem.*, **264**, 6730.

130. Akita, M., Nielsen, E., and Keegstra, K. (1997). Identification of protein transport complexes in the chloroplastic envelope membranes via chemical cross-linking. *J. Cell Biol.*, **136**, 983.

131. Kouranov, A., Chen, X., Fuks, B., and Schnell, D. J. (1998). Tic20 and Tic22 are new components of the protein import apparatus at the chloroplast inner envelope membrane. *J. Cell Biol.*, **143**, 991.

132. Jackson, D. T., Froehlich, J. E., and Keegstra, K. (1998). The hydrophilic domain of Tic110, an inner envelope membrane component of the chloroplastic protein translocation apparatus, faces the stromal compartment. *J. Biol. Chem.*, **273**, 16583.

133. Lübeck, J., Soll, J., Akita, M., Nielsen, E., and Keegstra, K. (1996). Topology of IEP110, a component of the chloroplastic protein import machinery present in the inner envelope membrane. *EMBO J.*, **15**, 4230.

134. Nielsen, E., Akita, M., Davila-Aponte, J., and Keegstra, K. (1997). Stable association of chloroplastic precursors with protein translocation complexes that contain proteins from both envelope membranes and a stromal Hsp100 molecular chaperone. *EMBO J.*, **16**, 935.

135. Kessler, F. and Blobel, G. (1996). Interaction of the protein import and folding machineries of the chloroplast. *Proc. Natl. Acad. Sci. USA*, **93**, 7684.

136. Caliebe, A., Grimm, R., Kaiser, G., Lübeck, J., Soll, J., and Heins, L. (1997). The chloroplastic protein import machinery contains a Rieske-type iron-sulfur cluster and a mononuclear iron-binding protein. *EMBO J.*, **16**, 7342.

137. Ko, K., Bornemisza, O., Kourtz, L., Ko, Z. W., Plaxton, W. C., and Cashmore, A. R. (1992). Isolation and characterization of a cDNA clone encoding a cognate 70-kDa heat shock protein of the chloroplast envelope. *J. Biol. Chem.*, **267**, 2986.

138. Kourtz, L. and Ko, K. (1997). The early stage of chloroplast protein import involves Com70. *J. Biol. Chem.*, **272**, 2808.

139. Wu, C., Seibert, F. S., and Ko, K. (1994). Identification of chloroplast envelope proteins in close physical proximity to a partially translocated chimeric precursor protein. *J. Biol. Chem.*, **269**, 32264.

140. Moore, T. and Keegstra, K. (1993). Characterization of a cDNA clone encoding a chloroplast-targeted Clp homologue. *Plant Mol. Biol.*, **21**, 525.

141. Shanklin, J., DeWitt, N. D., and Flanagan, J. M. (1995). The stroma of higher plant plastids contain ClpP and ClpC, functional homologs of Escherichia coli ClpP and ClpA: an archetypal two-component ATP-dependent protease. *Plant Cell*, **7**, 1713.

142. Marshall, J. S., DeRocher, A. E., Keegstra, K., and Vierling, E. (1990). Identification of heat shock protein hsp70 homologues in chloroplasts. *Proc. Natl. Acad. Sci. USA*, **87**, 374.

143. Tsugeki, R. and Nishimura, M. (1993). Interaction of homologues of Hsp 70 and Cpn 60 with ferredoxin-NADP+ reductase upon its import into chloroplasts. *FEBS Lett.*, **320**, 198.

144. Schlicher, T. and Soll, J. (1997). Chloroplastic isoforms of DnaJ and GrpE in pea. *Plant Mol. Biol.*, **33**, 181.

145. Yuan, J., Henry, R., and Cline, K. (1993). Stromal factor plays an essential role in protein integration into thylakoids that cannot be replaced by unfolding or by heat shock protein Hsp70. *Proc. Natl. Acad. Sci. USA*, **90**, 8552.

146. Yalovsky, S., Paulsen, H., Michaeli, D., Chitnis, P. G., and Nechustai, R. (1992). Involvement of a chloroplast HSP70 heatshock protein in the integration of a protein (light-harvesting complex protein precursor) into the thylakoid membrane. *Proc. Natl. Acad. Sci. USA*, **89**, 5616.

147. Hemmingsen, S. M., Woolford, C., van der Vies, S. M., Tilly, K., Dennis, D. T., Georgopoulos, C. P., *et al.* (1988). Homologues plant and bacterial proteins chaperone oligomeric protein assembly. *Nature*, **333**, 330.

148. Martel, R., Cloney, L. P., Pelcher, L. E., and Hemmingsen, S. M. (1990). Unique composition of plastid chaperonin-60: alpha and beta polypeptide-encoding genes are highly divergent. *Gene*, **94**, 181.

149. Viitanen, P. V., Schmidt, M., Buchner, J., Suzuki, T., Vierling, E., Dickson, R., *et al.* (1995). Functional characterization of the higher plant chloroplast chaperonins. *J. Biol. Chem.,* **270**, 18158.

150. Bertsch, U., Soll, J., Seetharam, R., and Viitanen, P. V. (1992). Identification, characterization, and DNA sequence of a functional 'double' groES-like chaperonin from chloroplasts of higher plants. *Proc. Natl. Acad. Sci USA,* **89**, 8696.

151. Baneyx, F., Bertsch, U., Kalbach, C. E., van der Vies, S. M., Soll, J., and Gatenby, A. A. (1995). Spinach chloroplast cpn21 co-chaperonin possesses two functional domains fused together in a toroidal structure and exhibits nucleotide-dependent binding to plastid chaperonin 60. *J. Biol. Chem.,* **270**, 10695.

152. Hachiya, N., Alam, R., Sakasegawa, Y., Sakaguchi, M., Mihara, K., and Omura, T. (1993). A mitochondrial import factor purified from rat liver cytosol is an ATP-dependent conformational modulator for precursor proteins. *EMBO J.,* **12**, 1579.

153. Alam, R., Hachiya, N., Sakaguchi, M., Kawabata, S., Iwanaga, S., Kitajima, M., *et al.* (1994). cDNA cloning and characterization of mitochondrial import stimulation factor (MSF) purified from rat liver cytosol. *J. Biochem.,* **116**, 416.

154. Leonhard, K., Stiegler, A., Neupert, W., and Langer, T. (1999). Chaperone-like activity of the AAA domain of the yeast Yme1 AAA protease. *Nature,* **398**, 348.

155. Tauer, R., Mannhaupt, G., Schnall, R., Pajic, A., Langer, T., and Feldmann, H. (1994). Yta10p, a member of a novel ATPase family in yeast, is essential for mitochondrial function. *FEBS Lett.,* **353**, 197.

4 | The roles of the cytosolic chaperonin, CCT, in normal eukaryotic cell growth

KEITH R. WILLISON and JULIE GRANTHAM

1. Introduction

How do polypeptide chains synthesized on soluble polysomes in eukaryotic cytosol proceed to their native states? Is it reasonable to ask such questions of the folding pathways of tens of thousands of different proteins and expect to derive general answers or rules? We suspect that many proteins have sampled enough folding pathways during evolution to have had selected within their sequences chemical properties that allow them to fold up rapidly and concomitantly with chain elongation. The study of proteins which require the cooperation of molecular chaperones to assist their successful folding has provided major insights into understanding some of the problems which proteins face in proceeding to their native states.

The molecular chaperones of eukaryotic cytosol include the major heat shock protein families such as Hsc/Hsp70 and Hsp90. Under normal growth conditions, members of the Hsc70 proteins are mainly involved in interactions with ribosome-bound nascent polypeptide chains; Ssb1, 2 being the most abundant members of this family of molecular chaperones in *Saccharomyces cerevisiae* (1). In contrast, Hsp90 interacts in the maturation pathways of proteins such as the progesterone receptor and various protein kinases. In the case of the progesterone receptor, an important function of Hsp90 is to facilitate loading of steroid hormone.

The chaperonin containing TCP-1 (CCT) is member of the double-ring group of molecular chaperones called the chaperonins. CCT is the only chaperonin found in eukaryotic cytosol. This statement excludes the chaperonins of endosymbiotic organelles, such as mitochondria and chloroplasts, which are found in the cytosolic compartment but which contain GroEL-like eubacterial chaperonins. The true function of CCT is not well understood and there is a considerable divergence of opinion as to its role in general protein folding in cytosol. Its mechanism of action and other properties are also poorly understood, and this review attempts to outline the background of this system and indicate some future directions for experimentation.

It is clear that CCT is involved in the folding of the actins and tubulins *in vivo* and this is why we have concentrated on trying to understand how CCT acts upon these particular substrate proteins. Actins and tubulins are vital structural proteins in cells and are involved in the formation of dynamic polymer systems. Although they are completely unrelated in sequence and evolutionary origin, they have one common property which we believe lies at the core of their folding mechanisms. They both can bind and hydrolyze nucleotide and they use this activity to regulate their filament formation properties and behaviour. Actins bind adenine nucleotides and tubulins bind guanine nucleotides. We are interested in understanding how CCT facilitates the folding of the actin and tubulin polypeptide chains, and when and how the respective nucleotide cofactor is loaded onto the folding molecule. We believe that the binding of specific cofactors to folding proteins may be a frequently encountered problem during evolution of many diverse proteins, and that the study of CCT may provide general insights into how the process is controlled.

2. The CCT complex

2.1 Composition of CCT

Double-ring chaperonins can have seven-, eight-, or nine-fold symmetries constructed from one to eight subunit types (2). CCT is composed of 16 subunits arranged in two back-to-back eightfold pseudo-symmetrical rings. Our present model for the structural composition and arrangement of the rings of hexadecameric CCT is that the eightfold pseudo-symmetrical ring has each position occupied by a particular CCT subunit species (3), and that each of the eight core CCT polypeptides is encoded by an unique gene in all eukaryotes (4). This view derives firstly from characterization of the eight members of the core CCT gene family; first in *Mus musculus* (5–7) and subsequently through the completed *Saccharomyces cerevisiae* (8) and *Caenorhabditis elegans* genome projects (Table 1). The biochemical analysis of CCT by a combination of protein sequencing (5, 9), mass spectrometry (10), and antibody characterization (11) consistently identifies eight CCT polypeptides comprising this chaperonin. (reviewed in (12), see Table 1). The single exception is the tissue-specific CCTζ2 subunit, so far only found expressed in mammalian testis (13).

The evidence for the two rings of CCT having the same arrangement of CCT subunits derives from Liou and Willison (3), who found CCT microcomplexes which contained two or three different CCT subunits. They assumed that the interactions they observed between CCT subunits in microcomplexes reflected the interactions between adjacent CCT subunits within an 8-mer ring, and built a matrix of CCT subunit interactions which yielded to a unique solution from 5040 possibilities (Fig. 1A). Subsequently, Liou *et al.* (14) discovered a single-ring mediated assembly and disassembly cycle of CCT which can be utilized experimentally to incorporate *in vitro* translated CCT subunits into endogenous CCT in rabbit reticulocyte lysate. They incorporated one mouse CCTα subunit, mutated such that it was unable to bind to a monoclonal antibody (mAb 23C) recognizing the C-terminus of CCTα, into CCT 16-

Table 1 Eukaryotic CCT subunits

Protein		Null mutants		
Mammal	**Invertebrate**	**Mouse**	**Yeast[2]**	**Drosophila**
CCTα	Cct1$_p$		Δ	
CCTβ	Cct2$_p$		Δ	
CCTγ	Cct3$_p$	$\Delta/+$[1]	Δ	
CCTδ	Cct4$_p$		Δ	
CCTϵ	Cct5$_p$		Δ	Δ/Δ[3]
CCTζ_1	Cct6$_p$		Δ	
CCTζ_2	–[4]		–	
CCTη	Cct7$_p$		Δ	
CCTθ	Cct8$_p$		Δ	

1. CCTγ inactivated heterozygously ES cells by insertion but the homozygous phenotype is undetermined: (clone WO13B04). Visit the GeneTrap Project for Gene Trap Sequence Tags (http://tikus.gsf.de).
2. CCT genes *of Saccharomyces cerevisiae* are essential (8). The Stanford University 'Function Junction' has annotated information on all CCT genes of yeast and links to the worm CCT genes (http://genome-www.stanford.edu/cgi-bin/SGD/functionJunction).
3. CCTϵ inactivated in *Drosophila embryogenesis* by P-element insertion to give lethality (95).
4. CCTζ_2 is found only in vertebrates and is expressed in a tissue specific manner (13).

mers. Antibody binding measurements suggested that the CCT hexadecamer contained two copies of CCTα, presumably one copy per ring (14). Recently we (15) tested this model directly through immuno-decorating mouse testis CCT with mAb 23C, followed by negative staining and visualization in the electron microscope. In processed images it could be seen that only one antibody molecule was able to bind inside each ring. Furthermore, in all CCT-23C immuno-complexes containing two bound antibody molecules, each antibody was bound to opposite rings. This data supports the idea of a single CCTα subunit being present in each ring, although it does not prove this to be the case, because if there were, say, two CCTα subunits adjacent to each other within the ring, perhaps stearic hindrance could prevent binding of two antibody molecules.

However, we have also observed top views of negatively stained CCT bound by two other anti-CCT subunit monoclonal antibodies; one recognizing an epitope located on the exterior face of the apical domain of the CCTδ subunit (mAb 8G), and the other an epitope located on the exterior face of the equatorial domain of the CCTα subunit (mAb 91A). The antibodies bind to the outside of the rings and very clearly mark their respective subunits (16). Only one bound antibody molecule per ring is observed, supporting the idea that single copies of CCTα and CCTδ are present per ring. CCT-α-actin complexes bound by antibody were examined and, using the α-actin position as a reference, it was found that CCTα and CCTδ occupy the slots in the ring predicted by Liou and Willison's model (Fig. 1A). One can also see, in com-

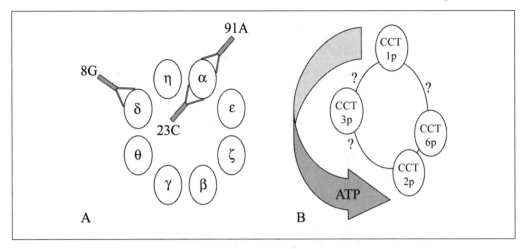

Fig. 1 CCT subunit order. (A) Topological arrangement of the eight CCT subunits within a ring according to model of Liou and Willison (1997) (3). Monoclonal antibodies 91A, 23C, and 8G bind to CCTα subunits (91A, 23C) and CCTδ subunits (8G), respectively, as shown by Llorca *et al.* (1999) (16), Grantham *et al.* (2000) (15). (B) Genetic hierarchy of CCT subunit ATP-binding/hydrolysis activity as proposed by Lin and Sherman (1997) (53).

parisons of the CCT-antibody images, that any potential effects of stearic hindrance are likely to be less of a problem with those antibodies which bind to the outside of the CCT ring than are those binding the inside of the cavity (15, 16). Clearly, the use of further monoclonal antibodies recognizing not only the other CCT subunits, but also different domains of CCT subunits, will be useful in solving the topology of the subunit arrangement and also the phasing between the two rings, and this work is in progress. Until a high-resolution X-ray structure of CCT is solved, this immuno-decoration approach promises to be a productive way forward in verifying our model. Furthermore, it is clear that even if an atomic structure of 'core' CCT were determined, outstanding problems would remain in some systems. Such questions as 'Where are the tissue-specific CCTζ2 subunits (13) located in the rings of mouse testis CCT?' spring to mind. Monoclonal antibodies which discriminate between CCTζ1 and CCTζ2 subunits could help solve the problem of whether these subunits replace each other in order to occupy the same position in the ring, or not. Analogous problems exist in the study of the four-ringed eukaryotic proteasome composed of two copies of each of 14 subunit species; an atomic structure has been obtained of the yeast proteasome 28-mer (17), but not of the mammalian immune proteasome complex, which has tissue-specific γ-interferon-inducible subunits predicted to exchange positions by substitution with their constitutively expressed homologues.

2.2 Structure of CCT

Substantial advances in structural analysis of Type II chaperonins have occurred over the past 3 years and these have been well reviewed recently by Gutsche *et al.* (18). The first high-resolution structure determined was that of an isolated apical domain of the α-subunit of the thermosome of *Thermoplasma acidophilum* (19). The core of the domain resembles the apical domain fold of GroEL (20) but a novel, and mainly α-helical, structure erupts from its upper surface and is called the helical protrusion. In the X-ray structure of the α/β thermosome hexadecamer (21) the eight apical protrusions in each ring point towards the central, pseudo-eightfold axis of the chaperonin and seem to cap the cavities. Thus it seems that Type II chaperonins have their own inbuilt lid and perhaps this explains why a GroES-like co-chaperonin has never been observed in this system. Indeed, because phylogenetic analysis shows the Type I chaperonins to be a much more homogeneous group than the Type II group (7), and since all Type I chaperonins studied also possess GroES co-chaperonins, we argue that the GroES co-chaperonin system represents a later evolutionary development in chaperonin function specific to eubacteria and the endosymbiotic organelles of eukaryotes (12). Since one presumes that the original chaperonin evolved its functions in archaebacterial-like species using a single polypeptide species, it also seems feasible that the evolution of the Type I chaperonin system involved loss of the helical protrusions and subsequent rescue by GroES of the lost function, giving rise to an assembly constructed from two separate gene products. The one common feature of the Type II apical protrusions and GroES which we can discern is the ring of conserved glutamate residues in the opening at the top of the roof formed by each structure (see fig. 1 (21), and fig. 7 (22)). Perhaps these negatively charged residues facilitate the exit of water molecules through the orifice as the lids close up their respective cavities and enclose substrate.

The structural studies of the Madrid group, who are in collaboration with us, have shed significant light on several structural aspects of CCT behaviour (16, 23).

These studies have analysed highly purified CCT isolated from mouse testis cells, in the transmission electron microscope, using the techniques of single-particle analysis and image reconstruction of both negatively stained and vitrified samples. Chaperonins are ideally suited for this approach due to their large size and inherent symmetry (24). Initially, the behaviour of CCT in adenine nucleotides was examined by negative staining and image processing of both top and side views in two dimensions only (25). CCT takes up stain readily because it has two openings at its ends and many around its circumference. Side views of both the nucleotide-free apo-CCT and ADP-CCT structures appeared symmetric in respect of the two rings (Fig. 2A) whereas the rings of ATP-CCT and AMP-PNP-CCT showed asymmetry between each other (Fig. 2B). One of the rings of ATP-CCT showed upward and outward movements of the apical domains whereas the opposite ring resembled an ADP-like ring. Subsequently, side views and top views of apo-CCT and ATP-CCT particles were collected in cryo-electron microscopical analysis of vitrified particles (23). Three-dimensional reconstructions were obtained at 28Å resolution after image processing of several thousand particles (Figs 2C and 2D). The ATP-bound form of CCT revealed the asymmetric folding conformation of this chaperonin and, furthermore, the enclosed hemispheric cavity of the ATP-ring fitted very well the rings of the X-ray structure of the thermosome (21). The protrusions of the eight apical domains appear like fingers capping the cavity to produce a flimsily covered dome structure.

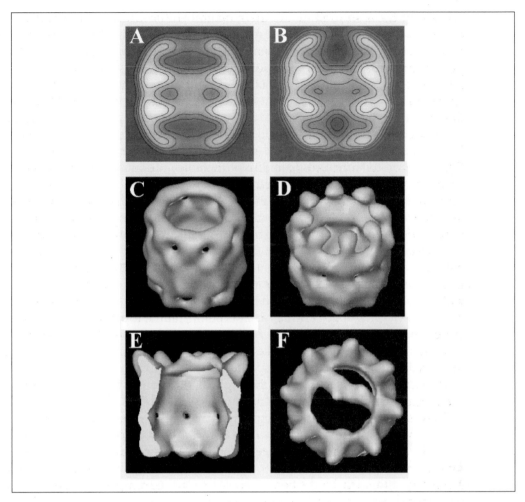

Fig. 2 Two-dimensional and three-dimensional reconstructions of mouse testis CCT. Two-dimensional projection of negatively stained images of apo-CCT (A) and ATP-CCT (B), showing striking upward movements of both the equatorial and apical domains of the CCT subunits upon incubation in ATP (25). Three-dimensional reconstruction of cryo-electron microscopical images of apo-CCT (C) and ATP-CCT (D) (23). (E, F) Three-dimensional reconstruction of cryo-electron microscopical images of apo-CCT-α-actin complexes (16). E shows cut-away section of complex. F shows top view of complex. The α-actin is the density seen spanning the top of the cavity, resting between two CCT subunits in the 1#:#4 position.

The fingers are probably very flexible and we have, in fact, obtained images of the monoclonal antibody 23C binding from outside across the upper rim to the inside of the cavity of an apo-CCT ring in which distortions of the rings seem to be induced by the antibody binding (15). Remarkably, these antibody-bound CCT complexes can fold actin and tubulin at normal rates in rabbit reticulocyte lysate. Klumpp *et al.* (1997) found two conformations of the apical protrusion in crystals of the α-thermosome isolated apical domain (19). It will be interesting to learn if any functionality in folding is due to apical domain conformational changes during the folding cycle of

Type II chaperonins. All the many structural studies carried out so far with GroEL show its apical domain to be homotypic in structure and that the folding cycle consists of rigid body movements between the three major domains (26, 27). Furthermore, evidence suggests that the substrate is not completely sequestered by the helical protrusions, since, in reticulocyte lysate, CCT-bound actin remains highly susceptible to proteolysis by trypsin during the folding cycle (15). In contrast, in GroES capped GroEL-substrate complexes the enclosed substrate is highly resistant to proteases, consistent with the fact that GroES seals the cavity completely (26).

Certain features of the holo-thermosome X-ray structure determination may be explained by the inherent flexibility of the apical domains. The conditions used for crystallization favoured a symmetrical, doubly closed hexadecamer form resembling a soccer ball (21). Addition of Mg-ADP and Mg-ADP-AlF$_3$ to such crystals allowed assignment of residues involved in nucleotide interactions and suggested a mode of nucleotide binding highly analogous to GroEL (21). Nevertheless, the closed Mg-ATP-thermosome form may not resemble the Mg-ADP bound solution state and seems, in fact, to be induced by sulfate ions (28). Small-angle neutron scattering is a technique which provides information about the overall structure of populations of macromolecules in solution and it has been used recently to probe the conformations of the *Thermoplasma acidophilum* chaperonin upon ATPase cycling (28). This study found that not only is apo-chaperonin open in solution, but also that ATP binding induces its further expansion. The closure of the lid seems actually to occur during ATP hydrolysis, but before phosphate release, and may represent the rate-limiting step of the nucleotide cycle. This step appears to reflect a transition from a tight state to a loose one. In the electron microscopical studies on CCT the exact nature of ATP-CCT was not measured and, indeed, we specifically stated that one could only presume the hemispherical ring was ATP-bound, based on comparisons with apo-CCT and ADP-CCT forms (23, 25). Recently we have examined the stability of nucleotide bound forms of CCT in urea (29). We found that incubation of CCT in aluminium fluoride to mimic the transition state of the CCT ATPase reaction was not destabilizing, but that when conditions were such that ATP-CCT was able to hydrolyse ATP, it became extremely unstable, even at low urea concentrations. It is possible that a loose state of CCT is an ADP-P$_i$ intermediate, as observed for the thermosome.

In broad terms, the Type II chaperonins have much in common with the binding of GroES and nucleotide to GroEL (24, 26). In the case of CCT, we suggested a GroEL-compatible mechanism whereby apo-CCT complex resembles the substrate acceptor form (presumably a structure having at least one ADP-CCT ring *in vivo*) and, upon ATP-binding, the cavity of the ATP ring expands and is enclosed by the helical protrusions (23). Clearly, much more work is required to fully understand the allosteric regulation of CCT by nucleotide and the contribution of substrate binding to the process.

We have begun to understand why CCT-bound substrate can be protease-accessible through our determination of a 3D-reconstruction of an apo-CCT-α-actin complex by cryo-electron microscopy (16). The bound actin is suspended between the apical domains of a pair of subunits and can be seen as a bar of density resting across an open cavity (Figs 2E, 2F), and the actin folding intermediate would be expected to be freely accessible to the bulk cytosol and its constituent proteins, assuming such CCT-actin complexes exist for an appreciable lifetime *in vivo*. The CCT-α-actin complex shows that actin can bind to apo-CCT in two different ways. The interaction is CCT subunit-specific and geometry-specific since α-actin binds through CCTδ and CCTβ, or CCTδ and CCTε. The two binding conformations differ by rotation of 135° and share the same 1:4 arrangement with respect to the CCT subunit positions (see Fig. 1A).

This 30 Å-resolution 3D-reconstruction of CCT-α-actin complex further shows that actin can interact with a similar region of the CCTδ, β, and ε apical domains located well below the helical protrusion. Several authors argue that the helical protrusions of the α/β-thermosome must contain the sites for substrate binding because they contain hydrophobic residues (19, 21, 30). However, Willison (1999) suggested that the substrate binding sites of CCT could reside in the region of the CCT apical domains corresponding to the known substrate binding region of GroEL, even though the regions are rather different in structure to each other (12). This proposal is based on bioinformatic and phylogenetic analysis of Type II sequences by Counsell and Willison (unpublished results). There is an abundance of charged residues located in these regions of eight CCT apical domains, and we suggest that CCT recognizes charged regions in its substrates. We have recently proposed that charged residues in β-actin are involved in CCT binding, presumably through electrostatic interactions (31). Lin *et al.* (32) show severe effects of mutation of lysine residues in the apical domain of CCT6p on tubulin folding in *Saccharomyces cerevisiae in vivo*. Hynes and Willison (2000) have shown that several interactions between CCT and actin folding intermediates are stable in mixed micelle detergent buffers. Such buffer conditions preserve most known antibody-antigen interactions, for example, but are known to disrupt protein-protein interactions mediated through hydrophobic interactions (31). Hynes and Willison (2000) propose that certain loops and strands found on the surface of native actin are involved in mediating interactions with CCT, based upon analysis of actin derived oligopeptide sequences which bind avidly to CCT. Structural analysis of holo-CCT-actin complexes, or of CCT apical domains bound by substrates or fragments thereof, are now required to identify the precise location of the substrate binding residues located on CCT apical domains.

2.3 Substrates of CCT *in vitro* and *in vivo*

The actins and tubulins are substrates for CCT *in vivo* in mammalian cells and in yeast (8, 33) and together represent the major substrates folded by CCT. Recently, however, several more CCT binding proteins have been identified (Table 2). It is not yet clear whether all of these proteins themselves require interactions with CCT to reach their native state, or whether some are regulatory proteins which modulate the action of CCT during the folding of actin and tubulin. Studies directed at determining the ability of other proteins to form binary complexes with CCT (34) showed that, although proteins aside from actin and tubulin (Cap-binding protein, cyclin B, H-ras, c-myc and TCP-1) were shown to bind to CCT, they did so with a much lower

Table 2 CCT substrates and interacting proteins

Substrate	Binds to CCT or CCT subunit	CCT binding assay				Folded by CCT	In vivo substrate	Reference(s)
		In vivo	In vitro	Rabbit reticulocyte lysate	Yeast 2-hybrid			
Actin	Yes	Yes	Yes	Yes	–	Yes	Yes	(57, 96)
Tubulin	Yes	Yes	Yes	Yes	–	Yes	Yes	(57, 96, 97)
Luciferase	Yes	No	Yes	Yes	–	Yes	?	(59, 60)
Gα transducin	Yes	Yes	–	Yes		Yes	Yes	(63)
Cyclin E	Yes	Yes	–	Yes	CCT β,δ η	Yes ?	?	(47, 98)
EBNA 3	Yes	No	No	–	CCTε	?	?	(99)
Myosin II HMM	Yes	No	No	Yes	–	?	?	(71)
VHL tumour suppressor	Yes (transfection)	Yes	No	Yes	–	?	?	(100)
NF-H fragment	Yes	No	Yes	–	–	?	?	(101)
Hepatitis B virus capsid	Yes	–	Yes	–	–	Role in assembly	?	(102)
HIV TAR	CCTδ	–	–	–	–	Co-purification	?	(103)
Microtubules	Individual subunits	–	Yes	–	–	Role in assembly?	–	(104)
F-actin	Yes	–	Yes	–	–	Role in assembly?	–	(105)
Hsc70	Yes	Yes	–	Yes	–		–	(37, 47)
Prefoldin	Yes	Yes	Yes	No	–	?	Cofactor?	(68, 73)
Hop/p60	Yes	Yes	Yes	Yes	–	–	–	(106)
Range of denatured proteins	Yes	–	Yes	–	–	–	–	(34, 35)

Dashes indicate 'not determined'.

affinity than did actin and tubulin. Additionally, these unfolded test proteins bound much more rapidly to CCT than actin and tubulin. This may be due to rapid, non-specific interactions occurring between CCT and early folding intermediates, rather than the specific interactions which occur between CCT and actin (16). A further study of CCT binding proteins (35) showed that the previously studied group of denatured binding proteins bound to CCT when added as a mixture, but another series of test proteins when purified showed selective binding to CCT. It is possible that fully denatured proteins, and especially, mixtures of proteins, have a tendency to bind to CCT, but this activity may not reflect genuine interactions occurring between CCT and its bona fide substrates *in vivo*.

Thulasiraman *et al.* (1999) suggest that between 15 and 22% of newly synthesized proteins interact with Hsc70, and between 9 and 15% interact with CCT in mammalian cell lines (36). The latter is a surprisingly high number when considering the limited range of CCT substrates which have been identified to date (Table 2). These measurements also diverge from the previous estimates of Sternlicht *et al.* (1993) *in vivo* in cell lines, showing actin and tubulin as the predominant substrates for CCT in pulse chase experiments after recovery of CCT from cell extracts using monoclonal antibody. However, Thulasiraman *et al.* suggest that, as other studies do not utilize an ATP-depleting step to stabilize CCT/substrate interactions, these may have been missed in previous studies and that CCT, indeed, may have a broader range of folding substrates than previously suggested (36). Further analysis in various cell lines and tissues will be required to gain a clearer view of the number of CCT substrates *in vivo*. Furthermore, once we begin to understand the specificity of binding by these candidate substrates, as is occurring for actin (16), a firmer base for assigning whether a CCT-interacting protein is a bona fide substrate, or not, will emerge.

2.4 Regulation of CCT expression

Testis cells and rapidly growing embryonic cells have between 10^5 and 3×10^5 CCT hexadecamers per cell (5, 37). Lewis *et al.* (1992) found that CCT subunits were not induced by heat shock in mammalian cells and Tcp-1/Cct1 mRNA is not heat induced in *Saccharomyces cerevisiae* (38) or *Tetrahymena* (39). This is a slightly surprising result, since GroE and thermosome are highly up-regulated by heat shock treatment; the thermosome can comprise 50% of total cell protein after heat shock (40, 41). CCTα mRNA has been shown to be induced by two- to threefold upon heat shock in human Jurkat cells by array analysis (42). Recently, Kubota's laboratory found that, in various mammalian cell lines, CCT subunit levels can be induced by the chemical stressors, sodium arsenite and L-azetedine-2-carboxylic acid; 1.6–3.1-fold induction during continuous stress and 1.6–3.7-fold during recovery from stress (43). Interestingly, CCT was induced transiently by continuous chemical stress, peaking at 6 hours of continuous treatment, whereas Hsp70 was still rising beyond 10-fold induction after 12 hours' treatment (43). Perhaps CCT is recovering a specific subset of unfolded substrates early during stress, whereas Hsp70 continues to deal with a larger group of unfolded polypeptide chains remaining in the cell for a longer time. Kubota and colleagues have analysed gene regulatory sequences of the nine mouse CCT genes in order to understand how the CCT genes are transcriptionally regulated (44). The production of CCT seems predominantly to be coordinately regulated at the RNA level in order to provide equivalent levels of the eight protein subunits. Each CCT gene has consensus heat shock transcription factor (HSF) binding sites which appear functional in transient transfection assays because they respond to raising the levels of the HSF-1 and HSF-2 transcription factors (44). Heat shock elements may be involved in constitutive expression of CCT genes. Also, perhaps the high expression of CCT in the mammalian testis, and the induction of the testis expressed CCTz2 gene, are mediated through the HSF-2 activity known to be abundantly expressed in the testis.

Studies on the CCT gene family in *Tetrahymena* species, which are unicellular ciliates, have demonstrated a strong correlation between CCT and tubulin gene expression during re-growth of cilia after experimental deciliation (39, 45). One also supposes that the reason for the huge abundance of CCT in mammalian testis is to cope with the large amount of tubulin required for synthesis of the sperm tail flagellum (2, 46). However, this simple model that supposes CCT levels should correlate with tubulin and actin abundance is not generally applicable, since, for example, CCT levels differ one hundred-fold between mouse testis and heart extracts. Heart contains large amounts of actin and it is puzzling that it has little CCT, as is also the case with skeletal muscle. In fact, it seems that CCT mRNA and protein levels in mouse tissues and mammalian cell lines actually correlate most tightly with growth rate (47). In various diverse organisms, treatments which cause differentiation and cell growth arrest generally down-regulate CCT protein (47, 48) and RNA levels (47, 49). Kubota's laboratory have performed a careful study on CCT expression in seven mammalian cell lines (47) and find that CCT subunit expression is directly correlated ($p > 0.01$) with growth rate as measured by [^{3}H] thymidine incorporation. Furthermore, maximum CCT protein and mRNA levels are observed at the G_1/S transition through early S-phase and, remarkably, tubulin seems to be maximally associated with CCT at early S-phase, not G_0/G_1 phase. In this study the correlation between CCT, actin, and tubulin levels was not tight, and the authors suggest that CCT may be mainly involved in folding non-cytoskeletal substrates during rapid growth (47).

We now suggest another hypothesis which supposes that CCT is more than just a folding machine. Perhaps the actual function of CCT is not just to fold actin and tubulin, but also to count or measure in some way newly synthesized actin and tubulin proteins. Through CCT's ability to capture folding intermediates, it can discriminate new actin and tubulin monomers from the bulk population. We can estimate crudely that in mammalian cells, at any moment in time, there is a pool of around 10^4 newly synthesized, actively folding tubulin molecules compared with the total cell tubulin population of around 10^7 α/β-tubulin heterodimers (33), and similar numbers probably pertain to actin. In yeast, doubling once every 120 minutes at 30 °C, there are 54 molecules of mRNA and 205 000 molecules of Act1p (actin) per cell, and an estimated 540 nascent actin chains per cell (1). Thus, the potential exists for CCT to act as a counting device which can examine the rates of actin and tubulin synthesis. Rather than have a mechanism to count each member of the entire actin and tubulin population, CCT could be measuring the new polypeptides passing into the pool (i.e. sample 0.1% of the total population). The output of the counting process could be a qualitatively different actin or tubulin monomer, perhaps post-translationally modified, or in some different nucleotide state to the bulk population. Indeed, the substrate folding intermediates trapped on CCT (16) probably assume unique conformations not attained again during the life cycle of the protein, making this a possible target conformation for modification enzymes. This counting device idea has many attractive possibilities for adding information to the actin and tubulin systems by extracting information from folding pathways using the nucleotide binding, exchange, and hydrolysis activities of the eight CCT subunits. Thermo-dynamic considerations show that one can 'purchase' much information from eight ATP molecules and, in combination with a trapped folding intermediate which can be captured and released in an ordered mechanism, information is being added to the system (see also ref. 12). The fact that CCT is the first machine identified which can bind both actins and tubulins also seeds in us the idea that it could be a component of a system able to integrate measurements of the fluxes of actin and tubulin synthesis with respect to each other. Perhaps this could be carried out through direct com-petition for binding sites, or using inter-ring cooperativity to affect substrate binding and folding rates. In the case of actin, it is established that ATP-actin polymerizes much more rapidly than ADP-actin (50). Were CCT to be producing ATP-actin, for example, such an activity might ensure the incorporation of newly synthesized actin monomers at sites of microfilament assembly.

These ideas that CCT is some sort of counting device are necessarily, at present, rough, but it is clear that the actin and tubulin folding pathways are genetically programmed by CCT and these proteins seem unable to fold unaided *in vivo*. As this CCT-dependence of actins and tubulins became fixed during evolution, the oppor-tunity probably arose for selection processes to build extra features into actin and tubulin folding intermediates, and hence into the cytoskeletal cycle. Being able to measure and/or record rates of entry of new actin and tubulin chains into the bulk monomer pools seems to us to be a useful sophistication of the cytoskeletal cycles, which are generally considered thermodynamically closed systems by most workers, that is, the input and output rates of monomers are in general not thought relevant to the functional polymerization cycles.

2.5 Analysis of CCT in yeast

Early studies on CCT genes in *Saccharomyces cerevisiae* indicated their importance in cell growth and cytoskeletal mediated processes. All eight CCT genes are essential for cell viability (8) and many conditional mutations in CCT subunits have severe effects on both actin and tubulin mediated processes (Table 3). All four mutations in Cct1p so far sequenced alter amino acid residues involved in ATP binding and/or hydrolysis (51, 52), based upon analysis of the homologous residues in the atomic structure of the ATP-thermosome structure (12, 21). It is likely that mutations in ATP

Table 3 Mutants of CCT subunits in *Saccharomyces cerevisiae*

Subunit species	Mutant	Mutation	Tubulin effect	Actin effect	Other phenotype	Reference(s)
Cct1p	*tcp1-1*	$D_{96} \rightarrow E$	+	+		(52, 107)
	tcp1-2	$G_{423} \rightarrow D$	+	+		(52)
	tcp1-3	$G_{45} \rightarrow S$	+	+		(52)
	tcp1α-245	$G_{48} \rightarrow E$	+	−		(51)
Cct2p	*tcp1β-270*	ND	+	−		(51)
	tcp1β-326	ND	+	−		(51)
	bin3-1	ND	+	+		(108)
	bin3-2	ND	+	+		(108)
	bin3-3	ND	+	+		(108)
	bin3-4	ND	+	+		(108)
Cct3p	*bin2-1*	ND	+	+		(110)
Cct4p	*anc2-1*	$G_{345} \rightarrow D$	−	+		(16, 54)
Cct6p	90 mutants[b]	Alanine scan	See ref. 32	See ref. 32	See ref. 32	(32)
		Dosage ↑	ND	+	tor2[kin-] rescue	(109)
Cct8p	−	Dosage ↑	ND	ND	Morphogenesis[c] suppression	(55)
	−	Dosage ↓ 50%	ND	ND	Defective morphogenesis	(55)
	CaCct8-Δ1p	Δ 1-95	ND	+	Morphogenesis suppression of Ras2[Val14] and EFG1	

[a] ND, not determined. The symbols + and − denote presence and absence of abnormal organization of tubulin actin in the temperature sensitive mutants, respectively. Modified from ref. 6. Mutations in subunit genes are listed by their mutant allele names as described in the primary papers.

[b] A wide range of effects on CCT activity were observed in this study of mutations, introduced throughout Cct6p; some affected both tubulin and actin phenotypes and others, particularly mutations in the apical domain, were highly tubulin specific.

[c] Cct8p study mainly performed in yeast, *Candida albicans*. Effects also observed on pseudo-hyphae formation in *Saccharomyces cerevisiae*.

binding regions have general effects on the activity of CCT, as indicated by the work of Lin and Sherman (53) who showed a hierarchical order of severity of such mutations on CCT function. In this genetic study of four subunits, Cct1p sits at the top of the pathway and Cct6p at the bottom. Some mutations which would be expected to compromise severely the ATP binding of Cct6p and/or hydrolysis, nevertheless, have few observable phenotypic effects *in vivo*. Lin and Sherman interpreted their data in terms of cooperative utility of ATP binding sites around the CCT ring (Fig. 1B) and possibly Cct1p is lying near the site of initiation of the cooperative process (see ref. 12 for further discussion).

Obviously there is now great interest in trying to identify the regions of CCT subunit apical domains which are involved in interactions directly with substrates.

The *anc2-1* mutant allele of the *Saccharomyces cerevisiae* *CCT4* gene encodes the yeast orthologue of the δ subunit of mammalian CCT. We surmised that *anc2-1* might have suffered an alteration in its apical domain due to the design of the genetic screen in which this mutant was isolated (54). Briefly, *anc2-1* was recovered in a search for extragenic mutations that fail to complement temperature-sensitive alleles of the single yeast actin gene ACT1, hence, **a**ctin-**n**on-**c**omplementing. *Anc2-1* exacerbates the phenotype of the semi-dominant, temperature-sensitive allele *act1-4* (Glu259Val) and other mutations within subdomain 4 of actin. However, *anc2-1* also *complements* five actin alleles; *act-1-1*, *act1-2*, *act1-122*, *act1-124* and *act1-125*, whose mutations lie in subdomain 2 of *ACT1* (fig. 6 of (16)). The wild-type *CCT4* and mutant *anc2-1* genes were found to differ by a single nucleotide change (G-A) which changes Gly345 to Asp. This glycine residue lies on a β-strand found on the outside surface of the globular part of the Type II thermosome apical domain and is absolutely conserved in all chaperonins. Since substrate is thought to interact with the surface of the apical domain facing the cavity, it is likely that *anc2-1* (Gly345Asp) has altered properties in apical domain movements required for binding and/or release of substrates, rather than in direct interaction with substrate. This interpretation is consistent with the phenotypes of diploid yeast strains which contain wild-type and mutant copies of both actin and Cct4p proteins CCT, since the chaperonin complexes containing a mutant Cct4p subunit must still be able to provide effective interaction with wild-type actin folding intermediates in addition to increasing the yield of functional mutant actins; nevertheless, the properties of this mutant strongly support the picture from the 3D-reconstruction of CCT-α-actin, (16), that CCTδ interacts directly with the subdomain 2 of actin. Hynes and Willison (2000) suggest that a main binding region in human β-actin, called β-Actin Site I, is the long, extended loop at the tip of the subdomain 2 (31). Ursic *et al.* (1994) examined synthetic lethality effects between CCT subunit ATP-site mutations and actin and tubulin alleles, and found some strong effects which, again, indicate specificity of interaction between chaperonin and its substrates (52). These studies mostly examined effects on vegetative growth, but other studies have begun to examine effects of CCT subunit mutation of dosage levels on morphogenesis in yeast (Table 3). Particularly striking is the involvement of Cct8p in regulation of hyphae formation in *Candida albicans* (55). It seems that full wild-type activity of CCT may be absolutely required when cells are undergoing some of the drastic cytoskeletal remodelling events which are required for differentiation.

An interesting comparison of Cct RNA levels during vegetative growth and sporulation can be carried out using the latest available array data (http:cmgm. stanford.edu/pbrown/explore/index.html). All eight Cct subunit RNAs are strongly co-regulated during logarithmic phase growth, but the co-regulation seems much less tight during sporulation. Perhaps specific Cct subunits have special roles in the sporulation process in addition to forming the 'core' CCT hexadecamer. Fascinatingly, examination of the global response of *Saccharomyces cerevisiae* to the alkylating agent, methylmethanesulfonate, shows Cct8 mRNA to be induced 4.9-fold, but none of the other seven subunits are up-regulated or down-regulated (56). This is consistent with the suggestion that some CCT subunits are markers or regulators of CCT activity (3). Clearly, further analysis of Cct in yeasts will be a powerful tool in coming to a true understanding of the role of CCT in growth and growth control *in vivo*.

2.6   Mechanism of action

The function and mechanism of action of CCT in folding of substrates is controversial and there are several competing models containing mechanistic components which must be mutually exclusive if a single chaperone-mediated folding pathway exists. Cowan's group first showed that cytosolic chaperonin could capture chemically denatured recombinant β-actin and that, in the presence of ATP, it could then convey a proportion of the bound actin to the native state (57). Hartl and colleagues repeated these initial observations (58), and proceeded to suggest that actin is captured by chaperonin co-translationally during its synthesis on the ribosome, and folded in a single round of interaction (36, 59–62). Horwich and colleagues challenged the single round of interaction model by suggesting that actin undergoes multiple rounds of interaction with CCT until native state is attained (63), analogous to their general model for the action of GroEL in protein folding. Analysis of tubulin folding by CCT supports this concept of multiple rounds of substrate interaction with CCT, because tubulin seems to be able to cycle between chaperonins during folding *in vitro* (64, 65). When either α- or β-tubulin are sequestered from dimer by the requisite cofactor, cofactor D in the case of β-tubulin and cofactor E in that of α-tubulin, the remaining subunit decays to a non-native state within seconds and is capturable by chaperonin (66). The requirement of interactions with CCT for the production of native actin and tubulin represents one step on an apparently complex pathway, in which interactions with other pre-CCT chaperones (and in the case of tubulin, post-CCT cofactors) will occur; see model of Cowan and Lewis, 1999 (66).

There have been other studies directed at determining the folding pathways of the major CCT substrates, actin and tubulin. Welch's laboratory utilized an anti-puromycin antibody to identify nascent polypeptides in cell lysates and to investigate the associated chaperones (67). This study identified Hsp70 and Hsc70 as the major chaperones associated with truncated, puromycin-released actin chains, while CCT was predominantly associated with full-length polypeptides. Thus, this is in contrast to the studies by Frydman *et al.*, which used firefly luciferase as a model protein and found CCT in a co-translational complex which included the chaperones Hsp70 and Hsp40 (59, 60). However, in the case of actin it was also observed that, in immunoprecipitation experiments using antibodies to Hsc70 and CCT, Hsc70 interacted with short nascent actin chains in cell lysates, while CCT interacted in a co-translational manner with longer nascent chains (60). Recently this laboratory has performed photo cross-linking of nascent actin chains to CCT and proposes that actin polypeptides as short as 133 amino acids initially interact with CCT and that CCT may directly interact with the ribosome (62). There is much more work to be done in order to understand the timing of substrate capture by CCT *in vivo* and the ATP-dependent folding cycle.

2.7   GIM/prefoldin

In addition to Hsp70, which has been shown in several studies to interact with nascent actin polypeptide chains, a recently discovered chaperone, named prefoldin, has been implicated in the folding pathway of actin and tubulin (68). Prefoldin, which was identified by its ability to bind unfolded actin, consists of six subunit polypeptides with a size range of 14–23 kDa. Prefoldin is the mammalian orthologue of the yeast Gim protein complex, identified through a set of **G**enes **i**nvolved in **m**icrotubule biogenesis in yeast (69). The observations of Vainberg *et al.* (68) are supported by Hansen *et al.* (70), who showed that prefoldin will bind co-translationally at 20 °C to actin following the synthesis of approximately the first 145 amino acids, and remained bound until transfer to CCT. Prefoldin interacts with tubulin following the synthesis of at least 250 residues. The study by Vainberg *et al.* (68) demonstrated that prefoldin could direct target substrate proteins to CCT, even in the presence of a potentially competing Type I chaperonin, in a nucleotide-independent manner. It was suggested that this transfer occurred via direct interactions between CCT and prefoldin, because prefoldin was shown to interact with immobilized CCT but not GroEL; an interaction which was maintained in the presence of ATP. The ability of CCT substrates to be targeted to CCT regardless of the nucleotide present is inconsistent with the observed nucleotide-induced conformational changes of CCT (23), in which CCT is shown to undergo asymmetric changes in response to ATP binding similar to those seen with GroEL (24). These changes presumably reflect different substrate acceptor states of CCT, so it is a little surprising that in the presence of different nucleotides there is no difference in the targeting of substrate proteins to CCT by prefoldin.

Recently, further substrates for CCT in addition to actin and tubulin have been identified both *in vitro*, in the case of myosin (71) and *in vivo*, in the case of G-α transducin (63), but it would still appear that the actin and tubulin represent the major folding substrates of CCT. This raises the question of whether prefoldin is a co-chaperone specific to actin and tubulin, or whether it is a general chaperone. Hansen *et al.* suggest that prefoldin may be a specific co-chaperone for actin and tubulin, based on the inability of both prefoldin and CCT to interact with cytosolic yeast invertase, green fluorescent protein, chloramphenicol acetyltransferase, and luciferase (70). However, the identification of prefoldin homologues in strains of archaeabacteria (for example, *Methanobacterium thermoautotrophicum*) has led to the suggestion that, as the archaea do not possess an Hsp70 chaperone system (12, 72), prefoldin (in this study, named MtGimC) is a general chaperone in these organisms (73). Leroux *et al.* identified the prefoldin homologue *in M. thermoautotrophicum* which was shown to exist as a hexamer formed from two α and four β subunits. Eukaryotic prefoldin subunits 2 and 5 correspond to α while prefoldin 1, 3, 4, and 6 correspond to the β subunit. The question of the substrate specificity of MtGimC was addressed by assaying for its ability to prevent aggregation of chemically denatured proteins. Unfolded hen lysosyme, bovine mitochondrial rhodanese, and glucose dehydrogenase were all shown to undergo suppressed aggregation in the presence of MtGimC. Many strains of archaea contain FtsZ, a tubulin homologue, but to date no homologues of actin have been identified. We propose that actin evolved in eukaryotes by horizontal gene transfer (See Fig. 4). Therefore, if eukaryotic prefoldin is, indeed, specific to actin and tubulin, perhaps this reflects different subunits being required for interactions between actin and tubulin, and eukaryotic prefoldin may have evolved a more complex subunit composition to accommodate interactions with actin as well as tubulin. It is suggested by Leroux *et al.* (73) that, as they were unable to show an interaction with this archaeal prefoldin complex and CCT, the more diverse subunit composition of eukaryotic prefoldin may be required for this interaction. However, it was not possible to show an interaction between MtGimC and the archaebacterial Type II chaperonin, the thermosome.

There are conflicting reports regarding whether there is a direct interaction between CCT and prefoldin, and the exact role of prefoldin in the actin folding pathway. The first view suggests that prefoldin binds to nascent actin chains, and upon completion of synthesis actin is targeted to CCT. The second view supports a role for Hsc70 in capturing nascent actin chains, and that it is the premature release of actin from CCT that is prevented from occurring by interactions between prefoldin, actin and CCT. Siegers *et al.* (74) showed that approximately 10% of cellular GimC could be co-immunoprecipitated with myc-tagged CCT and suggest that as the levels of GimC and CCT are relatively similar, this interaction may be real, although transient.

Although the exact nature of prefoldin interactions with actin and tubulin during their biogenesis are not yet known, increasing evidence is emerging to suggest a role for prefoldin in the production of native actin and tubulin. Therefore, one asks whether interactions with prefoldin are an obligate step in the biogenesis of these proteins as well as the demonstrably required interactions with CCT. Vainberg *et al.* (1998) showed that deletion of PFD5 in yeast which encodes the prefoldin 5 subunit results in slow growth, but is not lethal, and that microtubule structures appeared to be identical to wild-type, suggesting that while prefoldin may not be essential, it may significantly contribute to levels of native tubulin available in the cell (68). Siegers *et al.* (74) reported that the deletion of different GIM genes in yeast had varying effects on the viability of conditional mutations of the CCTα subunit and that single deletions of Gim subunits resulted in a fivefold decrease in the folding of actin (an effect which was also seen with double and triple Gim deletions). Also, actin was found to be accessible to a GroEL trap in Gim1 deficient stains. Again, this suggests that although Gim/prefoldin is not required for the production of native actin, it does have a significant effect on the efficiency of actin folding in yeast.

2.8 Tubulin cofactors

Actin is released from CCT in either a native state, or committed to reaching its native state. However, interactions with a series of tubulin-specific cofactors (A, B, C, D, and E) are required to form assembly-competent α/β tubulin. Tubulin is released from CCT in a quasi-native form which is recognizable by tubulin cofactors (64, 75). Quasi-native tubulin has been shown to consist of three domains, only one of which interacts strongly with CCT, while another domain, which binds only weakly to CCT, was suggested to be involved in the release of tubulin from CCT (76). In support of tubulin being released from CCT in a quasi-native form, α-tubulin was shown to form its GTP binding pocket while bound to CCT, and CCT-bound α-tubulin can possess non-exchangeable GTP (77).

Both α-tubulin and β-tubulin require interactions with cofactors C, D, and E, while cofactor A is specific to β-tubulin and cofactor B is specific to α-tubulin (65, 78, 79). Interestingly, γ-tubulin requires interactions with CCT, but no subsequent interactions with tubulin cofactors are required for it to reach its native state. (80).

Purified bovine cofactor A was reported to increase the rate of ATP hydrolysis by CCT (81). However, a later study (75) determined that cofactor A does not bind to CCT or to nucleotides, and only increases the CCT ATPase activity when β-tubulin is the bound substrate. It was suggested, therefore, that cofactor A acts by stabilizing the β-tubulin released from CCT. The cofactor A homologue in *S. cerevisiae*, Rbl2p, has been shown to bind to tubulin *in vivo* and when over-expressed can rescue the lethal over-expression of β-tubulin (82). This supports the theory that Rbl2p/cofactor A acts by stabilizing monomeric β-tubulin released from CCT. The crystal structure of Rbl2p has recently been solved (83). Rbl2p forms an elongated structure comprised of three α-helices and which assembles as a dimer. Although it has previously been reported that cofactor A exists as a monomer (75), this study supports the view that Rbl2p exists as a dimer in solution, based upon gel filtration experiments. The surface of the Rbl2p dimer consists predominantly of charged and polar residues with no clustering of hydrophobic residues. This discovery is consistent with the view that Rbl2p/cofactor A binds to quasi-native forms of β-tubulin which will have been released from CCT, rather than being a general chaperone interacting with hydrophobic regions of unfolded substrate proteins in the early stages of protein folding.

Cofactor B forms a complex with α-tubulin intermediates in which α-tubulin is sensitive to subtilisin digestion and will not exchange into native α/β tubulin dimers. Deletions of the gene encoding yeast homologue of cofactor B (alf1p) result in the increased sensitivity of microtubules to benomyl (84). Tian *et al.* propose a model in which, following the interactions of α-tubulin with cofactors B or E, and β-tubulin with cofactors A or D, a supercomplex is formed containing α- and β-tubulin and cofactors C, D, and E (79). It is shown that these cofactors have a role as GTPase-activating proteins and that it is the hydrolysis of GTP by β-tubulin that leads to the release of the native tubulin heterodimer from the supercomplex.

Putative homologues of cofactors A, B, D, and E, but not C, have been identified in *S. cerevisiae*, but null mutants of these genes are not lethal. However, in *Schizosaccharomyces pombe*, the cofactor D homologue, Alp1[D] (85) and homologues to cofactors B and E are essential (86). The over-expression of Alp1[D] was shown to rescue the loss of Alp11[B] or Alp21[E] (cofactor B and E homologues), and it is suggested that these cofactors act in a linear manner in assembled microtubule biogenesis (Alp11[B] – Alp21[E] – Alp1[D]) (86). This study proposes a model in which cofactor D can remain associated with microtubules (an interaction which has been observed between cofactor D and purified brain tubulin by co-sedimentation (85)). It is suggested that this interaction may be involved in microtubule integrity, whereas interactions with cofactor B are required for the stabilization of α-tubulin upon release from CCT, and that cofactor E mediates interactions between α-tubulin and cofactors B and D (86). Cofactors B and E possess a CLIP 170 microtubule binding domain which is found in several microtubule associated proteins. Feierbach *et al.* (1999) showed that mutation of this domain prevented binding to α-tubulin (84). As Alf1$_p$ (*S. cerevisiae* cofactor B) can bind to both α-tubulin monomers and to microtubules, it is suggested that proteins which bind to microtubules via CLIP 170 domains do so via interactions with α-tubulin. Radcliffe *et al.* determined that fission yeast cofactor E, Alp21^E, does not have a CLIP 170 domain, suggesting that interactions via this domain may not be essential to the cofactor function (86).

What is the significance of the presence of the CLIP 170 domain in cofactors B and E, and are the tubulin cofactors simply MAPS? Cofactor D has been shown to co-sediment with bovine brain microtubules *in vitro* (85) and cofactors C and E cycle with microtubules during assembly *in vitro* (65). Feierbach *et al.* demonstrated that when the yeast cofactor B homologue Alf1$_p$ coupled to GFP was over-expressed at relatively low levels, it localized to cytoplasmic and nuclear microtubules, but when expressed at much higher levels, no microtubules were detectable, consistent with the activity of Alf1$_p$ sequestering α-tubulin (84). Radcliffe *et al.* showed that the cofactor B homologue in *S. pombe* is localized in the cytoplasm rather than bound to microtubules via its CLIP 170 domain (86).

It remains unclear as to whether the main function of these tubulin cofactors is predominantly as classical MAPS, or whether their major function is in their role as tubulin cofactors required for the biogenesis of microtubules. Unlike CCT, which is required for the production of native tubulin *in vivo*, there appear to be discrepancies in the absolute requirements of tubulin cofactors across different species. It may be significant that γ-tubulin, which forms a ring-like structure in the centrosome involved in microtubule nucleation, does not appear to require the assistance of the tubulin cofactors (80), yet interacts with CCT. It can be predicted that the requirement for cofactor interactions lies with the formation of assembly competent α-and β-tubulin, and that assembled microtubules themselves may utilize the cofactors as MAPS. Hirata *et al.* (1998) and Tian *et al.* (1996) both suggest that the cofactors may have dual functions involving interactions with assembling tubulin and with existing microtubules (65, 85). Therefore, the levels of free and microtubule-bound cofactors may act as a way in which microtubule assembly can be regulated.

2.9 Evolution of CCT and the eukaryotic cytoskeleton

Here we outline our present views concerning the co-evolution of the CCT system and the modern actins and tubulins of eukaryotes. Kubota *et al.* (1994) showed that all of the eight CCT subunits diverged from each other approximately two billion years ago (5). This places the genetic events which caused the amplification and divergence of the CCT family in the same time-frame as the period when it is believed that the first eukaryote was evolving. It has long been argued that the development of a more sophisticated cytoskeleton allowed the transition from prokaryotes to eukaryotes (87). We have speculated that the evolution of successful efficient folding pathways for the eukaryotic actins and tubulins may have allowed for selection of cytoskeletal proteins having more sophisticated functions than their immediate precursors (2, 12, 88). Did the presence of a multi-subunit CCT-like chaperonin facilitate this process by assisting the folding of evolving actin and tubulin proteins in new ways? If this was the case, can we surmise what that facilitation was from examination of the present function of CCT found in all extant eukaryotes so far examined?

It is now generally believed that most components of the eukaryotic cytoplasm evolved from archaebacterial precursors. CCT subunits are 40% identical in amino acid sequence to subunits of archaebacterial thermosomes, and together they comprise the Type II chaperonin family (7). The archaebacterial members come in two sorts, having eight or nine subunits per ring. The eightfold rings may have one or two subunit species and, in a couple of chaperonins which have two subunit species, it is known from direct structural analysis that the subunits alternate their positions in the rings (21, 89). The atomic structure of the α/β chaperonin from *Thermoplasma acidophilum* (21) fits well onto the three-dimensional 30Å structure of CCT determined by cryo-electron microscopy (23), since both chaperonins have eightfold pseudo-symmetry in each ring. It is likely that CCT evolved from a thermosome-like precursor found in the archaebacterial parent cell which gave rise to the eukaryotic cytoplasm (2). Recently, Archibald *et al.* (1999) sequenced several additional chaperonin subunit genes obtained from various species in different archaeal lineages (90). In conjunction with data obtained from archaebacterial genome sequencing projects, they were able to analyse the phylogeny and evolution of 30 chaperonin subunit amino acid sequences in Euryarchaeotes and Crenarchaeotes, the two recognized kingdoms within the Archaea. In addition to showing that the ninefold symmetrical chaperonins are probably restricted to only some species of *Sulfolobus* they showed that the eightfold symmetrical chaperonins can gain and lose subunit species (Fig. 3). Since some lineages contain both α/β and α-only chaperonins, the essential functions of these chaperonins cannot be related to whether or not they are comprised of one or two subunit types. Archibald (1999) argued that amplification of chaperonin subunit number leads to co-evolved interdependence between subunits, possibly reflecting the consequence of duplicated subunits diverging in sequence, and evolving to become interacting neighbours in the complex. They suggest that once preferred arrangements and specific intermolecular interactions are made between pairs of subunits, it may subsequently be difficult to revert to a monomeric state. Thus hetero-oligomerism is viewed as a consequence of an evolutionary ratchet effect, and is considered to be selectively neutral since it seems not to change the overall function of the hetero-oligomer, in Archaea at least. It is further suggested that the completely heteromeric CCT is an example of such a process taken to completion (90). This view discounts the attribution of any functional significance to the presence of eight CCT subunits in CCT. However, in the case of CCT, this model ignores an important observation concerning the sequence conservation in the apical domains of the eight CCT polypeptides. In apical domains of chaperonin subunits are located the sites on chaperonin subunits known to be directly involved in substrate binding, and this fact is well established in the GroEL Type I chaperonin system (27, 91). The amino acid sequences

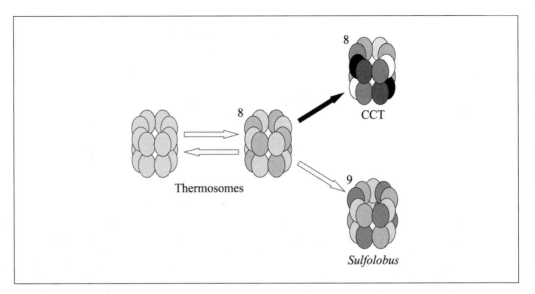

Fig. 3 Sketch of Group II chaperonin evolution. Thermosomes exist as eightfold in euryarchaeotes symmetrical objects, and eightfold or ninefold in crenarchaeotes symmetrical objects (90). CCT is proposed to have evolved from an α/β thermosome-like chaperonin (2). The scheme does NOT imply evolutionary lineage. Figure adapted from ref. 90.

of the apical domains of the eight different CCT subunits are highly divergent from each other within a species, but each individual apical domain is highly conserved across species (92). This result suggests that there has been continuous selection for function in each of the eight apical domains, at least since the yeast divergence one thousand million years ago. This data, and other genetic and biochemical evidence, had persuaded us for some years that CCT might interact with actins and tubulins in a sequence-specific fashion. By this we meant that specific regions of substrates interacted with particular CCT apical domains (6, 12). These ideas are supported by the recent structural determination of a complex between CCT and α-actin (16) in which actin interacts with two CCT apical domains in two different ways using a subunit and geometry-specific binding mechanism.

Figure 4 shows a model for the order of evolution of CCT and its two cytoskeletal substrates based upon analysis of complete genome sequences (see ref. 12). We propose that in the originating archaebacterial-like cells, chaperonin and tubulin precursors were present, and that these components co-evolved during the early evolution of eukaryotic precursor organisms. Because actin homologues are demonstrably absent in several Archael genomes, our novel suggestion is that the actin-fold entered the early eukaryotic lineage by horizontal gene transfer, as may also be the case for Hsc70 introgression (93, 94). Thus the co-evolution of CCT and its substrates may have been a step-wise process, first occurring between FtsZ-like tubulin precursors and a thermosome-like chaperonin, and subsequently involving actin-like precursors and the partially evolved chaperonin.

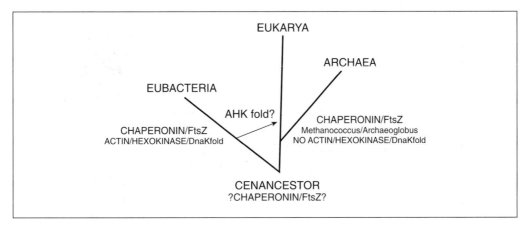

Fig. 4 Horizontal transmission of the AHK fold. Archaea do not appear to contain DNA sequences homologous to actin, hexokinase, or DnaK (AHK fold) (Willison (1996), unpublished results; 12). Most Archael species contain two copies of FtsZ genes (Faguy, D. M. and Doolittle, W. F. (1998). Cytoskeletal proteins: the evolution of cell division. *Curr. Biol.*, **8**, 338) although *Aeropyrum pernix* apparently has none (Faguy, D. M., and Doolittle, W. F. (1999). Genomics: lessons for the Aeropyrum pernix genome. *Curr. Biol.*, **9**, 883). We propose that the AHK fold entered the early eukaryotic cell lineage by horizontal gene transfer.

3. Conclusions

- Chaperonin containing TCP-1 (CCT) is found in the cytosol of eukaryotes. One of its main functions *in vivo* is to assist the folding of the actins and tubulins and this activity is essential.

- CCT is a hexadecamer composed of two copies each of eight subunit types encoded by a family of related genes. The subunits appear to have different roles in substrate interactions.

- The spectrum of substrates for CCT is a controversial area. Some investigators suggest that as many as 10% of eukaryotic proteins are folded by CCT, but others think that CCT is highly specific for actins and tubulins.

- CCT is an ATPase. Its nucleotide-dependent allosteric cycle is still poorly understood. The relationship between the nucleotide cycle and substrate binding and folding is only just beginning to be revealed.

- CCT is very old; it evolved into its eight subunit types around two billion years ago at about the same time the first eukaryotic cell was evolving. We believe that CCT helped drive the successful evolution of the modern cytoskeleton, since it is the presence of a sophisticated cytoskeleton which is the main distinguishing feature between prokaryotes and eukaryotes.

- We suggest that CCT has evolved to become more than solely a protein folding machine in modern eukaryotic cells. There are indications that it may actually be a complicated counting device which helps the cell measure and integrate the rates of synthesis of the actins and tubulins.

We have outlined some of the properties of CCT and its substrates and have drawn attention to some of the areas of controversy surrounding the study of this complicated folding machinery. Recently, substantial progress has been made in understanding the specificity of the interaction between specific CCT subunits and actin (16). We predict that further progress on this front with other CCT binding proteins will illuminate the chemical nature of the interactions between CCT and its spectrum of substrates. Only with this data to hand will we then be able to understand whether CCT is particularly concerned with actin and tubulin, or whether it is in fact a broad-specificity folding machine for a host of cytosolic proteins.

We incline strongly to the view that CCT is highly specialized, and propose that it may be using properties of folding intermediates in the actin and tubulin folding pathways to extract useful information for the cell concerning the flux and synthesis of these proteins. The CCT and cytoskeletal systems seem co-dependent in modern cells, and we suggest that the evolution of the cytoskeletal system in early eukaryotes may have depended upon the activity of CCT and/or its precursors.

Acknowledgements

Our laboratory is funded by the Cancer Research Campaign. We thank Sylvia Holt for helping to prepare the manuscript.

References

1. Futcher, B., Latter, G. I., Monardo, P., McLaughlin, C. S., and Garrels, J. I. (1999). A sampling of the yeast Proteome. *Molec. and Cellular Biol.*, **11**, 7357.
2. Willison, K. R. and Horwich, A. L. (1996). Structure and function of chaperonins in archaebacteria and eukaryotic cytosol. In *The chaperonins* (ed. R. J. Ellis), p. 107–36. Academic Press,
3. Liou, A. K. F. and Willison, K. R. (1997). Elucidation of the subunit orientation in CCT (Chaperonin Containing TCP1) from the subunit composition of CCT. micro-complexes. *EMBO J.*, **16**, 4311.
4. Kubota, H. and Willison, K. R. (1997). Cytosolic chaperonins—an overview. In *Guidebook to molecular chaperones and protein-folding catalysts* (ed. M.-J. Gething), pp. 207–11. Sambrook/Tooze publications at Oxford University Press, Oxford.
5. Kubota, H., Hynes, G., Carne, A., Ashworth, A., and Willison, K. (1994). Identification of six Tcp-1-related genes encoding divergent subunits of the TCP-1-containing chaperonin. *Curr. Biol.*, **4**, 89.
6. Kubota, H., Hynes, G., and Willison, K. (1995). The eighth Cct gene, Cctq, encoding the theta subunit of the cytosolic chaperonin containing TCP-1. *Gene*, **154**, 231.
7. Kubota, H., Hynes, G., and Willison, K. (1995). The chaperonin containing t-complex polypeptide 1 (TCP-1). Multisubunit machinery assisting in protein folding and assembly in the eukaryotic cytosol. *Eur. J. Biochem.*, **230**, 3.
8. Stoldt, V., Rademacher, F., Hehren, V., Ernst, J. F., Pearce, D. A., and Sherman, F. (1996). Review: the Cct eukaryotic chaperonin subunits of Saccharomyces cerevisiae and other yeasts. *Yeast*, **12**, 523.

9. Rommelaere, H., van Troys, M., Gao, Y., Melki, R., Cowan, N. J., Vandekerckhove, J., and Ampe, C. (1993). Eukaryotic cytosolic chaperonin contains t-complex polypeptide 1 and seven related subunits. *Proc. Natl. Acad. Sci. USA*, **90**, 11975.

10. Hynes, G., Sutton, C. W., U, S., and Willison, K. R. (1996). Peptide mass fingerprinting of chaperonin-containing TCP-1 (CCT) and copurifying proteins. *FASEB J.*, **10**, 137.

11. Hynes, G., Kubota, H., and Willison, K. R. (1995). Antibody characterisation of two distinct conformations of the chaperonin-containing TCP-1 from mouse testis. *FEBS Lett.*, **358**, 129.

12. Willison, K. R. (1999). Composition and function of the eukaryotic cytosolic chaperonin containing TCP-1. In *Molecular chaperones and folding catalysts. Regulation, cellular function and mechanisms* (ed. B. Bukau), pp. 555–71. Harwood Academic,

13. Kubota, H., Hynes, G. M., Kerr, S. M., and Willison, K. R. (1997). Tissue-specific subunit of the mouse cytosolic chaperonin-containing TCP-1. *FEBS Lett.*, **402**, 53.

14. Liou, K. F., McCormack, E. A., and Willison, K. R. (1998). The chaperonin containing TCP-1 (CCT) displays a single-ring mediated disassembly and reassembly cycle. *Biological Chem. Hoppe-Seyler*, **379**, 311.

15. Grantham, J., Llorca, O., Valpuesta, J. M., and Willison, K. R. (2000). Partial occlusion of both cavities of the eukaryotic chaperonin with antibody has no effect upon the rates of beta-actin or alpha-tubulin folding. *J. Biol. Chem.*, **275**, 4587.

16. Llorca, O., McCormack, E. A., Hynes, G., Grantham, J., Cordell, J., Carrascosa, J. L., Willison, K. R., Fernandez, J. J., and Valpuesta, J. M. (1999). Eukaryotic Type II chaperonin CCT interacts with actin through specific subunits. *Nature*, **402**, 693.

17. Groll, M., Ditzel, L., Löwe, J., Stock, D., Bochtler, M., Bartunik, H. D., and Huber, R. (1997). Structure of 20S proteasome from yeast at 2.4 Å resolution. *Nature*, **386**, 463.

18. Gutsche, I., Essen, L. O., and Baumeister, W. (1999). Group II chaperonins: new TRiC(k)s and turns of a protein folding machine. *J. Mol. Biol.*, **293**, 295.

19. Klumpp, M., Baumeister, W., and Essen, L.-O. (1997). Structure of the substrate binding domain of the thermosome, an archaeal group II chaperonin. *Cell*, **91**, 263.

20. Braig, K., Otwinowski, Z., Hegde, R., Boisvert, D. C., Joachimiak, A., Horwich, A. L., and Sigler, P. B. (1994). The crystal structure of the bacterial chaperonin GroEL at 2.8 Å. *Nature*, **371**, 578.

21. Ditzel, L., Lowe, J., Stock, D., Stetter, K.-O., Huber, H., Huber, R., and Steinbacher, S. (1998). Crystal structure of the thermosome, the archaeal chaperonin and homolog of CCT. *Cell*, **93**, 125.

22. Burston, S. G. and Saibil, H. R. (1999). The relationship between chaperonin structure and function. In *Molecular chaperones and folding catalysts. Regulation, cellular function and mechanisms* (ed. B. Bukau), pp. 523–53. Harwood Academic,

23. Llorca, O., Smyth, M. G., Carrascosa, J. L., Willison, K. R., Radermacher, M., Steinbacher, S., and Valpuesta, J. M. (1999). 3D reconstruction of the ATP-bound form of CCT reveals the asymmetric folding conformation of a Type II chaperonin. *Nat. Struct. Biol.*, **6**, 639.

24. Roseman, A. M., Chen, S., White, H., Braig, K., and Saibil, H. R. (1996). The chaperonin ATPase cycle: mechanism of allosteric switching and movements of substrate-binding domains in GroEL. *Cell*, **87**, 241.

25. Llorca, O., Smyth, M. G., Marco, S., Carrascosa, J. L., Willison, K. R., and Valpuesta, J. M. (1998). ATP binding induces large conformational changes in the apical and equatorial domains of the eukaryotic chaperonin containing TCP-1 complex. *J. Biol. Chem.*, **273**, 10091.

26. Xu, X., Horwich, A. L., and Sigler, P. B. (1997). The crystal structure of the asymmetric GroEL-GroES-(ADP)$_7$ chaperonin complex. *Nature*, **388**, 741.

27. Chen, L. and Sigler, P. B. (1999). The crystal structure of a GroEL/peptide complex: plasticity as a basis for substrate diversity. *Cell*, **99**, 757.

28. Gutsche, I., Holzinger, J., Röble, M., Heumann, H., Baumeister, W., and May, R. P. (2000). Conformational rearrangements of an archaeal chaperonin upon ATPase cycling. *Curr. Biol.*, **10**, 405.

29. Rohman, M. (1999). Biochemical characterisation of chaperonin containing TCP-1. PhD. thesis, University of London.

30. Horwich, A. L. and Saibil, H. R. (1998). The thermosome: chaperonin with a built-in lid. *Nat. Struct. Biol.*, **5**, 333.

31. Hynes, G. M. and Willison, K. R. (2000). Individual subunits of the eukaryotic cytosolic chaperonin mediate interactions with binding sites located on subdomains of β-actin. *J. Biol. Chem.*, **275**, 18985.

32. Lin, P., Cardillo, T. S., Richard, L. M., Segal, G. B., and Sherman, F. (1997). Analysis of mutationally altered forms of the Cct6 subunit of the chaperonin from *Saccharomyces cerevisiae*. *Genetics*, **147**, 1609.

33. Sternlicht, H., Farr, G. W., Sternlicht, M. L., Driscoll, J. K., Willison, K., and Yaffe, M. B. (1993). The t-complex polypeptide 1 complex is a chaperonin for tubulin and actin in vivo. *Proc. Natl. Acad. Sci. USA*, **90**, 9422.

34. Melki, R. and Cowan, N. J. (1994). Facilitated folding of actins and tubulins occurs via a nucleotide-dependent interaction between cytoplasmic chaperonin and distinctive folding intermediates. *Mol. Cell. Biol.*, **14**, 2895.

35. Melki, R., Batelier, G., Soulé, S., and Williams Jr, R. C. (1997). Cytoplasmic chaperonin containing TCP-1: structural and functional characterization. *Biochem.*, **36**, 5817.

36. Thulasiraman, V., Yang, C. F., and Frydman, J. (1999). In vivo newly translated polypeptides are sequestered in a protected folding environment. *EMBO J.*, **18**, 85.

37. Lewis, V. A., Hynes, G. M., Zheng, D., Saibil, H., and Willison, K. (1992). T-complex polypeptide-1 is a subunit of a heteromeric particle in the eukaryotic cytosol. *Nature*, **358**, 249.

38. Ursic, D. and Culbertson, M. R. (1992). Is yeast TCP1 a chaperonin? *Nature*, **356**, 392.

39. Soares, H., Penques, D., Mouta, C., and Rodrigues-Pousada, C. (1994). A tetrahymena orthologue of the mouse chaperonin subunit CCT gamma and its coexpression with tubulin during cilia recovery. *J. Biol. Chem.*, **269**, 29299.

40. Phipps, B., Hoffman, A., Stetter, K. O., and Baumeister, W. (1991). A novel ATPase complex selectively accumulated upon heat shock is a major cellular component of thermophilic archaebacteria. *EMBO J.*, **10**, 1711.

41. Trent, J. D., Osipik, J., and Pinkau, T. (1990). Acquired thermotolerance and heat shock in the extremely thermophilic archaebacterium Sulfolobus sp. strain B12. *J. Bacteriol.*, **172**, 1478.

42. Schena, M., Shalon, D., Heller, R., Chai, A., Brown, P. O., and Davis, R. W. (1996). Parallel human genome analysis: microarray-based expression monitoring of 1000 genes. *Biochem.*, **93**, 10614.

43. Yokota, S., Yanagi, H., Yura, T., and Kubota, H. (2000). Upregulation of cytosolic chaperonin CCT subunits during recovery from chemical stress that causes accumulation of unfolded proteins. *Eur. J. Biochem.*, **267**, 1658.

44. Kubota, H., Matsumoto, S., Yokota, S., Yanagi, H., and Yura, T. (1999). Transcriptional activation of mouse cytosolic chaperonin CCT subunit genes by heat shock factors HSF1 and HSF2. *FEBS Lett.*, **461**, 125.

45. Cyrne, L., Guerreiro, P., Cardoso, A. C., Rodrigues-Pousada, C., and Soares, H. (1996). The Tetrahymena chaperonin subunit CCT eta gene is coexpressed with CCT gamma gene during cilia biogenesis and cell sexual reproduction. *FEBS Lett.*, **383**, 277.

46. Willison, K., Hynes, G., Davies, P., Goldsborough, A., and Lewis, V. A. (1990). Expression of three t-complex genes, Tcp-1, D17 Leh117c3 and D17 Leh66, in purified murine spermatogenic cell populations. *Genet. Res.*, **56**, 193–201.

47. Yokota, S., Yanagi, H., Yura, T., and Kubota, H. (1999). Cytosolic chaperonin is up-regulated during cell growth. *J. Biol. Chem.*, **274**, 37070.

48. Hynes, G., Celis, J. E., Lewis, V. A., Carne, A., U, S., Lauridsen, J. B., and Willison, K. R. (1996). Analysis of chaperonin-containing TCP-1 subunits in the human keratinocyte two-dimensional protein database: further characterisation of antibodies to individual subunits. *Electrophoresis*, **17**, 1720.

49. Iijima, M., Shimizu, H., Tanaka, Y., and Urushihara, H. (1998). A Dictyostelium discoideum homologue to Tcp-1 is essential for growth and development. *Gene*, **213**, 101.

50. Sheterline, P., Clayton, J., and Sparrow, J. C. (1994). Actin. *Protein Profile*, **1**, 271.

51. Miklos, D., Caplan, S., Mertens, D., Hynes, G., Pitluk, Z., Kashi, Y., *et al.* (1994). Primary structure and function of a second essential member of the hetero-oligomeric TCP-1 chaperonin complex of yeast, TCP-1β. *Proc. Natl. Acad. Sci. USA*, **91**, 2742.

52. Ursic, D., Sedbrook, J. C., Himmel, K. L., and Culbertson, M. R. (1994). The essential yeast Tcp1 protein affects actin and microtubules. *Mol. Biol. of the Cell*, **5**, 1065.

53. Lin, P. and Sherman, F. (1997). The unique hetero-oligomeric nature of the subunits in the catalytic cooperativity of the yeast Cct chaperonin complex. *Proc. Natl. Acad. Sci. USA*, **94**, 10780.

54. Vinh, D. B. and Drubin, D. G. (1994). A yeast TCP-1-like protein is required for actin function in vivo. *Proc. Natl. Acad. Sci. USA*, **91**, 9116.

55. Rademacher, F., Kehren, F., Stoldt, V. R., and Ernst, J. F. (1998). A Candida albicans chaperonin subunit (CaCct8p) as a suppressor of morphogenesis and Ras phenotypes in C. albicans and Saccharomyces cerevisiae. *Microbiol.*, **144**, 2951.

56. Jelinsky, S. A. and Samson, L. D. (1999). Global response of *Saccharomyces cerevisiae* to an alkylating agent. *Proc. Natl. Acad. Sci. USA*, **96**, 1486.

57. Gao, Y., Thomas, J. O., Chow, R. L., Lee, G.-H., and Cowan, N. J. (1992). A cytoplasmic chaperonin that catalyzes β-actin folding. *Cell*, **69**, 1043.

58. Frydman, J., Nimmesgern, E., Erdjument-Bromage, H., Wall, J. S., Tempst, P., and Hartl, F.-U. (1992). Function in protein folding of TRiC, a cytosolic ring complex containing TCP-1 and structurally related subunits. *EMBO J.*, **11**, 4767.

59. Frydman, J., Nimmesgern, E., Ohtsuka, K., and Hartl, F.-U. (1994). Folding of nascent polypeptide chains in a high molecular mass assembly with molecular chaperones. *Nature*, **370**, 111.

60. Frydman, J. and Hartl, F. U. (1996). Principles of chaperone-assisted protein folding: differences between in vitro and in vivo mechanisms. *Science*, **272**, 1497.

61. Netzer, W. J. and Hartl, F. U. (1997). Recombination of protein domains facilitated by co-translational folding in eukaryotes. *Nature*, **388**, 343.

62. McCallum, C. D., Do, H., Johnson, A. E., and Frydman, J. (2000). The interaction of the chaperonin tailless complex polypeptide 1 (TCP1) ring complex (TRiC) with ribosome-bound nascent chains examined using photo-cross-linking. *J. Cell Biol.*, **149**, 591.

63. Farr, G. W., Schar, I. E.-C., Schumacher, R. J., Sondek, S., and Horwich, A. L. (1997). Chaperonin-mediated folding in the eukaryotic cytosol proceeds through rounds of release of native and nonnative forms. *Cell*, **89**, 927.

64. Tian, G., Vainberg, I. E., Tap, W. D., Lewis, S. A., and Cowan, N. J. (1995). Specificity in chaperonin-mediated protein folding. *Nature*, **375**, 250.

65. Tian, G., Huang, Y., Rommelaere, H., Vandekerckhove, J., Ampe, C., and Cowan, N. J. (1996). Pathway leading to correctly folded β-tubulin. *Cell*, **86**, 287.

66. Cowan, N. J. and Lewis, S. A. (1999). A chaperone with a hydrophilic surface. *Nat. Struct. Biol.*, **6**, 990.

67. Eggers, D. K., Welch, W. J., and Hansen, W. J. (1997). Complexes between nascent polypeptides and their molecular chaperones in the cytosol of mammalian cells. *Mol. Biol. of the Cell*, **8**, 1559.

68. Vainberg, I. E., Lewis, S. A., Rommelaere, H., Ampe, C., Vandekerckhove, J., Klein, H. L., and Cowan, N. J. (1998). Prefoldin, a chaperone that delivers unfolded proteins to cytosolic chaperonin. *Cell*, **93**, 863.

69. Geissler, S., Siegers, K., and Schiebel, E. (1998). A novel protein complex promoting formation of functional alpha- and gamma-tubulin. *EMBO J.*, **17**, 952.

70. Hansen, W. J., Cowan, N. J., and Welch, W. J. (1999). Prefoldin-nascent chain complexes in the folding of cytoskeletal proteins. *J. Cell Biol.*, **145**, 265.

71. Srikakulam, R. and Winkelmann, D. A. (1999). Myosin II folding is mediated by a molecular chaperonin. *J. Biol. Chem.*, **274**, 27265.

72. Macario, A. J., Lange, M., Ahring, B. K., and De Macario, E. C. (1999). Stress genes and proteins in the archaea. *Microbiol. Mol. Biol. Rev.*, **63**, 923.

73. Leroux, M. R., Fandrich, M., Klunker, D., Siegers, K., Lupas, A. N., Brown, J. R., *et al.* (1999). MtGimC, a novel archaeal chaperone related to the eukaryotic chaperonin cofactor GimC/prefoldin. *EMBO J.*, **18**, 6730.

74. Siegers, K., Waldmann, T., Leroux, M. R., Grein, K., Shevchenko, A., Schiebel, E., and Hartl, F. U. (1999). Compartmentation of protein folding in vivo: sequestration of non-native polypeptide by the chaperonin-GimC system. *EMBO J.*, **18**, 75.

75. Melki, R., Rommelaere, H., Leguy, R., Vandekerckhove, J., and Ampe, C. (1996). Cofactor A is a molecular chaperone required for β-tubulin folding: functional and structural characterization. *Biochem.*, **32**, 10422.

76. Dobrzynski, J. K., Sternlicht, M. L., Farr, G. W., and Sternlicht, H. (1996). Newly-synthesized β-tubulin demonstrates domain-specific interactions with the cytosolic chaperonin. *Biochem.*, **35**, 15870.

77. Tian, G., Vainberg, I. E., Tap, W. D., Lewis, S. A., and Cowan, N. J. (1995). Quasi-native chaperonin-bound intermediates in facilitated protein folding. *J. Biol. Chem.*, **270**, 23910.

78. Tian, G., Lewis, S. A., Feierbach, B., Stearns, T., Rommelaere, H., Ampe, C., and Cowan, N. J. (1997). Tubulin subunits exist in an activated conformational state generated and maintained by protein cofactors. *J. Cell Biol.*, **138**, 821.

79. Tian, G., Bhamidipati, A., Cowan, J. J., and Lewis, S. A. (1999). Tubulin folding cofactors as GTPase-activating proteins. GTP hydrolysis and the assembly of the alpha/beta-tubulin heterodimer. *J. Biol. Chem.*, **274**, 24054.

80. Melki, R., Vainberg, I. E., Chow, R. L., and Cowan, N. J. (1993). Chaperonin-mediated folding of vertebrate actin-related protein and γ-tubulin. *J. Cell Biol.*, **122**, 1301.

81. Gao, Y., Melki, R., Walden, P. D., Lewis, S. A., Ampe, C., Rommelaere, H., *et al.* (1994). A novel cochaperonin that modulates the ATPase activity of cytoplasmic chaperonin. *J. Cell Biol.*, **125**, 989.

82. Archer, J. E., Vega, L. R., and Solomon, F. (1995). Rbl2p, a yeast protein that binds to beta-tubulin and participates in microtubule in function in vivo. *Cell*, **82**, 425.

83. Steinbacher, S. (1999). Crystal structure of the post-chaperonin beta-tubulin binding cofactor Rbl2p. *Nat. Struct. Biol.*, **6**, 1029.

84. Feierbach, B., Nogales, E., Downing, K. H., and Stearns, T. (1999). Alf1p, a CLIP-170

domain-containing protein, is functionally and physically associated with alpha-tubulin. *J. Cell Biol.*, **144**, 113.

85. Hirata, D., Masuda, H., Eddison, M., and Toda, T. (1998). Essential role of tubulin-folding cofactor D in microtubule assembly and its association with microtubules in fission yeast. *EMBO J.*, **17**, 658.

86. Radcliffe, P. A., Hirata, D., Vardy, L., and Toda, T. (1999). Functional dissection and hierarchy of tubulin-folding cofactor homologues in fission yeast. *Mol. Biol. Cell.*, **10**, 2987.

87. Maynard Smith, J. and Szathmary, E. (1995). The major transitions in evolution. W. H. Freeman & Co. (eds R. I. Morimoto, A. Tissiéres, and C. Georgopoulos) pp. 299–312.

88. Willison, K. R. and Kubota, H. (1994). The structure, function and genetics of the chaperonin containing TCP-1 (CCT) in eukaryotic cytosol. *The biology of heat shock proteins and molecular chaperones.* (eds R. I. Morimoto, A. Tissiéres, and C. Georgopoulos) pp. 299–312. CSH Press, NY.

89. Nitsch, M., Klumpp, M., Lupas, A., and Baumeister, W. (1997). The thermosome: alternating α and β-subunits within the chaperonin of the archaeon *Thermoplasma acidophilum*. *J. Mol. Biol.*, **267**, 142.

90. Archibald, J. M., Logsdon, J. M., and Doolittle, W. F. (1999). Recurrent paralogy in the evolution of archaeal chaperonins. *Curr. Biol.*, **9**, 1053.

91. Fenton, W. A., Kashi, Y., Furtak, K., and Horwich, A. L. (1994). Residues in chaperonin GroEL required for polypeptide binding and release. *Nature*, **371**, 614.

92. Kim, S., Willison, K. R., and Horwich, A. L. (1994). Cytosolic chaperonin subunits have a conserved ATPase domain but diverged polypeptide-binding domains. *Trends Biochem. Sci.*, **19**, 543.

93. Gupta, R. S., Aitken, K., Falah, M., and Singh, B. (1994). Cloning of Giardia lamblia heat shock protein HSP70 homologs: implications regarding origin of eukaryotic cells and of endoplasmic reticulum. *Proc. Natl. Acad. Sci. USA*, **91**, 2895.

94. Gupta, R. S. (1995). Phylogenetic analysis of the 90 kD heat shock family of protein sequences and an examination of the relationship among animals, plants, and fungi species. *Mol. Biol. Evol.*, **12**, 1063.

95. Spradling, A. C., Stern, D., Beaton, A., Rehm, E. J., Laverty, T., Mozden, N., *et al.* (1999). The Berkeley Drosophila Genome Project Gene Disruption Project: single P-element insertions mutating 25% of vital Drosophila genes. *Genetics*, **153**, 135.

96. Sternlicht, H., Farr, G. W., Sternlicht, M. L., Driscoll, J. K., Willison, K., and Yaffe, M. B. (1993). The t-complex polypeptide 1 complex is a chaperonin for tubulin and actin in vivo. *Proc. Natl. Acad. Sci. USA.*, **90**, 9422.

97. Moudjou, M., Bordes, N., Paintrand, M., and Bornens, M. (1996). γ-tubulin in mammalian cells: the centrosomal and the cytosolic forms. *J. Cell Sci.*, **109**, 875.

98. Won, K.-A., Schumacher, R. J., Farr, G. W., Horwich, A. L., and Reed, S. I. (1998). Maturation of human cyclin E requires the function of eukaryotic chaperonin CCT. *Molec. Cell Biol.*, **18**, 7584.

99. Kashuba, E., Pokrovskaja, K., Klein, G., and Szekely, L. (1999). Epstein-Barr virus-encoded nuclear protein EBNA-3 interacts with the ε-subunit of the t-complex protein 1 chaperonin complex. *J. Human Virol.*, **2**, 33.

100. Feldman, D. E., Thulasiraman, V., Ferreyra, R. G., and Frydman, J. (1999). Formation of the VHL-elongin BC tumor suppressor complex is mediated by the chaperonin TRiC. *Molec. Cell*, **4**, 1051.

101. Roobol, A. and Carden, M. J. (1993). Identification of chaperonin particles in mammalian

brain cytosol and of t-complex polypeptide 1 as one of their components. *J. Neurochem.*, **60**, 2327.

102. Lingappa, J. R., Martin, R. L., Wong, M. L., Ganem, D., Welch, W. J., and Lingappa, V. R. (1994). A eukaryotic cytosolic chaperonin is associated with a high molecular weight intermediate in the assembly of hepatitis B virus capsid, a multimeric particle. *J. Cell Biol.*, **125**, 99.

103. Wu-Baer, F., Lane, W. S., and Gaynor, R. B. (1996). Identification of a group of cellular cofactors that stimulate the binding of RNA polymerase II and TRP-185 to human immunodeficiency virus 1 TAR RNA. *J. Biol. Chem.*, **271**, 4201.

104. Roobol, A., Sahyoun, Z. P., and Carden, M. J. (1999). Selected subunits of the cytosolic chaperonin associate with microtubules assembled in vitro. *J. Biol. Chem.*, **274**, 2408.

105. Grantham, J. (1998). Studies on the chaperonin containing TCP-1. PhD thesis, University of Kent.

106. Gebauer, M., Melki, R., and Gehring, U. (1998). The chaperone cofactor Hop/p60 interacts with the cytosolic chaperonin-containing TCP-1 and affects its nucleotide exchange and protein folding abilities. *J. Biol. Chem.*, **273**, 29475.

107. Ursic, D. and Culbertson, M. R. (1991). The yeast homolog to mouse Tcp-1 affects microtubule-mediated processes. *Mol. Cell. Biol.*, **11**, 2629.

108. Chen, X., Sullivan, D. S., and Huffaker, T. C. (1994). Two yeast genes with similarity to TCP-1 are required for microtubule and actin function *in vivo*. *Proc. Natl. Acad. Sci. USA*, **91**, 911.

109. Schmidt, A., Kunz, J., and Hall, M. N. (1996). TOR2 is required for organisation of the actin cytoskeleton in yeast. *Proc. Natl. Acad. Sci. USA*, **93**, 13780.

George Green Library - Issue Receipt

Customer name: Oluwaleye-Udom, Damilola-Folarin

Title: Molecular chaperones in the cell / edited by Peter Lund.

ID: 1002508304

Due: 15/01/2008 23:59

Total items: 1
20/11/2007 12:40

All items must be returned before the due date and time.

5 | The roles of the major cytoplasmic chaperones in normal eukaryotic growth: Hsc70 and its cofactors

CHRISTINE PFUND, WEI YAN, and ELIZABETH A. CRAIG

The role of chaperones in the prevention of protein aggregation in cells experiencing stress is well established (see Chapters 1 and 9); however, problems of protein misfolding and aggregation are not confined to the stressed cell. Even under optimal growth conditions, there is a need for chaperone action. In the past few years, this aspect of chaperone function has become increasingly obvious as researchers focus their attention on the processes of translation, translocation across membranes, and the assembly/disassembly of protein complexes. Within the crowded environment of the cell, each protein must be protected from misfolding and aggregation. In this chapter, we discuss the processes in the normal cell that require chaperone assistance and the evidence from studies using eukaryotes that support a role for both Hsp70s and their cohort proteins in these processes.

1. Chaperones involved in protein biogenesis

A newly translated protein contains all of the information within its primary amino acid sequence necessary for its final tertiary structure. In fact, many denatured proteins can fold spontaneously into their native conformations in a test tube (1). *In vivo*, however, folding of a newly translating polypeptide is not as simple. At the onset of translation, the first 30–40 amino acids are protected within the exit channel of the ribosome (2). Once the polypeptide progresses beyond this length, amino acids begin to be exposed to the cytosol. Until an entire domain is synthesized, the translating protein likely remains unfolded, thereby exposing hydrophobic patches to the crowded environment outside the ribosome. These exposed hydrophobic patches render the unfolded protein susceptible to aggregation.

Chaperones may function to protect the nascent chain from aggregation by binding to these exposed hydrophobic patches until the protein is fully translated and can fold correctly. This hypothesis predicts an increase in the number of aggregated nascent chains in the absence of chaperones. While there is substantial data to support this idea *in vitro* (3), evidence for increased aggregation in the absence of chaperones *in vivo* is more limited. A number of newly synthesized proteins aggregate after shift of a *groEL* temperature sensitive mutant to the nonpermissive temperature (4). Recently, the Bukau and Hartl laboratories have reported that deletion of two genes encoding chaperones of *E. coli*, DnaK, and trigger factor, is lethal (5, 6). Furthermore, in the absence of these two chaperones there is an increase in the aggregation of newly synthesized proteins (5). These analyses of aggregation have all been carried out using the prokaryote *E. coli*. At this point there is no data from studies using eukaryotes that address this question.

Another consideration is the constraints put upon the folding of a protein while still tethered to the ribosome as a peptidyl-tRNA. However, several proteins have been shown to achieve an active conformation while still bound to the ribosome in *in vitro* translation lysates. For example, heme can bind to globin while it is still attached to the ribosome, suggesting that globin can achieve its tertiary structure co-translationally (7). Both rhodenase and luciferase, when altered by the addition of 26 or more amino acids at their C-termini, have been found enzymatically active while bound to *E. coli* and wheat germ ribosomes, respectively (8, 9). In addition, the reovirus attachment protein s1 can form homotrimers while being synthesized, further supporting the hypothesis that proteins can obtain tertiary structure co-translationally (10). However, since the rate of translation in lysates is much slower than in the living cells, the biological significance of the co-translational folding observed in these situations is unclear. Recently, however, co-translational folding of the Semliki Forest virus (SFV) capsid protein was observed in both mammalian and bacterial cells (11). SFV capsid contains a chymotrypsin-like protease domain that must fold before it can autocatalytically cleave the capsid from the larger polyprotein. This cleavage, which does not require release from the ribosome, occurs prior to completion of synthesis of the polypeptide chain.

Somehow a balance between remaining unfolded until enough information is available to assure correct folding and protecting the unfolded polypeptide from aggregation must be achieved. Do molecular chaperones participate in these early steps of protein biogenesis? Although the exact mechanism by which chaperones function in protein biosynthesis in eukaryotes is unclear, a picture, which clearly indicates their involvement, is beginning to emerge. Molecular chaperones have been found associated with newly synthesized proteins, putting them on the frontlines of the folding of nascent polypeptides

1.1 Hsp / Hsc70 functions in protein synthesis

Over the past decade, Hsp70s have been implicated as chaperones for nascent polypeptides. There is evidence for an interaction between Hsp70s and newly synthesized proteins from studies in both mammalian and yeast cells.

1.1.1   Interaction of Hsp/Hsc70 of mammalian cells with nascent polypeptides

About 10 years ago William Welch's lab reported that Hsc70, the constitutively expressed Hsp70 in the cytosol of mammalian cells, interacts with newly synthesized proteins. Immunoprecipitation of Hsc70 from pulse-labelled HeLa cells resulted in immunoprecipitation of Hsc70 as well as a collection of newly synthesized proteins ranging in size from 20 to 200 kDa (12). When cells were subjected to a chase following the labelling period, the 'smear' of nascent chains which could be immunoprecipitated by Hsc70 antibodies was significantly reduced, suggesting that the interaction of Hsc70 with newly synthesized proteins is transient. However, if the amino acid analog, L-azetidine 2 carboxylic acid (Asc), whose incorporation results in the production of misfolded proteins, was added to cells the interaction of Hsc70 persisted. This persistence indicated a prolonged interaction of Hsc70 with misfolded nascent polypeptides. In addition, this Hsc70 interaction with newly synthesized proteins was shown to be sensitive to added ATP as would be expected for an Hsc70: substrate interaction. Hsc70 not only co-sedimented with polysomes, but also interacted with nascent chain released from ribosomes prematurely via treatment of cells with the amino-acyl tRNA analogue, puromycin (12, 13). These data suggest that Hsp70 functions as a chaperone co-translationally, but remains associated with misfolded nascent chains after release from the ribosome.

While these earlier studies focused on the Hsp70 class of molecular chaperones, the development of an antibody to puromycin itself allowed a more general investigation of nascent chain-associated chaperones. Eggers *et al.* (14) treated metabolically labelled cells with low concentrations of puromycin. This low concentration did not completely shut down protein synthesis, but promoted the release of a large number of translating proteins, 'tagging' them with puromycin (14). The puromycin-containing chains that were isolated by immunoprecipitation using antibodies against puromycin represented a random population of truncated proteins derived from all the actively translating mRNAs in the cell. In this unbiased approach the proteins associated with these prematurely released puromycin-tagged chains was analysed by two-dimensional gels; Hsp70 was shown to be the predominant species associated. The authors estimated that 15–20% of nascent polypeptides associate with Hsp70, in good agreement with other estimates (14–16).

While experiments with living cells offer the opportunity to analyse natural process, technical limitations abound. Reticulocyte *in vitro* translation systems have allowed analysis of selected proteins by a number of laboratories. For example, the interaction of proteins translated from both full-length and truncated CAT mRNAs with Hsp70 was analysed by native immunoprecipitations (16). A preferential association of Hsp70 with unfinished nascent chains compared to the full-length nascent CAT was found. In addition, the truncated CAT chains were protected from protease digestion, unlike the full-length chains. This protection is likely provided by Hsp70, which would protect incomplete and presumably misfolded nascent chain from aggregation. Interestingly, the CAT nascent chains released prematurely upon addition of puromycin were not only associated with Hsp70, but were also found in a high molecular weight complex of approximately 700 kDa (16) which is discussed later in Section 5.1.3 of this chapter.

Hartl and co-workers have also examined the interaction of chaperones and nascent chains *in vitro* using firefly luciferase in rabbit reticulocyte lysates. Nascent luciferase chains released prematurely from the ribosome with puromycin were analysed by gel filtration. Some 65% of these nascent chains were in a high molecular weight (HMW) complex of 1200 kDa. This complex was shown to contain Hsp70 (Hsc70/Hsp70 collectively) as well as other chaperones (17). Taking advantage of the ability to assay luciferase activity, Frydman *et al.* (17) went on to test the effects of Hsp70 depletion from *in vitro* translation reactions. When Hsp70 was depleted from the translation reaction, there was no inhibition of translation. There was, however, a 70% reduction in the amount of final luciferase specific activity. Little of this non-native luciferase was found in aggregates. Addition of purified Hsp70 to the depleted lysates prior to the initiation of translation resulted in the restoration of efficient folding of luciferase; yet addition of Hsp70 after completion of translation could not rescue the luciferase folding and activity, suggesting that Hsp70s must function co-translationally to facilitate proper folding (17).

1.1.2 Ribosome-associated Hsp70s of *Saccharomyces cerevisiae*

S. cerevisiae has 14 Hsp70s located in various compartments of the cell (Fig. 1). At least 8 of these, forming the Ssa, Ssb, and Sse subfamilies, are found in the cytosol. Cytosolic Hsp70s of the yeast *S. cerevisiae* have been analysed for roles in protein biogenesis. The Ssa and Ssb Hsp70s, which are about 60% identical in sequence, are the best-studied cytosolic Hsp70s. Hsp70s closely related to Ssbs are found in many fungi. Ssas, on the other hand, are closely related to cytosolic Hsp70s from all eukaryotes including the Hsp/Hsc70s of mammalian cells discussed above. Ribosome-associated Ssb has been implicated in the role of a chaperone for newly synthesizing proteins, while Ssa's involvement may be post-translational.

Ssb can be crosslinked to nascent polypeptide chains emerging from the ribosome (18). Using a yeast *in vitro* translation system, various truncated mRNAs were synthesized in the presence of radiolabel and a derivitized lysyl- tRNA bearing a photoactivatable crosslinker. After the translation reactions were complete, and the crosslinker was activated by light, a covalent interaction between the translated nascent chain and Ssb was detected. However, this interaction was not detectable if the nascent chain was first released from the ribosome using puromycin, suggesting that Ssb functions in the context of the ribosome. This ribosome-dependent interaction is different from the interaction with mammalian Hsp70, which maintains its interaction even after release from ribosomes. In addition, the Ssb-ribosome-nascent chain interaction was not sensitive to ATP. This lack of sensitivity is surprising in light of work from many laboratories indicating release of Hsp70s from polypeptide substrates in the presence of ATP.

Ssb's interaction with the ribosome does not appear to depend on its interaction with nascent chain. In cell lysates it is found associated with both translating and nontranslating ribosomes (18). However, this interaction is salt-sensitive when the

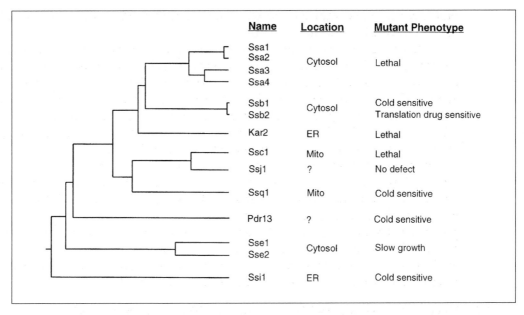

Name	Location	Mutant Phenotype
Ssa1 Ssa2 Ssa3 Ssa4	Cytosol	Lethal
Ssb1 Ssb2	Cytosol	Cold sensitive Translation drug sensitive
Kar2	ER	Lethal
Ssc1	Mito	Lethal
Ssj1	?	No defect
Ssq1	Mito	Cold sensitive
Pdr13	?	Cold sensitive
Sse1 Sse2	Cytosol	Slow growth
Ssi1	ER	Cold sensitive

Fig. 1 Polygenetic tree of the *S. cerevisiae* Hsp70s. The tree was constructed using the MegAlign software from DNASTAR (Madison, WI) by the clustal method. Amino acid sequences at the N-terminus were truncated before alignment to remove the leader sequences of Hsp70s, especially for those localized in the mitochondria (mito) and endoplasmic reticulum (ER). Therefore, the first amino acids used in the alignment are: 4 (Ssa1–4, Sse1–2), 9 (Ssb1–2), 32 (Ssc1), 52 (Kar2), 30 (Ssq1), 22 (Ssi1), 29 (Ssj1), and 39 (Pdr13). Subcellular location and phenotypes of deletion mutants are indicated for each Hsp70 family. This figure is adapted from the review by E. A. Craig, W. Yan, and P. James (35).

ribosome is not translating and lacks nascent chain, while resistant to 1.5 M KCl when on translating ribosomes. This higher affinity for the translating ribosome compared to the nontranslating ribosome may be due to the conformational differences between the ribosomes in the two states rather than the interaction with the nascent chain itself. These data suggest a model in which Ssb is localized to the ribosome prior to the emergence of the nascent chain. Once translation begins, Ssb may form a stable interaction with the translating ribosome, which allows it to function as a chaperone for translating polypeptides. It is unknown whether Ssb functions as a chaperone for a large population of nascent chains or has specific substrates; however, quantitation of the levels of Ssb in the cell demonstrated that there are enough molecules of Ssb for there to be two on every ribosome (18, 19).

Ssb proteins are not essential for cell viability. However, strains carrying deletions of the Ssb family are cold-sensitive, osmosensitive, and sensitive to translation-inhibiting drugs such as verrucarin A and the aminoglycoside, hygromycin B. This sensitivity to translation-inhibiting drugs provides *in vivo* evidence for a function of Ssb on the ribosome. In addition, *SSB* is regulated like a ribosomal protein gene suggesting a coordinate mode of regulation of this molecular chaperone and the protein synthetic machinery (20).

As discussed later in this chapter, Ssas clearly play a role in post-translational translocation of preproteins into the ER and mitochondria and other cellular processes. However, there is some suggestion of their involvement in the folding of newly synthesized proteins. One study using a temperature-sensitive *SSA1* mutant (*ssa1-45 ssa2 ssa3 ssa4*) found a lowered specific activity of a yeast cytosolic enzyme, ornithine transcabamylase (OTC), synthesized after temperature upshift, suggesting a role of Ssa in the folding of this protein (21). However, analysis of the folding of luciferase in yeast translation extracts failed to find a role for Ssas in folding, as depletion of Ssa from extracts had no affect on the specific activity of the synthesized protein (22). More analysis is required to assess the general requirement for Ssas in *in vivo* protein folding.

1.1.3 Hsp70s and initiation of translation

Although it is tempting to assign Hsp70s as having a role on the ribosome solely as chaperones for nascent chains, data suggest that Hsp70s may serve a more direct role in translation as well. Recently, Ssa of the yeast cytosol has been implicated in the initiation of translation. Polysome profiles from a temperature sensitive mutant of *SSA* (*ssa1-45 ssa2 ssa3 ssa4*) are altered when the strains are shifted to the nonpermissive temperature in a way reminiscent of mutants affecting translation initiation. 80S monosomes accumulate even though levels of mRNA remain abundant. Furthermore, Ssa, Sis1 (an Hsp40), and the poly-A binding protein, Pab1 were found to form a complex on the ribosome (L. Horton and J. Hensold, pers. comm.).

Studies in the rabbit reticulocyte lysate have demonstrated a relationship between Hsp70 levels and initiation of protein synthesis, as well. Hsp70 interacts with the heme-regulated eIF2α kinase, HRI. HRI specifically phosphorylates the α subunit of initiation factor eIF2, which results in the sequestration of eIF2B in an inactive form. Since eIF2B is needed for the recycling of eIF-2, its inactivation results in a block of translation initiation. It has been hypothesized that Hsp70 has two distinct roles in the maturation/regulation of HRI. Hsp70 (along with Hsp90) is needed for the folding of HRI into a mature form, which can then be activated by heme binding. In addition, Hsp70 can bind heme-containing HRI, inhibiting its activity. A model has been proposed that reduced levels of 'free' Hsp70 such as might occur under stress conditions would serve to eliminate Hsp70 binding to heme-activated HRI, thus leaving it active, which in turn would cause the inactivation of initiation factor 2. This increased level of active HRI might serve to reduce protein translation under stress conditions when Hsp70 may be unavailable to chaperone nascent chains (23–25).

1.2 Hsp40s involved in nascent chain synthesis and folding

It is widely believed that Hsp70s do not function alone. Another group of Hsps, the Hsp40s, function together with Hsp70s as molecular chaperones (for a review, see (26, 27)). Hsp40s regulate Hsp70's function by stimulating its ATPase activity. Hydrolysis of the Hsp70-bound ATP into ADP increases the affinity of the substrate binding to Hsp70 (28). Recent work by Rapoport and co-workers suggest that Hsp40 greatly increases the range of sequences with which an Hsp70 interacts (29). Additionally, some Hsp40s have been found to bind unfolded polypeptides, such as denatured rhodanese and luciferase, and prevent their aggregation (3, 17, 30–34), suggesting that they may function as molecular chaperones by themselves (35). However, a number of Hsp40s do not apparently bind substrate. These Hsp40s are thought to be involved in recruiting Hsp70 partners to their proper cellular localization (for a review, see (36)). Several Hsp40s are present in the eukaryotic cytosol. Work is ongoing to determine which, if any, Hsp40 is involved in early protein folding events.

1.2.1 Hdj-1 and Hdj-2 of mammalian cells

It has been hypothesized that an Hsp40 functions together with Hsp70 in the protection of newly synthesized proteins. Two of the Hsp40s of the mammalian cytosol, Hdj-1 and Hdj-2 are candidates for such a role. Both Hdj-1 (dj1) and Hdj-2 (dj2) contain the signature J-domain (J-domains are comprised of approximately 70 amino acids, including an HPD sequence), which plays an important role in Hsp40–Hsp70 interaction (26, 36) and an adjacent G/F-rich region; however, only Hdj-2 has a cysteine-rich domain (Fig. 2). In addition Hdj-2, which possesses a CaaX box, a site for prenylation, has been localized to the cytosolic side of the ER membrane (37). Previously, Hdj-1 was shown to function with Hsp70 to facilitate refolding of denatured luciferase (38); however, this activity has come into question as Terada et al. (71) have shown a refolding activity for Hdj-2 but not Hdj-1.

Evidence for an interaction of these two cytosolic Hsp40s with newly synthesized proteins is beginning to emerge; however, a consistent complete picture is still not evident. Frydman et al. (26) showed a Hsp40 in association with newly synthesized luciferase in reticulocyte lysates. These authors demonstrated that ribosomes translating luciferase had an Hsp40 bound in a HMW complex along with Hsc70. If this Hsp40 was removed from the lysate, there was a reduction in the amount of active luciferase synthesized, similar to the results described for depletion of Hsc70, thereby suggesting a role for both Hsp70 and Hsp40 in the folding of newly synthesized luciferase (17). In this study the identity of this Hsp40 was not determined, leaving a question as to the specific Hsp40 involved in this system. In contrast, Nagata et al. performed a more extensive investigation looking for a specific interaction of Hdj-1 and -2 with newly synthesized proteins in vivo, but were not able to detect an interaction for either Hsp40 (39). It should be noted that the antibodies against Hsp40 used by these authors differed and may in part explain the differences in the results. In addition, there may be other Hsp40 family members present that are capable of interaction with the antibodies used in the studies by Frydman's group. Clearly, further experiments need to be done to determine which, if any, Hsp40 is involved in the chaperoning of nascent polypeptides in mammalian cells.

More recently, Hdj-2 has been implicated in the folding of newly synthesized cystic fibrosis transmembrane conductance regulator (CFTR). Hdj-2 and Hsc70 were found to interact with early folding intermediates of CFTR on the cytosolic side of the ER membrane. It is suggested that Hdj-2 and Hsc70 function to prevent misfolding or aggregation of a cytosolic domain of CFTR until the protein is further synthesized

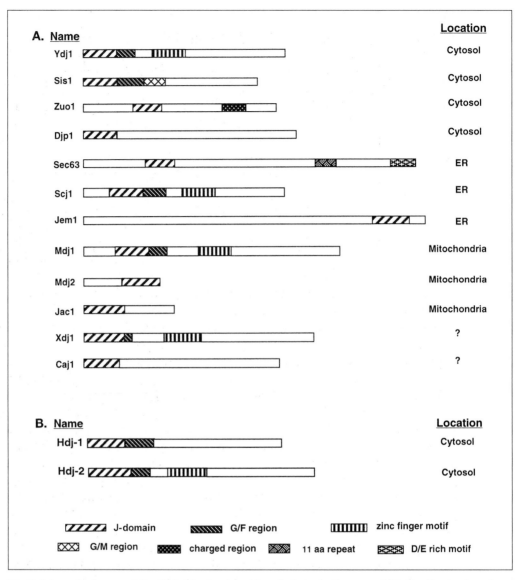

Fig. 2 Schematic diagram of Hsp40s of *S. cerevisiae* (A) and of mammalian cytosol (B). Hsp40s that have been studied to date are shown. Their subcellular location, if known, is indicated. This figure is adapted from the review by E. A. Craig, W. Yan, and P. James (35) and modified based on recent publications (41–43, 84).

and can undergo proper intramolecular interactions necessary for its proper folding (37). In these studies interactions with Hsc70 and Hdj-2 were observed in both living cells and reticulocyte lysates. Interaction of Hdj-1 was also observed in reticulocyte lysates, but little interaction was seen *in vivo*, leaving open the question of this Hsp40's involvement in CFTR maturation.

1.2.2 Hsp40s in *S. cerevisiae*

In *S. cerevisiae*, 16 Hsp40s have been identified by their sequence similarity to the *E. coli* Hsp40, DnaJ (Fig. 2). Although they have not all been analysed, some have been localized to major cellular compartments: three in the endoplasmic reticulum (Sec63, Scj1, and Jem1), three in the mitochondria (Mdj1, Mdj2, and Jac1) and at least four in the cytosol (Ydj1, Sis1, Zuo1, and Djp1) (35, 40–43) (C. Voisine and E. Craig, unpublished results). Of the cytosolic Hsp40s, Ydj1 and Sis1 have been studied most extensively. Ydj1 has the same domain structure as DnaJ and Hdj-2, having a J-domain, G/F region, and cysteine-rich region. Sis1 has a J-domain and G/F region, but lacks a cysteine-rich region. Rather, Sis1 has a G/M rich region immediately adjacent to the G/F region.

Sis1 and Ydj1 have similar biochemical properties *in vitro*. Both Ydj1 and Sis1 can stimulate the ATPase activity of the yeast cytosolic Hsp70 Ssa1, bind unfolded polypeptides, and function with Ssa1 to refold denatured luciferase (33). However, they appear to carry out different functions *in vivo*. Ydj1 has been implicated in the folding of proteins and the translocation of proteins into organelles as discussed in the next section (34, 44–46). Sis1, on the other hand, appears to be required for initiation of translation (47) making it a candidate chaperone for newly synthesized proteins. However, another yeast Hsp40, Zuo1, seems the best candidate for an Hsp40 involved in folding of nascent polypeptides, although Ydj1 and Sis1 remain viable candidates as players in the folding of newly synthesized proteins.

Zuo1 was initially identified *in vitro* as a tRNA and Z DNA-binding protein (48, 49). Like all DnaJ proteins, it contains a classic J-domain, but does not contain other domains typical of Hsp40s. However, Zuo1 has a domain rich in charged amino acids with an overall net positive charge. Zuo1 is a cytosolic protein found predominantly in association with ribosomes (40). Zuo1 associates with both translating and non-translating ribosomes, likely via an interaction with rRNA. Deletion of *ZUO1* results in sensitivity to translational inhibiting drugs such as hygromycin B, as well as cold-sensitivity, thereby supporting its role in translation *in vivo*. As with Ssb, the levels of Zuo1 in the cell are such that there is enough Zuo1 to accommodate one per ribosome (40).

Analysis of a series of deletions and truncations of Zuo1 indicated that the internal, highly charged region is required for both ribosome association and for RNA binding (Fig. 3) thereby favouring the idea that Zuo1 interacts with ribosomes, at least in part, by virtue of its association with ribosomal RNA. Deletion analysis of Zuo1 also indicates that regions other than the charged region are required for function *in vivo*. Not surprisingly, a deletion which encompasses the conserved J-domain is unable to rescue any of the Δ*zuo1* mutant phenotypes, suggesting that interaction with Hsp70 is important in Zuo1's *in vivo* role (40).

As mentioned, Zuo1 is not the only candidate Hsp40 in yeast associated with ribosomes; Sis1 is also ribosome-associated. A large fraction of Sis1 is in association with the 40S subunits and small polysomes (47). Interestingly, strains carrying a temperature-sensitive version of Sis1 accumulate 80S ribosomes and have decreased

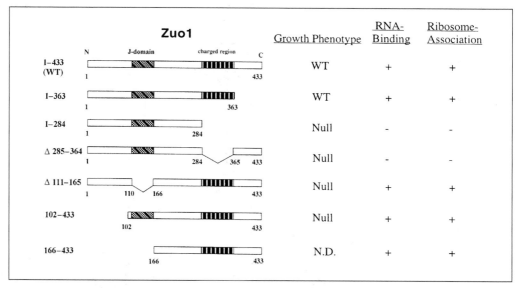

Fig. 3 Analysis summary of Zuotin deletion constructs. Zuo1 deletion constructs are diagrammed on the left (40). All mutants were tested in a strain lacking Zuo1 for their ability to rescue its cold-sensitivity and its hypersensitivity to the aminoglycoside, hygromycin B. N-terminal His-tagged wild-type and mutant proteins were partially purified from yeast and tested for their ability to bind to tRNA by northwestern blot analysis. Ribosome association was tested in extracts prepared from cells expressing wild-type or mutant Zuo1. Association was determined by western blot analysis of fractionated sucrose gradients and monitoring at OD_{260}. Summary of results from Yan *et al.* (40).

polysomes at the non-permissive temperature. This phenotype is suppressed by changes in a 60S ribosomal subunit protein. The authors proposed that Sis1 functions in the initiation of translation, possibly by helping to dissociate some part of the translation machinery. Recently, Sis1 has also been shown to be in a complex with Ssa1 and Pab1, again suggesting a role for Sis1 in initiation yet not clarifying if that role is in complex assembly or disassembly (L. Horton and J. Hensold, per. comm.). Yet the idea of Sis1 as an Hsp40 partner for an Hsp70 chaperoning nascent chains has not been excluded.

1.3 Chaperone machines involved in protein biogenesis

Based on the work from the Welch, Hartl, and Frydman laboratories, it seems likely that newly synthesized proteins associate with an HMW complex after release from ribosomes. This HMW complex is >700 kDa and has been found in association with C-terminal truncation products of CAT and luciferase using gel filtration. Hsp70 has been identified as one component of this HMW complex. Frydman *et al.* (17) have detected Hsp40 and TriC in this complex as well. It has been suggested that Hsp70, probably in cooperation with Hsp40, are components of a chaperone machine which recognizes and binds to nascent chains emerging from ribosomes as short as 40 amino acids. After about 250–300 amino acids are synthesized, the nascent chain con-

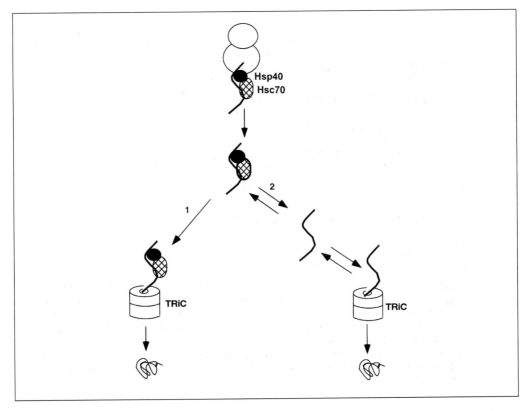

Fig. 4 Schematic model for a chaperone mode of action: sequential versus kinetic partitioning. Nascent poly-peptides emerging from the ribosome associate with chaperones. This association protects hydrophobic regions, which may be exposed in some nascent chains. This nascent chain-chaperone complex likely contains Hsp70 and Hsp40 but may contain other chaperones as well. As the nascent chain folds, it may follow pathway 1, a sequential pathway in which the nascent polypeptide is passed from one chaperone system to another, preventing any exposure to the cytosol prior to the completion of folding. Alternatively, the polypeptide may follow pathway 2, in which the nascent chain kinetically partitions among chaperones and the bulk cytosol until the final conformation is achieved. In this scenario, the newly synthesized protein may be exposed to the cytosol during its folding. It is important to note that certain proteins may require other chaperone systems for folding which are not depicted here. Additionally, some may fold independently of chaperone assistance.

tains enough information to be able to fold into its first domain (15–17). At this point the protein may proceed to fold on its own or could require the assistance of chaperones until the complete protein is synthesized. Even after release from the ribosome, many proteins may need assistance in folding which other cytosolic chaperones and/or chaperonins may supply.

Two models have been presented for the folding of newly synthesized proteins in the cell (Fig. 4). The kinetic partitioning model describes a scenario in which nascent chains could associate with Hsp/Hsc70 and Hsp40 as they are synthesized. Once synthesis was complete, these newly made proteins would be either folded by chaperones, such as TriC in the cytosol, or directed for translocation. At any point in time, these proteins could be bound to chaperones or free in the cytosol, either partially folded or in need of refolding. In this situation, there would be kinetic partitioning between chaperone systems and the bulk cytosol. Alternatively, proteins could follow a sequential pathway in which they are never accessible to the cytosol. In this type of pathway, unfolded protein would never have access to the cytosol, but would rather be handed from one chaperone system to another (50).

The Frydman lab recently reported evidence supporting a sequential mode of action. These authors over-expressed a chaperonin trap (that binds but cannot release unfolded protein) in cells and asked if newly synthesized proteins were accessible to this trap. Results demonstrated that the newly made proteins were never accessible to this trap, suggesting that nascent proteins are protected by chaperones throughout their biogenesis (15). These data imply that chaperones play an important role for many newly synthesized proteins from the time they emerge from the ribosome until they are folded into their final conformation. Some nascent polypeptides may require Hsp70 and Hsp40 for their proper folding, while others may require alternative chaperone systems. For example, nascent actin appears to need the assistance of prefoldin (PFD) in its early stages of folding and requires TRiC at a later stage (51) (see Chapter 4). Much investigation still needs to be done to identify all of the players in this folding pathway and the means by which they facilitate efficient protein biogenesis.

In yeast, the pathways of folding newly made proteins are still unclear, but data is beginning to establish which chaperones are involved and how these chaperones work together in protein biogenesis. As chaperones for nascent polypeptides, both Zuo1 and Ssb associate with ribosomes. Ssb has been shown to be two- to fourfold more abundant than ribosomes (18) and the ratio of Ssb to Zuo1 has been determined to be approximately 4:1 (40). Therefore, if Zuo1 and Ssb function together, there is sufficient Zuo1 within the cell to function with Ssb on every ribosome. Genetic data strongly suggest this possibility. Both Δ*zuo1* and Δ*ssb* strains are cold-, aminoglycoside-, and salt-sensitive. These similarities between the *ZUO1* and *SSB* mutant phenotypes are reminiscent of the similarity between *dnaK* and *dnaJ* mutant phenotypes (52).

Mechanistically, how Zuo1 and Ssb functionally interact remains unresolved. The simplistic idea that Zuo1 may target Ssb to the ribosome in the way Sec63 appears to target Kar2 to the ER membrane appears not to be true. Ssb binds both translating and nontranslating ribosomes with similar salt sensitivity in a Δ*zuo1* mutant and in wild-type cells (40). Zuo1 and Ssb may function together with Zuo1 regulating the interaction of Ssb with nascent chains either by targeting the nascent chain to Ssb or/and by stimulating its ATPase activity. However, an understanding of how Zuo1 and Ssb partner to facilitate protein synthesis and early stages of protein folding awaits further exploration, including a determination of whether Zuo1 binds directly to the nascent chain and an understanding of the effects of ribosome binding on biochemical properties of Ssb and Zuo1.

The Hsp70 partner of Sis1 is not established. Sis1 is ribosome associated and required for the initiation of protein translation (47). Because of their similar localization, it has been hypothesized that Sis1 and Ssb may function together in protein translation. These two chaperones may also function in other processes such as proteolysis as Ohba *et al.* (53, 54) have shown that over-expression of either *SSB1* or *SIS1* can rescue the growth defects of a proteasome mutation. However, in light of recent data showing localization of Ssa to polyribosomes and the interaction of Ssa, Sis1, and Pab1, it is possible that Sis1 and Ssa1 function in translation in a manner which requires the involvement of the poly (A) binding protein. This idea is consistent with the observed biochemical interaction between these two proteins (33).

In summary, Zuo1, Sis1, and Ssb are all ribosome-associated. A small amount of Ssa also appears to be ribosome-bound. It is unclear how Ssa and Sis1 function in translation, but based on genetic analysis it appears likely that Ssb and Zuo1 function in a similar capacity in protein biogenesis. Most of the Ssa and another cytosolic Hsp40, Ydj1, are found in the soluble fraction. Zuo1/Ssb may function in early stages of protein folding while Ydj1/Ssa function later, with some newly synthesized polypeptides being 'passed' from one class of Hsp70s to another. Dissection of the complex functional interactions between these and other chaperones in the cytosol will lead to a more realistic appreciation of the dynamic protein folding pathways in the cell and help to determine if nascent chains are protected and folded by an evolutionarily conserved chaperone machine.

2. Chaperones in post-translational translocation of proteins into organelles

There are fundamental differences in the mechanism of translocation into different cellular organelles. For example, the narrow channels into the endoplasmic reticulum (ER) and mitochondria demand that proteins be at least partially unfolded to enter, while folded proteins are imported into the nucleus and peroxisomes. However, cytosolic Hsp70s, and in some cases Hsp40s, have been implicated in post-translational translocation into all these organelles in both yeast and mammalian cells.

2.1 Protein import into the endoplasmic reticulum

Proteins are imported into the ER either co- or post-translationally. Evidence exists for a role of Hsp70s in post-translational translocation of proteins into the ER of yeast and mammalian cells. Hsp70 was identified as a factor in reticulocyte lysates which stimulated import of M13 coat protein into dog pancreas microsomes (55). Using yeast *in vitro* translocation assays, similar experiments identified one cytosolic class of Hsp70s, the Ssas, as proteins which stimulated post-translational transport of the secreted yeast pheromone prepro-α-factor across microsomal membranes (56). At the same time *in vivo* experiments demonstrated a role for Hsp70 in translocation. Cells depleted of Ssa accumulated prepro-α-factor, the precursor of a secreted pheromone. More extensive experiments with a temperature-sensitive *SSA1* mutant (*ssa1-45 ssa2 ssa3 ssa4*) revealed a defect in prepro-α-factor processing within 2 minutes of shift to the nonpermissive temperature, suggesting a direct involvement in the translocation process *in vivo* (46). Of six proteins destined for the ER, the translocation of only prepro-α-factor and proteinase A was inhibited. Whether the lack of precursor accumulation of the other four precursors is indicative of a lack of involvement of Ssa proteins in the post-translocational import of these proteins, or whether these proteins are completely translocated in a co-translational, Hsp70-independent pathway is not clear.

In yeast, the cytosolic Hsp40 Ydj1 has also been implicated in ER translocation. Like *SSA* mutants, a temperature-sensitive *YDJ1* mutant is defective in translocation of prepro-α-factor (45). Interestingly, Ydj1 is farnesylated. At least a portion is associated with the ER membrane, raising the possibility that Ssa and Ydj1 function together in post-translational translocation and that Ydj1 might function to target Ssa and bound preprotein to the membrane.

What is the role of cytosolic chaperones in post-translational translocation? The most popular hypothesis is based on the ability of Hsp70s and Hsp40s to bind unfolded proteins: Hsp70s and Hsp40s bind preprotein, delaying their folding, thus facilitating translocation through the channel into the ER lumen. This is a rationale hypothesis. However, at present there is little data to support it.

2.2 Protein import into mitochondria

As discussed in Chapter 3, both Hsp70s and Hsp40s have been shown to be important for the facilitation of import of at least some proteins into mitochondria, *in vivo* and *in vitro*. As in post-translational import into the ER, Ssa and Ydj1 types of Hsp70s and Hsp40s are important players. In yeast a temperature sensitive *SSA1* mutant (*ssa1-45 ssa2 ssa3 ssa4*) accumulates precursor forms of several proteins after a short shift to the nonpermissive temperature (46). However, of three proteins tested only one, the beta subunit of the F_1F_0 ATPase, showed such an effect suggesting that Hsp70 may be important for only a subset of proteins. The same set of proteins has not been analysed in *ydj1* mutants; however, such cells are also defective in import of the beta subunit of the F_1F_0 ATPase as well as citrate synthase (44, 57).

Roles of Hsp70 and Hsp40 have been observed in *in vitro* systems derived from both yeast and mammalian cells (56, 58). For example, yeast Ssa1/2 were isolated as proteins which stimulated import of proteins synthesized in wheat germ extracts (59). In addition, depletion of Hdj-2, but not Hdj-1 from reticulocyte lysate inhibited the translocation of preprotein into mammalian mitochondria, suggesting that Hdj-2 is the partner for Hsp70 in this process (60). All the data to date is consistent with the general model that Hsp70 and Hsp40 work together to maintain the import competence of proteins prior to their import. However, this idea has yet to be proven.

Proteins in addition to the Hsp40/Hsp70 chaperones act in the cytosol to promote import of proteins into mitochondria. For example, a protein called MSF (mitochondrial stimulating factor) was isolated from rat liver cytosol based on its ability to maintain the import competence of precursor proteins synthesized in wheat germ extracts to be imported into mammalian mitochondria (61). Thus MSF has characteristics of a general chaperone, binding to proteins and preventing their aggregation.

However, the two subunits of MSF, which are members of the 14-3-3 protein family, appear to facilitate the transport of proteins by a pathway that is independent of Hsp70 (58).

Other cytosolic proteins have also been implicated in mitochondrial import. One, presequence binding protein (PBP) from reticulocyte lysate specifically recognizes the presequence. PBP may provide specificity for binding to mitochondrial receptors (59). In addition nascent chain associated complex (NAC), a ribosome-associated heterodimer which associates with many, if not all, nascent chains, has recently been implicated in mitochondrial translocation (62, 63). Thus many factors, in addition to established chaperones, function to allow efficient targeting and translocation into mitochondria.

2.3   Protein import into peroxisomes

Peroxisomes, organelles required for beta-oxidation of fatty acids and hydrogen peroxide degradation, contain a variety of matrix proteins which are imported post-translationally. Unlike proteins translocated across the ER and mitochondrial membranes, proteins need not be unfolded prior to import into peroxisomes (64). In fact, some proteins have been shown to oligomerize in the cytosol and be imported without subunit dissociation (65, 66). In a permeabilized cell assay, Hsp70 antibodies inhibited the import of proteins into peroxisomes, suggesting a role for these chaperones even though proteins need not be maintained in an unfolded conformation (67).

A study of the involvement of Hsp70s in peroxisomal translocation in yeast has not been reported. However, a specific involvement of the cytosol-localized Hsp40 Djp1 has been shown (42). In a deletion mutant import of several peroxisomal proteins is impaired, while nuclear, ER, and mitochondrial import process appear normal. In addition, phenotypic effects of the mutant are only observed under conditions where peroxisomal function is required. The identity of a presumed Hsp70 partner for Djp1 has yet to be reported.

The role of either Hsp40 or Hsp70 in peroxisomal import is unresolved. The involvement of chaperones appears to be entirely cytosolic as no chaperone in the matrix of the peroxisome has been identified. It is possible that chaperones function in a manner similar to that proposed for cytosolic chaperones in import into the ER and mitochondria: maintenance of peroxisomal proteins in a translocation competent state. Although it has been shown that subunits of an oligomeric protein can together during import, there is evidence of binding of Hsp70 to peroxisomal-destined proteins. In addition, a monomeric protein was shown to be imported more efficiently than an oligomeric one (68). Alternatively, chaperones may be involved in facilitating the interaction of peroxisomal proteins with receptors that mediate import.

2.4   Protein import into the lysosome

Selective import of proteins into both mammalian lysosomes and yeast vacuoles is induced under starvation conditions (69, 70). This pathway is different from other nonselective modes of internalization of proteins within lysosomes, which often depend on forms of endocytosis or autophagy. Proteins that contain identified targeting signals are selectively imported by this starvation-induced process. In the well-studied mammalian system, Hsp70s may be involved at more than one step of this process (69). Cytosolic Hsc73, which can bind to a peptide containing the KFEQ sequence found in imported proteins, is needed for efficient import into isolated lysosomes (71). In addition, an Hsp70 which appears to be the most acidic form of cytosolic Hsc73 has been found inside the lysosome. Specific inhibition of intralysomal Hsc73 by directing antibodies to the lysosome also inhibited this degradation pathway (72). These results suggest that there is a requirement for at least two forms of Hsp70 for efficient degradation by this lysosomal pathway, one in the cytosol and one acting within the lysosome.

2.5   Protein import into the nucleus

Studies in living mammalian cells and permeabilized cell free systems first indicated a role for Hsp70s in nuclear transport (reviewed in (73)). The dependence on Hsp70 may be substrate specific. In a system using permeabilized cells in which import into the nucleus of proteins is dependent on added cytosolic factors, import of simian virus large T antigen was found to be dependent on added cytosolic Hsp70, but import of glucocorticoid receptors was not (74). More recently a system was developed using T antigen linked to green fluorescent protein to monitor nuclear import in yeast cells (75). Increased levels of the Ssa Hsp70 increased the rate of import in wild-type cells and overcame the defect of import in certain mutants which are defective in established components of the nuclear import pathway. Together these studies suggest a conserved role for Hsp70s in nuclear import, but their mechanism of action is unresolved.

3.   Chaperones involved in the disassembly of protein complexes: uncoating of clathrin

The disassembly of protein complexes is one of the first functions that was assigned to Hsp70s. Three such reactions have been studies extensively: (1) dissociation of the λP protein from the replication initiation complex allowing initiation of phage λ DNA replication in *E. coli* (76); (2) dissociation of the inactive dimeric P1 plasmid replication protein RepA into the active monomeric form in *E. coli* (77); (3) uncoating of clathrin-coated vesicles in mammalian cells. These three processes remain the only well-established examples of facilitation of disassembly of complexes by Hsp70, although there are suggestions in the literature that Hsp70s may be involved in other cellular disassembly processes as well. This section focuses on the only eukaryotic example, uncoating of clathrin-coated vesicles.

Monomeric clathrin triskelions are formed from three clathrin heavy chains and three clathrin light chains. During receptor-mediated endocytosis these monomeric clathrin triskelions polymerize, forming coats on pits in the plasma membrane which then invaginate into the cells forming clathrin-coated vesicles. Purified monomeric clathrin triskelions can spontaneously polymerize *in vitro*, forming so-called clathrin cages. The disassembly of the clathrin coat of clathrin-coated vesicles is stimulated by Hsc70 in an ATP-dependent manner (for review see (78)). During this reaction Hsc70 forms a stable trimeric complex: Hsc70: ADP: clathrin. Therefore, Hsc70 directly interacts with clathrin and hydrolysis of ATP traps the clathrin substrate in a complex that is slow to dissociate, similar to Hsp70's interaction with other substrate proteins. Uncoating of clathrin-coated vesicles *in vivo* also seems to be an Hsc70-dependent process as microinjection of Hsc70 antibody into tissue culture cells inhibited receptor-mediated endocytosis (79).

However, Hsc70 is unable to disassemble clathrin baskets formed from highly purified components. The search for a factor which would enable Hsc70 to disassemble clathrin baskets identified a 100 kDa protein named auxilin (80). Interestingly auxilin was found to contain a 'J'-domain, although unlike most Hsp40s it is located near the C-terminus. However, auxilin appears to have characteristics like other Hsp40s, as it is able to stimulate the ATPase activity of Hsc70 at low concentrations (81). Auxilin forms a stable complex with clathrin (82). Analysis of fragments of auxilin (Fig. 5) revealed that a segment including amino acids 547 to 814 of the 910 amino acid protein were sufficient for this clathrin binding. However this fragment was not active in clathrin uncoating assays. Fragment 547 to 910 which also includes the 'J'-domain not only bound clathrin, but had uncoating activity as well.

These reported data are consistent with the idea that auxilin binds clathrin on coated vesicles and recruits Hsc70 which then facilitates the disassembly of the

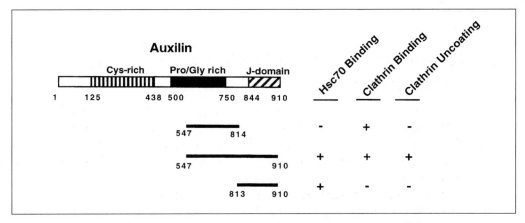

Fig. 5 Analysis summary of Auxilin fragments. Auxilin fragments are diagrammed on the left (82). The fragments were purified as GST fusion proteins from bacteria. The purified protein fragments were tested for binding to Hsc70 using a co-absorption assay to glutathione–sepharose beads. Clathrin binding was determined by co-sedimentation following incubation of each fragment with preassembled clathrin baskets. Clathrin uncoating was tested by quantitating the level of dissociated clathrin from preassembled baskets follow ultracentrifigation by SDS-PAGE. Summary of results from Holstein *et al.* (82).

complex. Consistent with this idea, fragments 547–814 and 813–910 were unable to function in *trans*, suggesting that the clathrin binding domain and the 'J'-domain must be linked in order to stimulate the uncoating reaction (82). According to this scenario, the high affinity of the clathrin binding domain of auxilin results in auxilin first binding to coated vesicles. Subsequently, the interaction of auxilin with Hsp70 via its 'J'-domain brings Hsp70 close to clathrin, facilitating binding not only because of its close proximity but also because of its ability to stimulate Hsp70s ATPase activity. These results are consistent with models proposed for other Hsp40/Hsp70 pairs in which the Hsp40 is thought to 'target' Hsp70 to its site of action as well as facilitate the productive interaction by stimulating the hydrolysis of ATP. A well-studied example of such an interaction is the targeting of the Hsp70 of the ER lumen (BiP/Kar2) to the translocation channel by the membrane bound Hsp40, Sec63 (83).

4. Summary

The research summarized in this chapter points to roles for cytosolic Hsp70s and their cohort proteins in the cellular processes of translation, translocation across organellar membranes, and assembly/disassembly of multimeric protein complexes. Current evidence places Hsp70s and Hsp40s in the proper location to participate in these functions. In addition, genetic evidence suggests that in the absence of these chaperones several of these processes are inefficient. However, many questions remain concerning the means by which these chaperones facilitate the processes in which they are involved.

There is clear evidence that Hsp70s associate with newly synthesized proteins and more recent data implicate Hsp70 and its co-chaperones in the proper folding of nascent polypeptides. However, many questions remain. For instance, can Hsp70 accomplish the folding job alone or does it hand proteins off to other chaperone systems? If so, how obligatory is a specific sequence for chaperone interactions? What percentage of newly synthesized proteins require Hsp70 assistance? Are certain types of proteins that are more prone to aggregation more apt to require chaperone action?

Some data suggests a role for Hsp70s in protecting translocating polypeptides prior to leaving the cytosol. However, it is unclear how Hsp70s accomplish this task and how many translocating proteins require Hsp70 assistance. Similarly, a role has been shown for Hsp70 and the Hsp40 auxilin in complex assembly/disassembly but it is unknown if this is a specialized role for cytosolic chaperones or if chaperone assistance is needed for formation of most complexes.

Many questions remain regarding Hsp70 and its co-chaperones. On one level we need to better understand the biochemical mechanisms by which Hsp70s and its co-chaperones function. On another level we need to establish the partnerships between different Hsp70s and Hsp40s, as well as the interaction between Hsp70s and other chaperone systems. In the years to come, a clearer picture of the pathways and networks of chaperones which play important roles in normal cellular processes will certainly be generated.

References

1. Anfinsen, C. B. (1973). Principles that govern the folding of protein chains. *Science*, **181**, 223–30.
2. Malkin, L. I. and Rich, A. (1967). Partial resistance of nascent polypeptide chains to proteolytic digestion due to ribosomal shielding. *J. Mol. Biol.*, **26**, 329–46.
3. Hartl, F. U. (1996). Molecular chaperones in cellular protein folding. *Nature*, **381**, 571–80.
4. Horwich, A. L., Low, K. B., Fenton, W. A., Hirshfield, I. N., and Furtak, K. (1993). Folding *in vivo* of bacterial cytoplasmic proteins: role of GroEL. *Cell*, **74**, 909–17.
5. Deuerling, E., Schulze-Specking, A., Tomoyasu, T., Mogk, A., and Bukau, B. (1999). Trigger factor and DnaK cooperate in folding of newly synthesized proteins. *Nature*, **400**, 693–6.
6. Teter, S. A., Houry, W. A., Ang, D., Tradler, T., Rockabrand, D., Fischer, G., *et al.* (1999). Polypeptide flux through bacterial Hsp70: Dnak cooperates with trigger factor in chaperoning nascent chains. *Cell*, **97**, 755–65.
7. Komar, A. A., Kommer, A., Krasheninnikov, I. A., and Spirin, A. S. (1993). Cotranslational heme binding to nascent globin chains. *FEBS Lett.*, **326**, 261–3.
8. Makeyev, E. V., Kolb, V. A., and Spirin, A. S. (1996). Enzymatic activity of the ribosome-bound nascent polypeptide. *FEBS Lett.*, **378**, 166–70.
9. Kudlicki, W., Chirgwin, J., Kramer, G., and Hardesty, B. (1995). Folding of an enzyme into an active conformation while bound as peptidyl-tRNA to the ribosome. *Biochem.*, **34**, 14284–7.
10. Gilmore, R., Coffey, M. C., Leone, G., McLure, K., and Lee, P. W. K. (1996). Co-translational trimerization of the reovirus cell attachment protein. *EMBO J.*, **15**, 2651–8.
11. Nicola, A., Chen, W., and Helenius, A. (1999). Co-translational folding of an alphavirus capsid protein in the cytosol of living cells. *Nature Cell Biol.*, **1**, 341–5.
12. Beckmann, R. P., Mizzen, L., and Welch, W. (1990). Interaction of Hsp70 with newly synthesized proteins: implications for protein folding and assembly. *Science*, **248**, 850–6.
13. Beckmann, R. P., Lovett, M., and Welch, W. J. (1992). Examining the function and regulation of hsp70 in cells subjected to metabolic stress. *J. Cell Biol.*, **117**, 1137–50.
14. Eggers, D. K., Welch, W. J., and Hansen, W. J. (1997). Complexes between nascent polypeptides and their molecular chaperones in the cytosol of mammalian cells. *Mol. Biol. Cell*, **8**, 1559–73.
15. Thulasiraman, V., Yang, C.-F., and Frydman, J. (1999). *In vivo* newly translated polypeptides are sequestered in a protected folding environment. *EMBO J.*, **18**, 85–95.
16. Hansen, W. J., Lingappa, V. R., and Welch, W. J. (1994). Complex environment of nascent polypeptide chains. *J. Biol. Chem.*, **269**, 26610–13.
17. Frydman, J., Nimmesgern, E., Ohtsuka, K., and Hartl, F. U. (1994). Folding of nascent polypeptide chains in a high molecular mass assembly with molecular chaperones. *Nature*, **370**, 111–17.
18. Pfund, C., Lopez-Hoyo, N., Ziegelhoffer, T., Schilke, B. A., Lopez-Buesa, P., Walter, W. *et al.* (1998). The molecular chaperone *SSB* from *S. cerevisiae* is a component of the ribosome-nascent chain complex. *EMBO J.*, **17**, 3981–9.
19. Nelson, R. J., Ziegelhoffer, T., Nicolet, C., Werner-Washburne, M., and Craig, E. A. (1992). The translation machinery and seventy kilodalton heat shock protein cooperate in protein synthesis. *Cell*, **71**, 97–105.

20. Lopez, N., Halladay, J., Walter, W., and Craig, E. A. (1999). SSB, encoding a ribosome-associated chaperone, is coordinately regulated with ribosomal protein genes. *J. Bacteriol.*, **181**, 3136–43.
21. Kim, S., Schilke, B., Craig, E., and Horwich, A. (1998). Folding *in vivo* of a newly translated yeast cytosolic enzyme is mediated by the SSA class of cytosolic yeast Hsp70 proteins. *Proc. Natl. Acad. Sci. USA*, **95**, 12860–5.
22. Bush, G. L. and Meyer, D. I. (1996). The refolding activity of the yeast heat shock proteins Ssa1 and Ssa2 defines their role in protein translocation. *J. Cell Biol.*, **135**, 1229–37.
23. Uma, S., Thulasiraman, V., and Matts, R. L. (1999). Dual role for Hsc70 in the biogenesis and regulation of the heme-regulated kinase of the α subunit of eukaryotic translation initiation factor 2. *Mol. Cell. Biol.*, **19**, 5861–71.
24. Matts, R. L., Hurst, R., and Xu, Z. (1993). Denatured proteins inhibit translation in hemin-supplemented rabbit reticulocyte lysate by inducing the activation of the heme-regulated eIF-2α kinase. *Biochem.*, **32**, 7321–8.
25. Matts, R. L. and Hurst, R. (1992). The relationship between protein synthesis and heat shock proteins levels in rabbit reticulocyte lysates. *J. Biol. Chem.*, **267**, 18168–74.
26. Cheetham, M. E. and Caplan, A. J. (1998). Structure, function and evolution of DnaJ: conservation and adaption of chaperone function. *Cell Stress and Chaperones*, **3**, 28–36.
27. Caplan, A. J., Cyr, D. M., and Douglas, M. G. (1993). Eukaryotic homologues of *Escherichia coli dnaJ*: a diverse protein family that functions with Hsp70 stress proteins. *Mol. Biol. Cell*, **4**, 555–63.
28. Bukau, B. and Horwich, A. L. (1998). The Hsp70 and Hsp60 chaperone machines. *Cell*, **92**, 351–66.
29. Misselwitz, B., Staeck, O., and Rapoport, T. (1998). J proteins catalytically activate Hsp70 molecules to trap a wide range of peptide sequences. *Molecular Cell*, **2**, 593–603.
30. Langer, T., Lu, C., Echols, H., Flanagan, J., Hayer, M. K., and Hartl, F. U. (1992). Successive action of DnaK, DnaJ, and GroEL along the pathway of chaperone-mediated protein folding *Nature*, **356**, 683–9.
31. Cyr, D. M. (1995). Cooperation of the molecular chaperone Ydj1 with specific Hsp70 homologs to suppress protein aggregation. *FEBS Lett.*, **359**, 129–32.
32. Prip-Buus, C., Westermann, B., Schmitt, M., Langer, T., Neipert, W., and Schwarz, E. (1996). Role of the mitochondrial DnaJ homologue, Mdj1p, in the prevention of heat- induced protein aggregation. *FEBS Lett.*, **390**, 142–6.
33. Lu, Z. and Cyr, D. M. (1998). Protein folding activity of Hsp70 is modified differentially by the Hsp40 co-chaperones Sis1 and Ydj1. *J. Biol. Chem.*, **273**, 27824–30.
34. Lu, Z. and Cyr, D. M. (1998). The conserved carboxyl terminus and zinc finger-like domain of the co-chaperone Ydj1 assist Hsp70 in protein folding. *J. Biol. Chem.*, **273**, 5970–8.
35. Craig, E., Yan, W., and James, P. (1999). Genetic dissection of the Hsp70 chaperone system of yeast. In *Molecular chaperones and folding catalysts: regulation, cellular function and mechanisms* (ed. B. Bukau), pp. 139–62. Harwood Academic, Amsterdam.
36. Kelley, W. L. (1998). The J-domain family and the recruitment of chaperone power. *Trends in Biochemistry*, **23**, 222–7.
37. Meacham, G. C., Lu, Z., King, S., Sorscher, E., Tousson, A., and Cyr, D. M. (1999). The Hdj-2/Hsc70 chaperone pair facilitates early steps in CFTR biogenesis. *EMBO J.*, **18**, 1492–1505.
38. Freeman, B. C. and Morimoto, R. I. (1996). The human cytosolic molecular chaperone hsp90, hsp70(hsc70) and hdj-1 have distinct roles in recognition of a non-native protein and proteinrefolding. *EMBO J.*, **15**, 2969–79.

39. Nagata, H., Hansen, W., Freeman, B., and Welch, W. (1998). Mammalian cytosolic DnaJ homologues affect the hsp70 chaperone-substrate reaction cycle, but do not interact directly with nascent or newly synthesized proteins. *Biochem.*, **37**, 6924–38.

40. Yan, W., Schilke, B., Pfund, C., Walter, W., Kim, S., and Craig, E. A. (1998). Zuotin, a ribosome-associated DnaJ molecular chaperone. *EMBO J.*, **17**, 4809–17.

41. Westermann, B. and Neupert, W. (1997). Mdj2p, a novel DnaJ homolog in the mitochondrial inner membrane of the yeast. *J. Mol. Biol.*, **272**, 477–83.

42. Hettema, E. H., Ruigrok, C. C. M., Koerkamp, M. G., Berg, M. v. d., Tabak, H. F., Distel, B., and Braakman, I. (1998). The cytosolic DnaJ-like protein Djp1p is involved specifically in peroxisomal protein import. *J. Cell Biol.*, **142**, 421–34.

43. Strain, J., Lorenz, C. R., Bode, J., Garland, S., Smolen, G. A., Ta, D. T., *et al.* (1998). Suppressors of superoxide dismutase (SOD1) deficiency in *Saccharomyces cerevisiae*. *J. Biol. Chem.*, **273**, 31138–44.

44. Atencio, D. P. and Yaffe, M. P. (1992). *MAS5*, a yeast homolog of DnaJ involved in mitochondrial protein import. *Mol. Cell. Biol.*, **12**, 283–91.

45. Caplan, A. J., Cyr, D. M. and Douglas, M. G. (1992). YDJ1p facilitates polypeptide translocation across different intracellular membranes by a conserved mechanism. *Cell*, **71**, 1143–55.

46. Becker, J., Walter, W., Yan, W., and Craig, E. A. (1996). Functional interaction of cytosolic Hsp70 and DnaJ-related protein, Ydj1p, in protein translocation in vivo. *Mol. Cell. Biol.*, **16**, 4378–86.

47. Zhong, T. and Arndt, K. T. (1993). The yeast *SIS1* protein, a DnaJ homolog, is required for initiation of translation. *Cell*, **73**, 1175–86.

48. Zhang, S., Lockshin, C., Herbert, A., Winter, E., and Rich, A. (1992). Zoutin, a putative Z-DNA binding protein in *Saccharomyces cerevisiae*. *EMBO J.*, **11**, 3787–96.

49. Wilhelm, M. L., Reinholt, J., Gangloff, J., Dirheimer, G., and Wilhelm, F. X. (1994). Transfer RNA binding protein in the nucleus of *Saccharomyces cerevisiae*. *FEBS Lett.*, **349**, 260–4.

50. Frydman, J. and Hartl, F. U. (1996). Principles of chaperone-assisted protein folding: differences between *in vitro* and *in vivo* mechanisms. *Science*, **272**, 1497–1502.

51. Hansen, W. J., Cowan, N. J., and Welch, W. J. (1999). Prefoldin-nascent chain complexes in the folding of cytoskeletal proteins. *J. Cell Biol.*, **145**, 265–77.

52. Sell, S. M., Eisen, C., Ang, D., Zylicz, M., and Georgopoulos, C. (1990). Isolation and characterization of dnaJ null mutants of *Escherichia coli*. *J. Bacteriol.*, **172**, 4827–35.

53. Ohba, M. (1997). Modulation of intracellular protein degradation by SSB1-SIS1 chaperon system in yeast *S. cerevisiae*. *FEBS Lett.*, **409**, 307–11.

54. Ohba, M. (1994). A 70-kDa heat shock cognate protein suppresses the defects caused by a proteasome mutation in *Saccharomyces cerevisiae*. *FEBS Lett.*, **351**, 263–6.

55. Zimmerman, R., Sagstetter, M., Lewis, M. J., and Pelham, H. R. B. (1988). Seventy-kilodalton heat shock proteins and an additional component from reticulocyte lysate stimulate import of M13 procoat protein into microsomes. *EMBO J.*, **7**, 2875–80.

56. Chirico, W., Waters, M. G., and Blobel, G. (1988). 70K heat shock related proteins stimulate protein translocation into microsomes. *Nature*, **332**, 805–10.

57. Deshaies, R., Koch, B., Werner-Washburne, M., Craig, E., and Schekman, R. (1988). A subfamily of stress proteins facilitates translocation of secretory and mitochondrial precursor polypeptides. *Nature*, **332**, 800–5.

58. Komiya, T., Sakaguchi, M., and Mihara, K. (1996). Cytoplasmic chaperones determine the targeting pathway of precursor proteins to mitochondria. *EMBO J.*, **15**, 399–407.

59. Murakami, K., Tanase, S., Morino, Y. and Mori, M. (1992). Presequence binding factor-dependent and -independent import of proteins into mitochondria. *J. Biol. Chem.*, **267**, 13119–22.

60. Terada, K., Kanazawa, M., Bukau, B., and Mori, M. (1997). The human DnaJ homologue dj2 facilitates mitochondrial protein import and luciferase refolding. *J. Cell Biol.*, **139**, 1089–95.

61. Hachiya, N., Komiya, T., Alam, R., Iwahashi, J., Sakaguchi, M., Omura, T., and Mihara, K. (1994). MSF, a novel cytoplasmic chaperone which functions in precursor targeting to mitochondria. *EMBO J.*, **13**, 5146–54.

62. George, R., Beddoe, T., Landl, K. and Lithgow, T. (1998). The yeast nascent polypeptide associated complex initiates protein targeting to mitochondria *in vivo*. *Proc. Natl. Acad. Sci. USA*, **95**, 2296–301.

63. Funfschilling, U. and Rospert, S. (1999). Nascent polypeptide-associated complex stimulates protein import into yeast mitochondria. *Mol. Biol. Cell*, **20**, 3289–99.

64. Walton, P., Hill, P., and Subramani, S. (1995). Import of stably folded proteins into peroxisomes. *Mol. Cell. Biol.*, **6**, 675–83.

65. Glover, J., Andrews, D., and Rachubinski, R. (1994). *Saccharomyces cerevisiae* peroxisomal thiolase is imported as a dimer. *Proc. Natl. Acad. Sci. USA*, **91**, 10541–5.

66. McNew, J. and Goodman, J. (1994). An oligomeric protein is imported into peroxisomes *in vitro*. *J. Cell Biol.*, **127**, 1245–57.

67. Walton, P., Wendland, M., Subramani, S., Rachubinski, R., and Welch, W. (1994). Involvement of 70-kD heat-shock proteins in peroxisomal import. *J. Cell Biol.*, **125**, 1037–46.

68. Crookes, W. and Olsen, L. (1998). The effects of chaperones and the influence of protein assembly on peroxisomal protein import. *J. Biol. Chem.*, **273**, 17236–42.

69. Dice, J. F., Agarraberes, F., Kirven-Brooks, M., Terlecky, L. J., and Terlecky, S. R. (1994). Heat Shock 70-kD proteins and lysosomal proteolysis. In *The biology of heat shock proteins and molecular chaperones* (ed. R. I. Morimoto *et al.*), pp. 137–51. Cold Spring Harbor Laboratory Press, Plainview, NY.

70. Horst, M., Knecht, E., and Schu, P. (1999). Import into and degradation of cytosolic proteins by isolated yeast vacuoles. *J. Cell Biol.*, **10**, 2879–89.

71. Cuervo, A. and Dice, J. (1996). A receptor for the selective uptake and degradation of proteins by lysosomes. *Science*, **273**, 5011–503.

72. Agarraberes, F., Terlecky, S., and Dice, J. (1997). An intralysosomal hsp70 is required for a selective pathway of lysosomal protein degradation. *J. Cell Biol.*, **137**, 825–34.

73. Melchior, F. and Gerace, L. (1995). Mechanisms of nuclear protein import. *Curr. Op. in Cell Biol.*, **7**, 310–18.

74. Yang, J. and DeFranco, D. (1994). Differential roles of heat shock protein 70 in the *in vitro* nuclear import of glucocorticoid receptor and simian virus 70 large tumor antigen. *Mol. Cell. Biol.*, **14**, 5088–98.

75. Shulga, N., Roberts, P., Gu, Z., Spitz, L., Tabb, M., Nomura, M., and Goldfarb, D. (1996). *In vivo* transport kinetics in *Saccharomyces cerevisiae*: a role for heat shock protein 70 during targeting and translocation. *J. Cell Biol.*, **135**, 329–39.

76. Georgopoulos, C., Liberek, K., Zylicz, M. and Ang, D. (1994). Properties of the heat shock proteins of *Escherichia coli* and the autoregulation of the heat shock response. In *The biology of heat shock proteins and molecular chaperones* (ed. R. I. Morimoto *et al.*), pp. 209–50. Cold Spring Harbor Laboratory Press, Plainview, NY.

77. Wickner, S., Skowyra, D., Hoskins, J., and McKenney, K. (1992). DnaJ, DnaK and GrpE heat shock proteins are required in *ori*P1 replication solely at the repA monomerization step. *Proc. Natl. Acad. Sci. USA*, **89**, 10345–9.

78. Eisenberg, E. and Greene, L. (1999). Role of Chaperones in uncoating of clathrin coated vesicles. In *Molecular chaperones and folding catalysts* (ed. B. Bukau), pp. 329–45. Harwood Academic, Amsterdam.

79. Honig, S., Kreimer, G., Robenek, H., and Jockusch, B. (1994). Receptor-mediated endocytosis is sensitive to antibodies. *Journal of Cell Science*, **107**, 1185–6.

80. Ungewickell, E., Ungewickell, H., Holstein, S. E. H., Lindner, R., Prasad, K., Barouch, W., *et al.* (1995). Role of auxilin in uncoating clathrin-coated vesicles. *Nature*, **378**, 632–5.

81. Jiang, R.-F., Greener, T., Barouch, W., Greene, L., and Eisenberg, E. (1997). Interaction of auxilin with the molecular chaperone, Hsc70. *J. Biol. Chem.*, **272**, 6141–5.

82. Holstein, S. E. H., Ungewickell, H., and Ungewickell, E. (1996). Mechanism of clathrin basket dissociation: separate functions of protein domains of the DnaJ homologue auxilin. *J. Cell Biol.*, **135**, 925–37.

83. Cyr, D. M., Langer, T., and Douglas, M. G. (1994). DnaJ-like proteins: molecular chaperones and specific regulators of Hsp70. *TIBS*, **19**, 176–81.

84. Nishikawa, S. and Endo, T. (1997). The yeast JEM1p is a DnaJ-like protein of the endoplasmic reticulum membrane required for nuclear fusion. *J. Biol. Chem.*, **272**, 12889–92.

6 | Hsp70 chaperone networks: the role of regulatory co-chaperones in coordinating stress responses with cell growth and death

JAEWHAN SONG and RICHARD I. MORIMOTO

1. Introduction

Among the fascinating problems in biology is the process by which proteins acquire and maintain their folded state. Interest in this problem has intensified with the recognition that folding intermediates adopt alternative conformations rich in β-sheets that can self-associate to form protein aggregates, protofilaments, and fibrils as associated with a class of diseases known as protein conformational or aggregation diseases, including Alzheimer's Disease, Parkinson's Disease, ALS, and Huntington's Disease (1, and see Chapter 11). As environmental and physiological stress and genetic mutations are known to negatively influence the stability of proteins and result in protein misfolding, it would not be surprising that the cellular stress response has an essential role in protein folding quality control.

A prominent biochemical pathway implicated in protein misfolding and aggregation is that involving protein homeostasis, the integration of folding, misfolding, refolding, and degradation of proteins. Specifically, what are the roles of molecular chaperones in altering the events of protein misfolding and aggregation and how does acute and chronic stress influence these events? Molecular chaperones also function during protein biogenesis in diverse events involving the folding and translocation of proteins and assembly into macromolecular complexes (2–7). The family of molecular chaperones is large and diverse and includes members of the Hsp104, Hsp90, Hsp70, dnaJ (Hsp40), immunophilins (Cyp40, FKBP), Hsp60 (chaperonins), the small heat shock proteins, and components of the steroid aporeceptor complex (p23, Bag1, Hip, Hop) (3, 8). Collectively, these proteins have been implicated in a growing number of diverse protein biosynthetic reactions, including elongation of the nascent polypeptide, co-translational folding, translocation and assembly into multi-meric complexes, regulation of protein function, and protein degradation. During stress, they prevent the appearance and accumulation of misfolded proteins by capturing unfolded non-native intermediates to prevent hydrophobic surfaces from self-associating. Upon recovery from stress, these intermediates can be refolded or degraded (9, 10). This ensures that non-native proteins, and in particular off-pathway intermediates, do not persist to form protein aggregates, which, in time, may lead to cell death.

Our current understanding of the function of molecular chaperones has benefited from *in vitro* biochemical studies complemented by genetic and cell biological observations. *In vitro* studies on Hsp70 and Hsp90 have provided insights on the biochemical function of chaperones and their interactions with unfolded polypeptide substrates as 'holders' or 'folders' (11–14). The 'holding' state corresponds to chaperone-dependent intermediates in which the unfolded polypeptide has acquired significant folded structure, however is non-native. Many chaperones, including the immunophilins (CyP40, FKBP52), cdc37, Hsp104, Hsp90, and Hsp70 interact with the unfolded substrate to generate a relatively protease-resistant non-native inter-mediate (11, 12). An analysis of the folded state of the non-native intermediate reveals at least two distinct intermediate states. The chaperones p23 (an Hsp90 co-chaperone and component of the steroid aporeceptor) and Hip (an Hsp70 co-chaperone) interact with unfolded polypeptides to generate a protease sensitive unfolded intermediate, consistent with the properties of a molten globule state (12). The 'folding' state is defined by the conversion of the non-native intermediate (the 'holding' state) to the native state, which in chaperone-dependent reactions requires the Hsp70 (Hsc70) chaperones, a member of the dnaJ family, and ATP (11). The re-lationship between 'holding' and 'folding' therefore establishes an equilibrium that determines the fate of proteins and affords regulatory possibilities to the substrates in heteromeric chaperone complexes.

This review will examine the biochemical properties of the Hsp70 molecular chaperone and the role of co-chaperones to regulate the chaperone and to coordinate cellular activities during stress.

2. The Hsp70 family of molecular chaperones and co-chaperones

Members of the Hsp70 family are found in most but not all prokaryotic genomes and in all eukaryotes as a multi-gene family with additional copies targeted to the endo-plasmic reticulum, mitochondria, and chloroplasts. Hsp70s are abundantly expressed during normal cell growth; they are required for *de novo* protein synthesis, to assist in the folding of nascent polypeptides, for the assembly and disassembly of macro-molecular complexes, and for the translocation of polypeptides across membranes (15–23; see also Chapters 1, 3, 5, and 8).

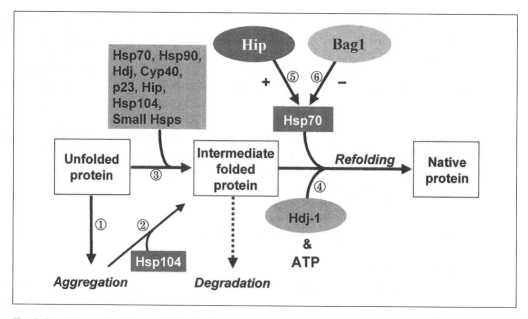

Fig. 1 Chaperone networks: regulating the fate of unfolded polypeptides. Upon heat shock or other forms of stress, proteins populate the unfolded state. ① In the absence of chaperones, unfolded proteins misfold and aggregate. ② In the presence of Hsp104, protein aggregates can be resolubilized into an intermediate folded state and subsequently refolded by Hsp70 and Hdj-1 to the native protein. ③ Alternatively, in the presence of chaperones, unfolded proteins are maintained as intermediates or partially folded states, which are substrates for refolding or degradation. ④ Upon addition of the co-chaperone Hdj-1 and ATP, the partially folded intermediates are refolded through the Hsp70-refolding process. Protein refolding can either be enhanced (⑤) or inhibited (⑥) by co-chaperones that interact directly with the Hsp70 ATPase domain.

Under conditions of physiological stress, Hsp70 family members associate with unfolded or partially folded intermediates to guide folding towards stable on-pathway states and to prevent the accumulation of misfolded species that can form protein aggregates. These partially folded intermediates, upon interaction with Hsp70 and co-chaperones, can be subsequently refolded or directed towards the degradative machinery (24–26) (Fig. 1). Which of these fates occurs for any particular substrate is likely to be dependent upon whether the Hsp70-substrate complex associates with a particular co-chaperone or accessory protein that modulates Hsp70 function. For example, DnaK, the *E. coli* orthologue of Hsp70, does not efficiently refold non-native proteins in the absence of the co-chaperones DnaJ and GrpE whose functions are to modulate ATP hydrolysis and ADP/ATP exchange rates of DnaK, respectively. This ensures that ATP hydrolysis by DnaK is tightly linked with substrate release (10, 27, 28).

Likewise, for the cytosolic Hsp70s of eukaryotes, the DnaJ proteins are necessary and sufficient for refolding activity *in vitro* (24). Orthologues for GrpE are not found in the cytosol or nucleus of eukaryotes; however, *in vivo* there are a number of novel co-chaperones that may regulate Hsp70, for example the expanded family of DnaJ proteins, Bag1, Hip, CHIP, and Hop (29–34). The analysis of genomic sequences from diverse species has revealed that DnaJ and GrpE co-exist together with DnaK in most prokaryotes. This is in contrast to *S. cerevisiae* and other eukaryotes where GrpE is restricted to the mitochondria, and the DnaJ family has expanded to a large multi-gene family encoding compartment-specific members. In the multicellular eukaryotes *C. elegans* and *D. melanogaster*, the DnaJ family has expanded in size and complexity and a novel collection of Hsp70 binding proteins including Hip, Hop, and Bag1 have emerged. These co-chaperones enhance or inhibit Hsp70 activities (Fig. 1); for example, Hdj-1/Hdj-2 and Hip stimulate refolding, while Bag1 inhibits this process (24, 29, 32, 35). Hop (p60), an adaptor protein that functions to stabilize Hsp70 and Hsp90 complexes, inhibits or activates Hsp70-mediated refolding process depending on the concentration of Hop (30, 31).

3. The family of Hsp70 proteins

3.1 Hsp70 Family

Members of the Hsp70 family of molecular chaperones in *E. coli* are referred to as DnaK, HscA (also known as Hsc66), and Hsc62 (36, and see Chapter 1). Hsc66 is 41% identical with DnaK and induced by cold shock rather than by heat shock (37). In vertebrates, the Hsp70 family includes Hsc70 and Hsp70 that are localized to the cytosol and nucleus, mtHsp70/Grp74 is localized to the mitochondria, and Grp78/Bip is targeted to the lumen of the endoplasmic reticulum (Table 1). Human mito-

Table 1 Components of the Hsp70 co-chaperones that regulate Hsp70 functions in *E. coli*, *S. cerevisiae*, and mammals

Organism	Chaperone	Co-chaperones	Subcellular localization
E. coli	DnaK	DnaJ/GrpE Cbpa	
S. cerevisiae	Ssa1-4	Ydj-1 Djp-1	Cytosol
	Ssb1-2	Zuotin Sis-1	
	Ssc1	Mdj-1/Mge1 (GrpE), Mdj-2	Mitochondria
	Kar2	Sec63, Scj1, Jem1	ER
Mammals	Hsp/Hsc70	Hdj-1, Hdj-2, Hsj1a/b Auxilin Mida1 (Zuotin) Bag1 (50K, 46K, 33K, 29K) Hip Hop/Hsp90	Cytosol
	mtHsp70 (Grp74)	Tid1/mt-GrpE	Mitochondria
	Bip(Grp78)	Sec63	ER

chondrial Hsp70 (mtHsp70) is nuclear encoded and shows a high degree of identity with the orthologues *E. coli* DnaK (56%), *S. cerevisiae* Ssc1 (62%), *C. elegans* (75%), *D. melanogaster* (74%), and *M. musculus* (91%). Mitochondrial Hsp70s are more closely related to DnaK (55–57%) and less closely related (44–47%) to the cytosolic and endoplasmic reticulum Hsp70s, consistent with the endosymbiotic origin of mitochondria. In contrast, another mitochondrial Hsp70, *S. cerevisiae* Ssq1 does not exhibit significant homology to either mtHsp70 (42–47%) or other Hsp70 homologues (39%), suggesting a different point of origin than for Ssc1 (38). In contrast, the endoplasmic reticulum-localized Hsp70s share a high degree of identity with the cytosolic Hsp70s (56–63%), which suggests that these members of the Hsp70 family diverged more recently with the appearance of the lumen as a specialized compartment.

3.2 Hsp110 family

A distant subgroup of the Hsp70 family is the 110 kDa heat shock protein (Hsp110) in mammals or Sse1 and Sse2 in *S. cerevisiae* (39–46). The overall organization of Hsp110 homologues has features similar to Hsp70 including the ATPase domain (31% identical), the peptide binding domain, and the flexible region (20% identical) involved in intra-molecular domain cross-talk (41). The principal differences between Hsp110 and Hsp70 lies in a 100-amino acid loop that connects the flexible region and peptide binding domains. Hsp110 homologues function *in vitro* to prevent the aggregation of unfolded polypeptides and have been reported to associate with both Hsp70 and Hsp25 (47). A related member of the Hsp110 family is Hsp105, which is induced by heat shock, osmotic stress, low temperature heat shock, ischemia, and HPV16 E7 oncoprotein (43, 48–50).

3.3 Biochemical and biophysical features of Hsp70

Biochemical and biophysical studies on Hsp70 have characterized a 45 kDa amino-terminal ATPase domain, an 18 kDa polypeptide binding domain, and a less well characterized 10 kDa C-terminal domain (Fig. 2). Complete chaperone activity requires both the ATPase and polypeptide binding domains (24, 51, 52). The tertiary structure of the ATPase and the polypeptide binding domains has been determined by NMR and X-ray crystallography; however, the structure of the intact full length protein remains elusive (53–57). The bovine Hsc70 and human Hsp70 ATPase domains contain two lobes separated by a cleft where ATP is bound and are nearly identical with each other and to actin and hexokinase I (53, 54, 58, 59).

The three-dimensional structure of the substrate-binding domain of mammalian Hsc70 and *E. coli* DnaK was determined by NMR and X-ray crystallography, respectively, and this identified the substrate binding domain (residues 383–540 of Hsc70 and residues 386–561 of DnaK) and the importance of DnaK residue L539 in the β-sheet-rich hydrophobic pocket for substrate binding (55–57). Screens for mutations in DnaK using a bacteriophage lambda replication assay identified residues F426S, S427P, and N451K that influenced peptide binding (60). As the peptide binding domain of Hsp70 is also important for oligomerization, the oligomeric state of Hsp70 is affected by substrate or ATP (61). The 10 kDa C-terminal domain is composed of four rigid helices with the last 33 amino acids containing the highly conserved EEVD motif suggested to have a role in the intramolecular communication between the ATPase and substrate binding domains and with the co-chaperone Hdj-1 (24, 62). Other mutations that influence intramolecular communication between these two domains have also been identified (63).

4. DnaJ

4.1 General features of the DnaJ family

DnaJ (Hsp40) proteins contain the characteristic J-domain required to stimulate the ATPase activity of Hsp70 (21, 24, 25, 27, 64–66) (Fig 2). The structures of the J-domain of human Hsp40 (Hdj-1) and *E. coli* DnaJ contain four helices, each possessing a highly conserved tripeptide (HPD) that may be important for interaction with Hsp70 (67, 68). Disruption of the HPD tripeptide region in Ydj1 inhibits the stimulatory effect on the Hsp70 ATPase and substrate release (65). Consistent with this, a peptide containing the HPD tripeptide competes with DnaJ for interaction with Hsp70 (69). Adjacent to the J-domain is the G/F-rich region, a zinc finger-like domain, and the conserved carboxyl-terminal region. The G/F-rich region does not have a well-defined structure, however, and is thought to assist in the packing of the J-domain with four well-defined helical structures and two connecting loops. The necessity of this structure is revealed by the observation that without the G/F rich region, the J-domain alone cannot stimulate the ATPase activity of Hsp70 or the ATP-dependent substrate binding (70, 71).

A subset of DnaJ proteins, such as the prototypical DnaJ of *E. coli*, also function independently as a molecular chaperone to maintain unfolded proteins in a folding competent state (72). Other members of the DnaJ family retain the J-domain and have other diverse motifs, and have been implicated in protein degradation, signal transduction, apoptosis, karyogamy, and other diverse cellular functions (73–81).

How do DnaJ homologues regulate Hsp70 function? In *E. coli*, Hsp70 exists in a low affinity ATP-state or a high affinity ADP-state that regulates cycles of complex formation between Hsp70 and unfolded substrates and the release of folded proteins. DnaJ enhances the DnaK ATPase rate, which together with GrpE stimulates nucleotide exchange approximately 5000-fold (27). Mutational analyses of a DnaK and DnaJ complex indicate that the DnaJ tripeptide HPD binds to the flexible loop of Hsp70 in the ATPase domain (residues 167, 170, and 173) (82, 83). It has been proposed that Hsp70, in the ATP-state undergoes a conformational change that results in the exposure of sites on Hsp70 for DnaJ interaction that facilitates binding of substrates (83).

4.2 DnaJ homologues of *E. coli, S. cerevisiae,* and humans

E. coli DnaJ was discovered by characterization of a temperature-sensitive mutation that affects viability and DNA replication at the restrictive temperature and was

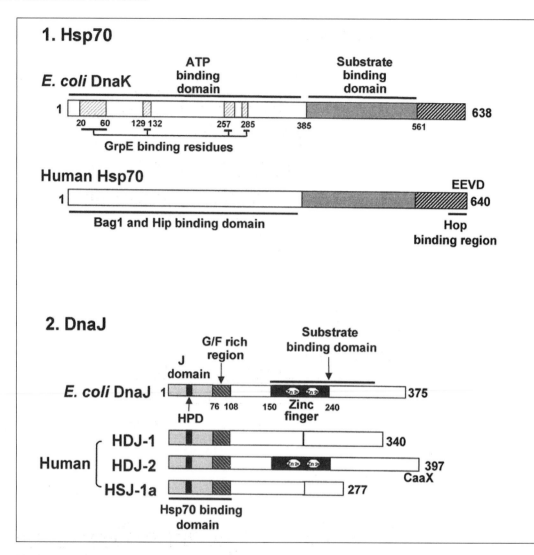

Fig. 2 Primary structure and functional domains of Hsp70 and the co-chaperones, DnaJ, Bag1, Hip, and Hop. The molecular size of each chaperone and co-chaperones are indicated by the amino acid residues flanking each model. The numbers indicated below the model for each chaperone and co-chaperone correspond to the boundaries of each functional motif. (1) Hsp70 is organized into three domains which are conserved among family members. The amino-terminal region constitutes the ATPase domain (residues 1–385), followed by the peptide binding domain (residues 385–561) and the regulatory domain containing the EEVD motif (residues 561–640). The ATPase domain of *E. coli* DnaK contains GrpE binding sites and the mammalian ATPase domain has Hip and Bag1 binding regions. The carboxy-terminus of eukaryotic Hsp70 corresponding to the EEVD motif has been identified as a binding site for Hop. (2) *E. coli* DnaJ is composed of four distinct domains. The amino-terminal region form the J-domain (residues 1–76), which is followed by a G/F rich domain (residues 76–108). The four CxxCxGxGx repeats (zinc finger domains) (residues 150–240), which are the third domain, are conserved among some of the family members. The fourth domain, the carboxy-terminal end (residues 240–375) is the least conserved domain among the family members. Human Hdj-2 that has the most similar structure to DnaJ contains a CxxA motif at the carboxy-terminal end for farnesylation. Hdj-1 and Hsj-1 do not have a zinc finger domain, yet they could still function as Hsp70 co-chaperones. (3) Human Bag1 has four different orthologues. All orthologues

3. Human Bag1

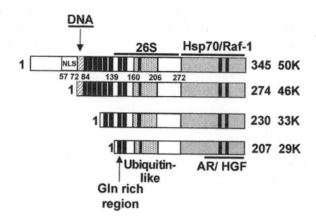

4. Human Hip

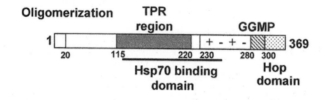

5. Human Hop

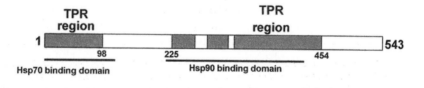

contains a 26S proteasome (residues 139–272 for 50K), androgen receptor (residues 260–345 for 50K), hepatocyte growth factor receptor (residues 260–345 for 50K), Hsp70 and Raf-1 binding sites (residues 225–345 for 50K). The Bag1 46K and 50K orthologues have DNA and binding sites (residues 72–84 for 50K). The Bag1 50K contains extra amino-terminal sequences, which have a nuclear localization signal (NLS) (residues 50–72 for 50K). (4) Human Hip is a TPR (tetratricopeptide) protein that binds to the Hsp70 ATPase domain. It is composed of an oligomerization domain at the N-terminal end (residues 1–20) followed by three TPR regions (residues 115–220), a highly charged region (residues 230–280), a GGMP region (residues 280–300), and a Hop homologous domain (residues 300–369). Hsp70 binds to the TPR region and highly charged region of Hip (residues 115–280). (5) Human Hop has two TPR regions. The TPRI, located at the N-terminal end (residues 1–98), binds to Hsp70, while the TPRII located in the middle (residues 224–454) of Hop, binds to Hsp90.

essential for lambda phage DNA replication (84–89). DnaJ and DnaK form a multi-protein complex together with the origin of lambda replication and lambda O and P proteins and DnaB. Partial disassembly of the nucleoprotein complex occurs in the presence of ATP and GrpE resulting in activation of DnaB and unwinding of lambda DNA (89, 90). The involvement of DnaJ as a DnaK co-chaperone to facilitate protein refolding in the presence of GroEL and GroES was observed using rhodanese as a model substrate (21). Subsequent studies with denatured luciferase as a substrate revealed that DnaJ, together with DnaK and GrpE allowed protein refolding (27). The DnaJ and DnaK system also cooperates *in vitro* with ClpB (*E. coli* homologue of *S. cerevisiae* Hsp104) to disaggregate protein aggregates (91, 92).

    E. coli CbpA (curved DNA-binding protein) was identified by its property to interact with a synthetic DNA sequence (93). CbpA contains a J-domain, G/F-rich region, the carboxyl-terminus-conserved region, and lacks the zinc finger-like domain. Expression of CbpA suppresses the biological defects of a DnaJ mutant for both lambda phage and host DNA replication (93, 94). Several *S. cerevisiae* homologues of CbpA including Djp1, Sis1, and Hlj1 have been identified. Djp1 is a cytosolic protein involved in peroxisome assembly (95). Sis1 is an essential nuclear enriched protein involved in nucleus migration from mother to daughter cell, a suppressor of the SIT4 gene with a slow growth phenotype, and is involved in translational initiation (96, 97). Sis1, together with Ssb1, suppresses the growth defect caused by a mutation in the proteasome subunits (98). The properties of Sis1 cannot be compensated by Ydj-1; recently it was shown that the specificity of Sis1 and Ydj1 could be due to the G/F region, since a chimeric protein comprised of the J-domain of Ydj-1 and the G/F region of Sis1 could suppress the lethality of Dsis1 (96, 99). Hdj-1, the human homologue of *E. coli* CbpA, was the first human DnaJ protein to be identified in Hsp70 mediated refolding process (24, 25). Hdj-1 alone does not have chaperone activity but functions together with Hsp70. Hdj-1 forms complexes with glucocorticoid receptor, p53, the human papillomavirus replication initiator, and E1 helicase (100–102). The binding of Hdj-1 (or Hdj-2) to E1 helicase together with Hsp70 suggests that the human Hsp70 chaperone network might also be involved in DNA synthesis.

    For DnaJ to interact with DnaK for protein refolding, both the cysteine rich zinc finger-like region containing four repeated sequences motif, CXXCXGXG, and the conserved carboxyl-terminus are required (72, 103). The zinc finger region, however, is not present in human Hdj-1, which nevertheless functions as a Hsp70 co-chaperone (24). Analysis of the motifs of *S. cerevisiae* Ydj-1 reveals that the conserved carboxyl terminus is responsible for prevention of protein aggregation (104). Another intriguing specificity domain of DnaJ is the G/F rich region that has been characterized in *S. cerevisiae* Sis1 (99). Yet, DnaJ proteins that lack the G/F region can interact with Hsp70 to activate the ATPase (105, 106).

    Of the eukaryotic DnaJ proteins, *S. cerevisiae* Ydj-1 is perhaps best characterized as a component of the nuclear matrix and involved in translocation across the mitochondria and the endoplasmic reticulum (103, 107–109). Ydj1 contains a CaaX box at the carboxyl-terminus where farnesylation occurs post-translationally for targeting to the endoplasmic reticulum. In the absence of this motif, Ydj1 acquires a temperature-sensitive growth phenotype (108). Likewise, human Hdj-2, which also has CaaX box and localizes to the golgi complexes and the endoplasmic reticulum (110). Ydj-1 exhibits specificity as a co-chaperone with the Ssa members of the *S. cerevisiae* Hsp70 family but not with the Ssb members (104, 109, 111) (Table 1). Ydj-1 is also involved in ubiquitin-dependent protein degradation revealing a dual role in protein folding and degradation (112).

The family of DnaJ proteins in humans is certain to be large and in excess of a dozen distinct members. As for *S. cerevisiae*, it seems likely that the properties of mammalian Hsp70s can be differentially regulated by association with specific DnaJ proteins. For example, Hdj-2, the human homologue of DnaJ, suppresses the aggregation of ΔF508 CFTR (cystic fibrosis transmembrane conductance regulator), a mutant form of CFTR, that misfolds in the early pathway of CFTR assembly in the endoplasmic reticulum (113). Likewise, over-expression of either Hdj-2 or Hdj-1 reduces the aggregation of polyglutamine-expansions associated with spinocerebellar ataxia type 3/Machado–Joseph disease and spinal bulbar muscular atrophy (114, 115).

4.3 Mitochondrial DnaJ proteins

Mitochondrial specific DnaJ homologues in *S. cerevisiae* are Mdj1 and Mdj2 (116) (Table 1). Mdj1 is necessary for the refolding of newly imported proteins together with Ssc1, the mitochondrial Hsp70, and for protein translocation across mitochondrial membranes. Mdj1 also serves to maintain an active form of mitochondrial DNA polymerase, and consequently is essential for the inheritance of the mitochondrial genome (116–118). In humans, two mitochondrial DnaJ homologues, corresponding to the splice variants Tid1L and Tid1S, have been identified (81) (Table 1). Tid1L activates apoptosis whereas Tid1S inhibits apoptosis. This link with cell growth control is also observed with the mitochondrial DnaJ homologue in *D. melanogaster*, Tid50, which suppresses malignant tumours growing on imaginal discs (119).

4.4 DnaJ endoplasmic reticulum homologues

DnaJ proteins in the endoplasmic reticulum have a characteristic structure that is distinct from cytosolic or mitochondrial homologues (Table 1). The *S. cerevisiae* orthologues Jem1, Sec63, and Scj1 contain transmembrane domains for attachment to the endoplasmic reticulum, such that the J-domain is oriented towards the lumen (78, 120, 121). All three proteins contain transmembrane domains for attachment to the endoplasmic reticulum membrane, with the J-domain facing the ER lumen (78, 120). In addition to the J-domain, Scj1 has a zinc finger-like domain, a G/F rich region, and a carboxyl-terminus-conserved region, while Jem1 and Sec63 lack all three. These DnaJ proteins interact with Kar2 (Bip) and are involved in protein biogenesis (80, 122, 123). Sec63 is an essential component of post-translational protein translocation across the ER membrane with Kar2, Sec62, Sec67, and Sec66 (124–127). Destruction of the kar2 gene prevents karyogamy during mating (80, 128). Mutations in Sec63 and Jem1 have the same phenotype as a Kar2 mutation. This indicates that Kar2 must interact with both Sec63 and Jem1 during ER biogenesis associated with nuclear fusion (78, 80). Cells with a null mutation in Scj1 exhibit a delay in maturation of glycosylated proteins which has led to the suggestion that Kar2 may also interact with Scj1 (129).

4.5 Other DnaJ homologues

Zuotin is a conserved DnaJ protein from yeast to human that was identified as a Z-DNA binding protein and ribosome associated protein (130–132) (Table 1). Zuotin has only a J-domain followed by an alanine/lysine rich region and interacts the *S. cerevisiae* Hsp70 member, Ssb, with ribosomes containing nascent polypeptide chains (132).

Cysteine string protein (Csp), a J-domain and cysteine-rich string protein, was identified from cholinergic synaptic vesicles (106). Mutation of Csp caused impaired presynaptic neuromuscular transmission in *D. melanogaster* consistent with a role in protein exocytosis (133). Csp interacts with Hsp70 and stimulates ATPase activity (134). Auxilin contains a J-domain in the carboxyl-terminus and LKD repeats followed by a tensin and proline-rich region and was identified as an integral component of clathrin-coated vesicles (135, 136) (Table 1). Uncoating of the clathrin coat by Hsp70 requires ATP and auxilin (137). It appears that auxilin interacts with Hsp70 and induces binding of Hsp70 to clathrin in the same way that Hsp40 affects the binding of substrate to Hsp70. The J-domain of auxilin inhibits the uncoating process of Hsp70 (138). However, auxilin cannot be replaced by other DnaJ proteins, suggesting specificity with its Hsp70 partner protein (139, 140).

5. Bag1

Bag1, further discussed in the next chapter in the context of signal transduction, is a multifunctional protein, originally identified as a Bcl-2 interacting protein that protects cells from apoptosis when over-expressed in mammalian cells (141) (Fig. 2). Bag1 is not present in prokaryotes or *S. cerevisiae* and is conserved in eukaryotes from *S. pombe, C. elegans* to humans. Bag1 contains a ubiquitin-like region in the amino-terminal region and sites in the carboxyl-terminus for interaction with Hsp70 and Raf-1 (29, 141). Subsequently, Bag1 was shown to interact *in vivo* with, and influence the activities of, Siah and the hormone receptors for androgen, HGF, retinoic acid, proteasome 26S, DNA, and glucocorticoid (142–148). Over-expression of Bag1 negatively regulates the chaperone activity of Hsp70 (149).

The interaction between Raf-1 and Bag1 was observed *in vivo* using a protein interaction assay and by immunoprecipitation studies, and through *in vitro* binding assays (150). The association of Bag1 with Raf-1 in the catalytic domain stimulates kinase activity in a Ras-independent mechanism. Bag1 interacts with the plasma membrane-associated tyrosine kinase domains of hepatocyte growth factor (HGF) and platelet-derived growth factor receptor which (144). Bag1 also interacts with and activates the HGF receptor independent of tyrosine phosphorylation, enhances HGF-

and PDGF-induced protection from apoptosis, and interacts with the retinoic acid receptor (RAR) to inhibit binding to RAR elements that suppress apoptosis (148).

The family of Bag1 proteins includes Bag1M (Rap46) and Bag1L. Androgen receptor (AR) interacts with testosterone and related androgens; association with Bag1L stimulates the ability of AR to transactivate the ARE (145). Another Bag1 homologue, Bag1M or Rap46, interacts with the glucocorticoid receptor (GR) and has diverse roles in development, differentiation, and cellular proliferation (147). The binding of Bag1M with GR suppresses the transcriptional activity of this receptor that mediates apoptosis. Another Bag1 interacting protein, Siah, a p53-induced gene, has a negative effect on cell proliferation in *Drosophila* (146). Over-expression of Bag1 suppressed Siah and p53 induced growth arrest. The observation of the complex formation between Bag1 and Hsp70 suggests a possible role for chaperone involvement in cell growth, death, and stress by regulating Bag1 and its substrates. The elucidation of interactions among these proteins would present a new mechanism in cellular processes.

6. Hip and Hop (p60)

Hip and Hop are TPR (tetratricopeptide) repeat-containing proteins that interact with Hsp70 and are conserved from *S. pombe* to human and are not in prokaryotes or *S. cerevisiae* (Fig. 2). Hip was identified as a component of aporeceptor complexes and as an Hsp70-binding protein that stimulates Hsp70 chaperone activity (32, 151). Hip contains an oligomerization domain (residues 1–100), a TPR repeat (residues 113–214), and a GGMP-rich region at the C-terminus (Fig. 2).

Hop is an abundant stress-induced protein that interacts with the carboxyl-terminus of Hsp70 to stimulate Hsp70-mediated refolding interacts with both Hsp90 and Hsp70 (30, 152). Hop contains two TPR regions; I (residues 1–115) and II (residues 222–431) (Fig. 2). TPR I region interacts primarily with Hsp70 while TPR II region is necessary for association with both Hsp70 and Hsp90 (153–155). The crystal structure of the TPR reveals that TPR I and II interacts with Hsp70 and Hsp90 via the EEVD motif located at the extreme carboxyl-terminus of both chaperones (154). Hop is also required for the formation of mature progesterone and glucocorticoid receptors that contain both Hsp70 and Hsp90 (156, 157).

7. Conclusion

It is intriguing that an increasing number of Hsp70 co-chaperones have roles as modulators of Hsp70 activity and separately in cell growth, cell signalling, and apoptosis. For nearly all co-chaperones, their cellular concentrations range from 0.3–5% of the levels of Hsp70, yet stoichiometric levels of chaperones and co-chaperones are required for protein refolding studies *in vitro* (C. Schmidt, unpublished observations). This suggests that co-chaperones are unlikely to be constitutively associated with chaperones, rather Hsp70 activities are regulated according to the specific co-chaperones complement available at each specific stage of the cell cycle or in response to stress. Indeed, we propose that the highly abundant class of molecular chaperones represented by Hsp70 may acquire exquisite specificity via interactions with a constellation of regulatory co-chaperones and furthermore that Hsp70 itself could function as a negative or positive regulator of the biological activities exhibited by co-chaperones as a means to coordinate cell stress with other cellular activities.

Acknowledgements

We thank Masahiro Takeda for discussion and assistance with the figures and Kate Veraldi and Ellen Nollen for assistance with the manuscript. R. M. was supported by grants from the National Institutes of Health, the Martin and Carol Golub Foundation, and the Hereditary Disease Foundation.

References

1. Tran, P. B. and Miller, R. J. (1999). Aggregates in neurodegenerative disease: crowds and power? *Trends Neurosci.*, **22**, 194.
2. Buchner, J. (1996). Supervising the fold: functional principles of molecular chaperones. *Faseb J.*, **10**, 10.
3. Bukau, B. (ed.) (1999). *Molecular chaperones and folding catalysts. Regulation, cellular function and mechanism*. Harwood Academic, Amsterdam.
4. Gething, M. J. and Sambrook, J. (1992). Protein folding in the cell. *Nature*, **355**, 33.
5. Hartl, F. U. (1996). Molecular chaperones in cellular protein folding. *Nature*, **381**, 571.
6. Horwich, A. L. and Weissman, J. S. (1997). Deadly conformations – protein misfolding in prion disease. *Cell*, **89**, 499.
7. Morimoto, R. I. and Santoro, M. G. (1998). Stress-inducible responses and heat shock proteins: new pharmacologic targets for cytoprotection. *Nat. Biotechnol.*, **16**, 833.
8. Gething, M. J. (1997). Protein folding: The difference with prokaryotes. *Nature*, **388**, 329.
9. Parsell, D. A., Kowal, A. S., Singer, M. A., and Lindquist, S. (1994). Protein disaggregation mediated by heat-shock protein Hsp104. *Nature*, **372**, 475.
10. Schroder, H., Langer, T., Hartl, F. U., and Bukau, B. (1993). DnaK, DnaJ and GrpE form a cellular chaperone machinery capable of repairing heat-induced protein damage. *EMBO J.*, **12**, 4137.
11. Freeman, B. C. and Morimoto, R. I. (1996). The human cytosolic molecular chaperones hsp90, hsp70 (hsc70) and hdj-1 have distinct roles in recognition of a non-native protein and protein refolding. *EMBO J.*, **15**, 2969.
12. Freeman, B. C., Toft, D. O., and Morimoto, R. I. (1996). Molecular chaperone machines: chaperone activities of the cyclophilin Cyp-40 and the steroid aporeceptor-associated protein p23 [see comments]. *Science*, **274**, 1718.
13. Jakob, U., Lilie, H., Meyer, I., and Buchner, J. (1995). Transient interaction of Hsp90 with early unfolding intermediates of citrate synthase. Implications for heat shock in vivo. *J. Biol. Chem.*, **270**, 7288.
14. Miyata, Y. and Yahara, I. (1992). The 90-kDa heat shock protein, HSP90, binds and protects casein kinase II from self-aggregation and enhances its kinase activity. *J. Biol. Chem.*, **267**, 7042.

15. Ungewickell, E. (1985). The 70-kd mammalian heat shock proteins are structurally and functionally related to the uncoating protein that releases clathrin triskelia from coated vesicles. *EMBO J.*, **4**, 3385.

16. Munro, S. and Pelham, H. R. (1986). An Hsp70-like protein in the ER: identity with the 78 kd glucose-regulated protein and immunoglobulin heavy chain binding protein. *Cell*, **46**, 291.

17. Chirico, W. J., Waters, M. G., and Blobel, G. (1988). 70K heat shock related proteins stimulate protein translocation into microsomes. *Nature*, **332**, 805.

18. Murakami, H., Pain, D., and Blobel, G. (1988). 70-kD heat shock-related protein is one of at least two distinct cytosolic factors stimulating protein import into mitochondria. *J. Cell Biol.*, **107**, 2051.

19. Beckmann, R. P., Mizzen, L. E., and Welch, W. J. (1990). Interaction of Hsp 70 with newly synthesized proteins: implications for protein folding and assembly. *Science*, **248**, 850.

20. Skowyra, D., Georgopoulos, C., and Zylicz, M. (1990). The *E. coli dnaK* gene product, the hsp70 homolog, can reactivate heat-inactivated RNA polymerase in an ATP hydrolysis-dependent manner. *Cell*, **62**, 939.

21. Langer, T., Lu, C., Echols, H., Flanagan, J., Hayer, M. K., and Hartl, F. U. (1992). Successive action of DnaK, DnaJ and GroEL along the pathway of chaperone-mediated protein folding. *Nature*, **356**, 683.

22. Nelson, R. J., Ziegelhoffer, T., Nicolet, C., Werner-Washburne, M., and Craig, E. A. (1992). The translation machinery and 70 kd heat shock protein cooperate in protein synthesis. *Cell*, **71**, 97.

23. Shi, Y. and Thomas, J. O. (1992). The transport of proteins into the nucleus requires the 70-kilodalton heat shock protein or its cytosolic cognate. *Mol. Cell Biol.*, **12**, 2186.

24. Freeman, B. C., Myers, M. P., Schumacher, R., and Morimoto, R. I. (1995). Identification of a regulatory motif in Hsp70 that affects ATPase activity, substrate binding and interaction with HDJ-1. *EMBO J.*, **14**, 2281.

25. Minami, Y., Hohfeld, J., Ohtsuka, K., and Hartl, F. U. (1996). Regulation of the heat-shock protein 70 reaction cycle by the mammalian DnaJ homolog, Hsp40. *J. Biol. Chem.*, **271**, 19617.

26. Bercovich, B., Stancovski, I., Mayer, A., Blumenfeld, N., Laszlo, A., Schwartz, A. L., and Ciechanover, A. (1997). Ubiquitin-dependent degradation of certain protein substrates in vitro requires the molecular chaperone Hsc70. *J. Biol. Chem.*, **272**, 9002.

27. Szabo, A., Langer, T., Schroder, H., Flanagan, J., Bukau, B., and Hartl, F. U. (1994). The ATP hydrolysis-dependent reaction cycle of the *Escherichia coli* Hsp70 system DnaK, DnaJ, and GrpE. *Proc. Natl. Acad. Sci. USA*, **91**, 10345.

28. Levy, E. J., McCarty, J., Bukau, B., and Chirico, W. J. (1995). Conserved ATPase and luciferase refolding activities between bacteria and yeast Hsp70 chaperones and modulators. *FEBS Lett.*, **368**, 435.

29. Takayama, S., Bimston, D. N., Matsuzawa, S., Freeman, B. C., Aime-Sempe, C., Xie, Z., Morimoto, R. I., and Reed, J. C. (1997). BAG-1 modulates the chaperone activity of Hsp70/Hsc70. *EMBO J.*, **16**, 4887.

30. Johnson, B. D., Schumacher, R. J., Ross, E. D., and Toft, D. O. (1998). Hop modulates Hsp70/Hsp90 interactions in protein folding. *J. Biol. Chem.* **273**, 3679.

31. Gebauer, M., Zeiner, M., and Gehring, U. (1997). Proteins interacting with the molecular chaperone hsp70/hsc70: physical associations and effects on refolding activity. *FEBS Lett.*, **417**, 109.

32. Hohfeld, J., Minami, Y., and Hartl, F. U. (1995). Hip, a novel cochaperone involved in the eukaryotic Hsc70/Hsp40 reaction cycle. *Cell*, **83**, 589.

33. Ballinger, C. A., Connell, P., Wu, Y., Hu, Z., Thompson, L. J., Yin, L. Y., and Patterson, C. (1999). Identification of CHIP, a novel tetratricopeptide repeat-containing protein that interacts with heat shock proteins and negatively regulates chaperone functions. *Mol. Cell Biol.* **19**, 4535.

34. Minami, Y. (1996). [Structure and function of the DnaJ/Hsp40 family]. *Tanpakushitsu Kakusan Koso.* **41**, 875.

35. Bimston, D., Song, J., Winchester, D., Takayama, S., Reed, J. C., and Morimoto, R. I. (1998). BAG-1, a negative regulator of Hsp70 chaperone activity, uncouples nucleotide hydrolysis from substrate release. *EMBO J.*, **17**, 6871.

36. Seaton, B. L. and Vickery, L. E. (1994). A gene encoding a DnaK/hsp70 homolog in Escherichia coli. *Proc. Natl. Acad. Sci. USA*, **91**, 2066.

37. Lelivelt, M. J. and Kawula, T. H. (1995). Hsc66, an Hsp70 homolog in *Escherichia coli*, is induced by cold shock but not by heat shock. *J. Bacteriol.*, **177**, 4900.

38. Voisine, C., Schilke, B., Ohlson, M., Beinert, H., Marszalek, J., and Craig, E. A. (2000). Role of the mitochondrial Hsp70s, Ssc1 and Ssq1, in the maturation of Yfh1. *Mol. Cell Biol.*, **20**, 3677.

39. Shyy, T. T., Subjeck, J. R., Heinaman, R., and Anderson, G. (1986). Effect of growth state and heat shock on nucleolar localization of the 110,000-Da heat shock protein in mouse embryo fibroblasts. *Cancer Res.*, **46**, 4738.

40. Mukai, H., Kuno, T., Tanaka, H., Hirata, D., Miyakawa, T., and Tanaka, C. (1993). Isolation and characterization of SSE1 and SSE2, new members of the yeast HSP70 multigene family. *Gene*, **132**, 57.

41. Lee-Yoon, D., Easton, D., Murawski, M., Burd, R., and Subjeck, J. R. (1995). Identification of a major subfamily of large hsp70-like proteins through the cloning of the mammalian 110-kDa heat shock protein. *J. Biol. Chem.*, **270**, 15725.

42. Chen, X., Easton, D., Oh, H. J., Lee-Yoon, D. S., Liu, X., and Subjeck, J. (1996). The 170 kDa glucose regulated stress protein is a large HSP70-, HSP110-like protein of the endoplasmic reticulum. *FEBS Lett.*, **380**, 68.

43. Kojima, R., Randall, J., Brenner, B. M., and Gullans, S. R. (1996). Osmotic stress protein 94 (Osp94). A new member of the Hsp110/SSE gene subfamily. *J. Biol. Chem.*, **271**, 12327.

44. Oh, H. J., Chen, X., and Subjeck, J. R. (1997). Hsp110 protects heat-denatured proteins and confers cellular thermoresistance. *J. Biol. Chem.*, **272**, 31636.

45. Santos, B. C., Chevaile, A., Kojima, R., and Gullans, S. R. (1998). Characterization of the Hsp110/SSE gene family response to hyperosmolality and other stresses. *Am. J. Physiol.*, **274**, F1054.

46. Nonoguchi, K., Itoh, K., Xue, J. H., Tokuchi, H., Nishiyama, H., Kaneko, Y., *et al.* (1999). Cloning of human cDNAs for Apg-1 and Apg-2, members of the Hsp110 family, and chromosomal assignment of their genes. *Gene*, **237**, 21.

47. Wang, X. Y., Chen, X., Oh, H. J., Repasky, E., Kazim, L., and Subjeck, J. (2000). Characterization of native interaction of hsp110 with hsp25 and hsc70. *FEBS Lett.*, **465**, 98.

48. Morozov, A., Subjeck, J., and Raychaudhuri, P. (1995). HPV16 E7 oncoprotein induces expression of a 110 kDa heat shock protein. *FEBS Lett.*, **371**, 214.

49. Kaneko, Y., Nishiyama, H., Nonoguchi, K., Higashitsuji, H., Kishishita, M., and Fujita, J. (1997). A novel hsp110-related gene, apg-1, that is abundantly expressed in the testis responds to a low temperature heat shock rather than the traditional elevated temperatures. *J. Biol. Chem.*, **272**, 2640.

50. Yagita, Y., Kitagawa, K., Taguchi, A., Ohtsuki, T., Kuwabara, K., Mabuchi, T., *et al.* (1999). Molecular cloning of a novel member of the HSP110 family of genes, ischemia-responsive

protein 94 kDa (irp94), expressed in rat brain after transient forebrain ischemia. *J. Neurochem.*, **72**, 1544.

51. Buchberger, A., Theyssen, H., Schroder, H., McCarty, J. S., Virgallita, G., Milkereit, P., *et al.* (1995). Nucleotide-induced conformational changes in the ATPase and substrate binding domains of the DnaK chaperone provide evidence for interdomain communication. *J. Biol. Chem.*, **270**, 16903.

52. McCarty, J. S., Buchberger, A., Reinstein, J., and Bukau, B. (1995). The role of ATP in the functional cycle of the DnaK chaperone system. *J. Mol. Biol.*, **249**, 126.

53. Flaherty, K. M., McKay, D. B., Kabsch, W., and Holmes, K. C. (1991). Similarity of the three-dimensional structures of actin and the ATPase fragment of a 70-kDa heat shock cognate protein. *Proc. Natl. Acad. Sci. USA*, **88**, 5041.

54. Sriram, M., Osipiuk, J., Freeman, B., Morimoto, R., and Joachimiak, A. (1997). Human Hsp70 molecular chaperone binds two calcium ions within the ATPase domain. *Structure*, **5**, 403.

55. Zhu, X., Zhao, X., Burkholder, W. F., Gragerov, A., Ogata, C. M., Gottesman, M. E., and Hendrickson, W. A. (1996). Structural analysis of substrate binding by the molecular chaperone DnaK. *Science*, **272**, 1606.

56. Wang, H., Kurochkin, A. V., Pang, Y., Hu, W., Flynn, G. C., and Zuiderweg, E. R. (1998). NMR solution structure of the 21 kDa chaperone protein DnaK substrate binding domain: a preview of chaperone-protein interaction. *Biochemistry.* **37**, 7929.

57. Morshauser, R. C., Hu, W., Wang, H., Pang, Y., Flynn, G. C., and Zuiderweg, E. R. (1999). High-resolution solution structure of the 18 kDa substrate-binding domain of the mammalian chaperone protein Hsc70. *J. Mol. Biol.*, **289**, 1387.

58. Fletterick, R. J., Bates, D. J., and Steitz, T. A. (1975). The structure of a yeast hexokinase monomer and its complexes with substrates at 2.7-A resolution. *Proc. Natl. Acad. Sci. USA*, **72**, 38.

59. Kabsch, W., Mannherz, H. G., Suck, D., Pai, E. F., and Holmes, K. C. (1990). Atomic structure of the actin:DNase I complex. *Nature*, **347**, 37.

60. Montgomery, D. L., Morimoto, R. I., and Gierasch, L. M. (1999). Mutations in the substrate binding domain of the Escherichia coli 70 kDa molecular chaperone, DnaK, which alter substrate affinity or interdomain coupling. *J. Mol. Biol.*, **286**, 915.

61. Fouchaq, B., Benaroudj, N., Ebel, C., and Ladjimi, M. M. (1999). Oligomerization of the 17-kDa peptide-binding domain of the molecular chaperone HSC70. *Eur. J. Biochem.*, **259**, 379.

62. Bertelsen, E. B., Zhou, H., Lowry, D. F., Flynn, G. C., and Dahlquist, F. W. (1999). Topology and dynamics of the 10 kDa C-terminal domain of DnaK in solution. *Protein Sci.*, **8**, 343.

63. Davis, J. E., Voisine, C., and Craig, E. A. (1999). Intragenic suppressors of Hsp70 mutants: interplay between the ATPase- and peptide-binding domains. *Proc. Natl. Acad. Sci. USA*, **96**, 9269.

64. Scidmore, M. A., Okamura, H. H., and Rose, M. D. (1993). Genetic interactions between KAR2 and SEC63, encoding eukaryotic homologues of DnaK and DnaJ in the endoplasmic reticulum. *Mol. Biol. Cell*, **4**, 1145.

65. Tsai, J. and Douglas, M. G. (1996). A conserved HPD sequence of the J-domain is necessary for YDJ1 stimulation of Hsp70 ATPase activity at a site distinct from substrate binding. *J. Biol. Chem.*, **271**, 9347.

66. Mayer, M. P., Laufen, T., Paal, K., McCarty, J. S., and Bukau, B. (1999). Investigation of the interaction between DnaK and DnaJ by surface plasmon resonance spectroscopy. *J. Mol. Biol.*, **289**, 1131.

67. Pellecchia, M., Szyperski, T., Wall, D., Georgopoulos, C., and Wuthrich, K. (1996). NMR structure of the J-domain and the Gly/Phe-rich region of the *Escherichia coli* DnaJ chaperone. *J. Mol. Biol.*, **260**, 236.

68. Qian, Y. Q., Patel, D., Hartl, F. U., and McColl, D. J. (1996). Nuclear magnetic resonance solution structure of the human Hsp40 (HDJ-1) J-domain. *J. Mol. Biol.*, **260**, 224.

69. Michels, A. A., Kanon, B., Bensaude, O., and Kampinga, H. H. (1999). Heat shock protein (Hsp) 40 mutants inhibit Hsp70 in mammalian cells. *J. Biol. Chem.*, **274**, 36757.

70. Karzai, A. W. and McMacken, R. (1996). A bipartite signaling mechanism involved in DnaJ-mediated activation of the Escherichia coli DnaK protein. *J. Biol. Chem.*, **271**, 11236.

71. Wall, D., Zylicz, M., and Georgopoulos, C. (1995). The conserved G/F motif of the DnaJ chaperone is necessary for the activation of the substrate binding properties of the DnaK chaperone. *J. Biol. Chem.*, **270**, 2139.

72. Szabo, A., Korszun, R., Hartl, F. U., and Flanagan, J. (1996). A zinc finger-like domain of the molecular chaperone DnaJ is involved in binding to denatured protein substrates. *EMBO J.*, **15**, 408.

73. Straus, D. B., Walter, W. A., and Gross, C. A. (1988). *Escherichia coli* heat shock gene mutants are defective in proteolysis. *Genes Dev.*, **2**, 1851.

74. Caplan, A. J., Langley, E., Wilson, E. M., and Vidal, J. (1995). Hormone-dependent trans-activation by the human androgen receptor is regulated by a dnaJ protein. *J. Biol. Chem.*, **270**, 5251.

75. Kimura, Y., Yahara, I., and Lindquist, S. (1995). Role of the protein chaperone YDJ1 in establishing Hsp90-mediated signal transduction pathways [see comments]. *Science*, **268**, 1362.

76. Liberek, K., Wall, D., and Georgopoulos, C. (1995). The DnaJ chaperone catalytically activates the DnaK chaperone to preferentially bind the sigma 32 heat shock transcriptional regulator. *Proc. Natl. Acad. Sci. USA*, **92**, 6224.

77. Yaglom, J. A., Goldberg, A. L., Finley, D., and Sherman, M. Y. (1996). The molecular chaperone Ydj1 is required for the p34CDC28-dependent phosphorylation of the cyclin Cln3 that signals its degradation. *Mol. Cell Biol.*, **16**, 3679.

78. Nishikawa, S. and Endo, T. (1997). The yeast JEM1p is a DnaJ-like protein of the endoplasmic reticulum membrane required for nuclear fusion. *J. Biol. Chem.*, **272**, 12889.

79. Blaszczak, A., Georgopoulos, C., and Liberek, K. (1999). On the mechanism of FtsH-dependent degradation of the sigma 32 transcriptional regulator of *Escherichia coli* and the role of the Dnak chaperone machine. *Mol. Microbiol.*, **31**, 157.

80. Brizzio, V., Khalfan, W., Huddler, D., Beh, C. T., Andersen, S. S., Latterich, M., and Rose, M. D. (1999). Genetic interactions between KAR7/SEC71, KAR8/JEM1, KAR5, and KAR2 during nuclear fusion in *Saccharomyces cerevisiae*. *Mol. Biol. Cell*, **10**, 609.

81. Syken, J., De-Medina, T., and Munger, K. (1999). TID1, a human homolog of the Drosophila tumor suppressor l(2)tid, encodes two mitochondrial modulators of apoptosis with opposing functions. *Proc. Natl. Acad. Sci. USA*, **96**, 8499.

82. Suh, W. C., Burkholder, W. F., Lu, C. Z., Zhao, X., Gottesman, M. E., and Gross, C. A. (1998). Interaction of the Hsp70 molecular chaperone, DnaK, with its cochaperone DnaJ. *Proc. Natl. Acad. Sci. USA*, **95**, 15223.

83. Gassler, C. S., Buchberger, A., Laufen, T., Mayer, M. P., Schroder, H., Valencia, A., and Bukau, B. (1998). Mutations in the DnaK chaperone affecting interaction with the DnaJ cochaperone. *Proc. Natl. Acad. Sci. USA*, **95**, 15229.

84. Sunshine, M., Feiss, M., Stuart, J., and Yochem, J. (1977). A new host gene (groPC) necessary for lambda DNA replication. *Mol. Gen. Genet.*, **151**, 27.

85. Saito, H. and Uchida, H. (1978). Organization and expression of the *dnaJ* and *dnaK* genes of *Escherichia coli* K12. *Mol. Gen. Genet.*, **164**, 1.

86. Yochem, J., Uchida, H., Sunshine, M., Saito, H., Georgopoulos, C. P., and Feiss, M. (1978). Genetic analysis of two genes, *dnaJ* and *dnaK*, necessary for *Escherichia coli* and bacteriophage lambda DNA replication. *Mol. Gen. Genet.*, **164**, 9.

87. Mensa-Wilmot, K., Seaby, R., Alfano, C., Wold, M. C., Gomes, B., and McMacken, R. (1989). Reconstitution of a nine-protein system that initiates bacteriophage lambda DNA replication. *J. Biol. Chem.*, **264**, 2853.

88. Zylicz, M., Ang, D., Liberek, K., and Georgopoulos, C. (1989). Initiation of lambda DNA replication with purified host- and bacteriophage-encoded proteins: the role of the dnaK, dnaJ and grpE heat shock proteins. *EMBO J.*, **8**, 1601.

89. Alfano, C. and McMacken, R. (1989). Heat shock protein-mediated disassembly of nucleoprotein structures is required for the initiation of bacteriophage lambda DNA replication. *J. Biol. Chem.*, **264**, 10709.

90. Liberek, K., Georgopoulos, C., and Zylicz, M. (1988). Role of the *Escherichia coli* DnaK and DnaJ heat shock proteins in the initiation of bacteriophage lambda DNA replication. *Proc. Natl. Acad. Sci. USA*, **85**, 6632.

91. Goloubinoff, P., Mogk, A., Zvi, A. P., Tomoyasu, T., and Bukau, B. (1999). Sequential mechanism of solubilization and refolding of stable protein aggregates by a bichaperone network. *Proc. Natl. Acad. Sci. USA*, **96**, 13732.

92. Zolkiewski, M. (1999). ClpB cooperates with DnaK, DnaJ, and GrpE in suppressing protein aggregation. A novel multi-chaperone system from *Escherichia coli*. *J. Biol. Chem.*, **274**, 28083.

93. Ueguchi, C., Kakeda, M., Yamada, H., and Mizuno, T. (1994). An analogue of the DnaJ molecular chaperone in *Escherichia coli*. *Proc. Natl. Acad. Sci. USA*, **91**, 1054.

94. Ueguchi, C., Shiozawa, T., Kakeda, M., Yamada, H., and Mizuno, T. (1995). A study of the double mutation of *dnaJ* and *cbpA*, whose gene products function as molecular chaperones in *Escherichia coli*. *J. Bacteriol.*, **177**, 3894.

95. Hettema, E. H., Ruigrok, C. C. M., Koerkamp, M. G., van den Berg, M., Tabak, H. F., Distel, B., and Braakman, I. (1998). The cytosolic DnaJ-like protein djp1p is involved specifically in peroxisomal protein import. *J. Cell Biol.*, **142**, 421.

96. Luke, M. M., Sutton, A., and Arndt, K. T. (1991). Characterization of SIS1, a *Saccharomyces cerevisiae* homologue of bacterial dnaJ proteins. *J. Cell Biol.*, **114**, 623.

97. Zhong, T. and Arndt, K. T. (1993). The yeast SIS1 protein, a DnaJ homolog, is required for the initiation of translation. *Cell*, **73**, 1175.

98. Ohba, M. (1997). Modulation of intracellular protein degradation by SSB1-SIS1 chaperon system in yeast S. cerevisiae. *FEBS Lett.*, **409**, 307.

99. Yan, W. and Craig, E. A. (1999). The glycine-phenylalanine-rich region determines the specificity of the yeast Hsp40 Sis1. *Mol. Cell Biol.*, **19**, 7751.

100. Liu, J. S., Kuo, S. R., Makhov, A. M., Cyr, D. M., Griffith, J. D., Broker, T. R., and Chow, L. T. (1998). Human Hsp70 and Hsp40 chaperone proteins facilitate human papillomavirus-11 E1 protein binding to the origin and stimulate cell-free DNA replication. *J. Biol. Chem.*, **273**, 30704.

101. Dittmar, K. D., Banach, M., Galigniana, M. D., and Pratt, W. B. (1998). The role of DnaJ-like proteins in glucocorticoid receptor.hsp90 heterocomplex assembly by the reconstituted hsp90.p60.hsp70 foldosome complex. *J. Biol. Chem.*, **273**, 7358.

102. Sugito, K., Yamane, M., Hattori, H., Hayashi, Y., Tohnai, I., Ueda, M., et al. (1995). Inter-

action between hsp70 and hsp40, eukaryotic homologues of DnaK and DnaJ, in human cells expressing mutant-type p53. *FEBS Lett.*, **358**, 161.

103. Caplan, A. J. and Douglas, M. G. (1991). Characterization of YDJ1: a yeast homologue of the bacterial dnaJ protein. *J. Cell Biol.*, **114**, 609.

104. Lu, Z. and Cyr, D. M. (1998). The conserved carboxyl terminus and zinc finger-like domain of the co-chaperone Ydj1 assist Hsp70 in protein folding. *J. Biol. Chem.*, **273**, 5970.

105. Jiang, R. F., Greener, T., Barouch, W., Greene, L., and Eisenberg, E. (1997). Interaction of auxilin with the molecular chaperone, Hsc70. *J. Biol. Chem.*, **272**, 6141.

106. Braun, J. E. and Scheller, R. H. (1995). Cysteine string protein, a DnaJ family member, is present on diverse secretory vesicles. *Neuropharmacology*, **34**, 1361.

107. Atencio, D. P. and Yaffe, M. P. (1992). MAS5, a yeast homolog of DnaJ involved in mitochondrial protein import. *Mol. Cell Biol.*, **12**, 283.

108. Caplan, A. J., Cyr, D. M., and Douglas, M. G. (1992). YDJ1p facilitates polypeptide translocation across different intracellular membranes by a conserved mechanism. *Cell*, **71**, 1143.

109. Becker, J., Walter, W., Yan, W., and Craig, E. A. (1996). Functional interaction of cytosolic hsp70 and a DnaJ-related protein, Ydj1p, in protein translocation in vivo. *Mol. Cell Biol.*, **16**, 4378.

110. Davis, A. R., Alevy, Y. G., Chellaiah, A., Quinn, M. T., and Mohanakumar, T. (1998). Characterization of HDJ-2, a human 40 kD heat shock protein. *Int. J. Biochem. Cell. Biol.*, **30**, 1203.

111. Cyr, D. M. (1995). Cooperation of the molecular chaperone Ydj1 with specific Hsp70 homologs to suppress protein aggregation. *FEBS Lett.*, **359**, 129.

112. Lee, D. H., Sherman, M. Y., and Goldberg, A. L. (1996). Involvement of the molecular chaperone Ydj1 in the ubiquitin-dependent degradation of short-lived and abnormal proteins in *Saccharomyces cerevisiae*. *Mol. Cell Biol.*, **16**, 4773.

113. Meacham, G. C., Lu, Z., King, S., Sorscher, E., Tousson, A., and Cyr, D. M. (1999). The Hdj-2/Hsc70 chaperone pair facilitates early steps in CFTR biogenesis. *EMBO J.*, **18**, 1492.

114. Chai, Y., Koppenhafer, S. L., Bonini, N. M., and Paulson, H. L. (1999). Analysis of the role of heat shock protein (Hsp) molecular chaperones in polyglutamine disease. *J. Neurosci.*, **19**, 10338.

115. Stenoien, D. L., Cummings, C. J., Adams, H. P., Mancini, M. G., Patel, K., DeMartino, G. N., *et al.* (1999). Polyglutamine-expanded androgen receptors form aggregates that sequester heat shock proteins, proteasome components and SRC-1, and are suppressed by the HDJ-2 chaperone. *Hum. Mol. Genet.*, **8**, 731.

116. Rowley, N., Prip-Buus, C., Westermann, B., Brown, C., Schwarz, E., Barrell, B., and Neupert, W. (1994). Mdj1p, a novel chaperone of the DnaJ family, is involved in mitochondrial biogenesis and protein folding. *Cell*, **77**, 249.

117. Westermann, B., Gaume, B., Herrmann, J. M., Neupert, W., and Schwarz, E. (1996). Role of the mitochondrial DnaJ homolog Mdj1p as a chaperone for mitochondrially synthesized and imported proteins. *Mol. Cell Biol.*, **16**, 7063.

118. Duchniewicz, M., Germaniuk, A., Westermann, B., Neupert, W., Schwarz, E., and Marszalek, J. (1999). Dual role of the mitochondrial chaperone Mdj1p in inheritance of mitochondrial DNA in yeast. *Mol. Cell Biol.*, **19**, 8201.

119. Kurzik-Dumke, U., Debes, A., Kaymer, M., and Dienes, P. (1998). Mitochondrial localization and temporal expression of the Drosophila melanogaster DnaJ homologous tumor suppressor Tid50. *Cell Stress Chaperones*, **3**, 12.

120. Feldheim, D., Rothblatt, J., and Schekman, R. (1992). Topology and functional domains of Sec63p, an endoplasmic reticulum membrane protein required for secretory protein translocation. *Mol. Cell Biol.*, **12**, 3288.

121. Schlenstedt, G., Harris, S., Risse, B., Lill, R., and Silver, P. A. (1995). A yeast DnaJ homologue, Scj1p, can function in the endoplasmic reticulum with BiP/Kar2p via a conserved domain that specifies interactions with Hsp70s. *J. Cell Biol.*, **129**, 979.

122. Brodsky, J. L., Bauerle, M., Horst, M., and McClellan, A. J. (1998). Mitochondrial Hsp70 cannot replace BiP in driving protein translocation into the yeast endoplasmic reticulum. *FEBS Lett.*, **435**, 183.

123. McClellan, A. J., Endres, J. B., Vogel, J. P., Palazzi, D., Rose, M. D., and Brodsky, J. L. (1998). Specific molecular chaperone interactions and an ATP-dependent conformational change are required during posttranslational protein translocation into the yeast ER. *Mol. Biol. Cell*, **9**, 3533.

124. Deshaies, R. J., Sanders, S. L., Feldheim, D. A., and Schekman, R. (1991). Assembly of yeast Sec proteins involved in translocation into the endoplasmic reticulum into a membrane-bound multisubunit complex. *Nature*, **349**, 806.

125. Brodsky, J. L. and Schekman, R. (1993). A Sec63p-BiP complex from yeast is required for protein translocation in a reconstituted proteoliposome. *J. Cell Biol.*, **123**, 1355.

126. Kurihara, T. and Silver, P. (1993). Suppression of a sec63 mutation identifies a novel component of the yeast endoplasmic reticulum translocation apparatus. *Mol. Biol. Cell*, **4**, 919.

127. Sanders, S. L., Whitfield, K. M., Vogel, J. P., Rose, M. D., and Schekman, R. W. (1992). Sec61p and BiP directly facilitate polypeptide translocation into the ER. *Cell*, **69**, 353.

128. Rose, M. D., Misra, L. M., and Vogel, J. P. (1989). KAR2, a karyogamy gene, is the yeast homolog of the mammalian BiP/GRP78 gene [published erratum appears in *Cell* 1989 Aug 25; 58(4): following 801]. *Cell*, **57**, 1211.

129. Silberstein, S., Schlenstedt, G., Silver, P. A., and Gilmore, R. (1998). A role for the DnaJ homologue Scj1p in protein folding in the yeast endoplasmic reticulum. *J. Cell Biol.*, **143**, 921.

130. Zhang, S., Lockshin, C., Herbert, A., Winter, E., and Rich, A. (1992). Zuotin, a putative Z-DNA binding protein in *Saccharomyces cerevisiae*. *EMBO J.*, **11**, 3787.

131. Hughes, R., Chan, F. Y., White, R. A., and Zon, L. I. (1995). Cloning and chromosomal localization of a mouse cDNA with homology to the *Saccharomyces cerevisiae* gene zuotin. *Genomics*, **29**, 546.

132. Yan, W., Schilke, B., Pfund, C., Walter, W., Kim, S., and Craig, E. A. (1998). Zuotin, a ribosome-associated DnaJ molecular chaperone. *EMBO J.*, **17**, 4809.

133. Heckmann, M., Adelsberger, H., and Dudel, J. (1997). Evoked transmitter release at neuromuscular junctions in wild type and cysteine string protein null mutant larvae of Drosophila. *Neurosci. Lett.*, **228**, 167.

134. Chamberlain, L. H. and Burgoyne, R. D. (2000). Cysteine-string protein: the chaperone at the synapse. *J. Neurochem.*, **74**, 1781.

135. Ahle, S. and Ungewickell, E. (1990). Auxilin, a newly identified clathrin-associated protein in coated vesicles from bovine brain. *J. Cell Biol.*, **111**, 19.

136. Schroder, S., Morris, S. A., Knorr, R., Plessmann, U., Weber, K., Nguyen, G. V., and Ungewickell, E. (1995). Primary structure of the neuronal clathrin-associated protein auxilin and its expression in bacteria. *Eur. J. Biochem.*, **228**, 297.

137. Ungewickell, E., Ungewickell, H., Holstein, S. E., Lindner, R., Prasad, K., Barouch, W., *et al.* (1995). Role of auxilin in uncoating clathrin-coated vesicles. *Nature*, **378**, 632.

138. Holstein, S. E., Ungewickell, H., and Ungewickell, E. (1996). Mechanism of clathrin basket dissociation: separate functions of protein domains of the DnaJ homologue auxilin. *J. Cell Biol.*, **135**, 925.

139. Cheetham, M. E., Anderton, B. H., and Jackson, A. P. (1996). Inhibition of hsc70-catalysed clathrin uncoating by HSJ1 proteins. *Biochem. J.*, **319**, 103.

140. King, C., Eisenberg, E., and Greene, L. (1997). Effect of yeast and human DnaJ homologs on clathrin uncoating by 70 kilodalton heat shock protein. *Biochemistry*, **36**, 4067.

141. Takayama, S., Sato, T., Krajewski, S., Kochel, K., Irie, S., Millan, J. A., and Reed, J. C. (1995). Cloning and functional analysis of BAG-1: a novel Bcl-2-binding protein with anti-cell death activity. *Cell*, **80**, 279.

142. Luders, J., Demand, J., and Hohfeld, J. (2000). The ubiquitin-related BAG-1 provides a link between the molecular chaperones Hsc70/Hsp70 and the proteasome. *J. Biol. Chem.*, **275**, 4613.

143. Zeiner, M., Niyaz, Y., and Gehring, U. (1999). The hsp70-associating protein Hap46 binds to DNA and stimulates transcription. *Proc. Natl. Acad. Sci. USA*, **96**, 10194.

144. Bardelli, A., Longati, P., Albero, D., Goruppi, S., Schneider, C., Ponzetto, C., and Comoglio, P. M. (1996). HGF receptor associates with the anti-apoptotic protein BAG-1 and prevents cell death. *EMBO J.*, **15**, 6205.

145. Froesch, B. A., Takayama, S., and Reed, J. C. (1998). BAG-1L protein enhances androgen receptor function. *J. Biol. Chem.*, **273**, 11660.

146. Matsuzawa, S., Takayama, S., Froesch, B. A., Zapata, J. M., and Reed, J. C. (1998). p53-inducible human homologue of Drosophila seven in absentia (Siah) inhibits cell growth: suppression by BAG-1. *EMBO J.*, **17**, 2736.

147. Kullmann, M., Schneikert, J., Moll, J., Heck, S., Zeiner, M., Gehring, U., and Cato, A. C. (1998). RAP46 is a negative regulator of glucocorticoid receptor action and hormone-induced apoptosis. *J. Biol. Chem.*, **273**, 14620.

148. Liu, R., Takayama, S., Zheng, Y., Froesch, B., Chen, G. Q., Zhang, X., *et al.* (1998). Interaction of BAG-1 with retinoic acid receptor and its inhibition of retinoic acid-induced apoptosis in cancer cells. *J. Biol. Chem.*, **273**, 16985.

149. Nollen, E. A., Brunsting, J. F., Song, J., Kampinga, H. H., and Morimoto, R. I. (2000). Bag1 functions in vivo as a negative regulator of Hsp70 chaperone activity. *Mol. Cell Biol.*, **20**, 1083.

150. Wang, H. G., Takayama, S., Rapp, U. R., and Reed, J. C. (1996). Bcl-2 interacting protein, BAG-1, binds to and activates the kinase Raf-1. *Proc. Natl. Acad. Sci. USA*, **93**, 7063.

151. Prapapanich, V., Chen, S., Nair, S. C., Rimerman, R. A., and Smith, D. F. (1996). Molecular cloning of human p48, a transient component of progesterone receptor complexes and an Hsp70-binding protein. *Mol. Endocrinol.*, **10**, 420.

152. Honore, B., Leffers, H., Madsen, P., Rasmussen, H. H., Vandekerckhove, J., and Celis, J. E. (1992). Molecular cloning and expression of a transformation-sensitive human protein containing the TPR motif and sharing identity to the stress-inducible yeast protein STI1. *J. Biol. Chem.*, **267**, 8485.

153. Chen, S. and Smith, D. F. (1998). Hop as an adaptor in the heat shock protein 70 (Hsp70) and hsp90 chaperone machinery. *J. Biol. Chem.*, **273**, 35194.

154. Scheufler, C., Brinker, A., Bourenkov, G., Pegoraro, S., Moroder, L., Bartunik, H., *et al.* (2000). Structure of TPR domain-peptide complexes: critical elements in the assembly of the Hsp70-Hsp90 multichaperone machine. *Cell*, **101**, 199.

155. Van Der Spuy, J., Kana, B. D., Dirr, H. W., and Blatch, G. L. (2000). Heat shock cognate

protein 70 chaperone-binding site in the co-chaperone murine stress-inducible protein 1 maps to within three consecutive tetratricopeptide repeat motifs. *Biochem. J.*, **345**, 645.

156. Chen, S., Prapapanich, V., Rimerman, R. A., Honore, B., and Smith, D. F. (1996). Interactions of p60, a mediator of progesterone receptor assembly, with heat shock proteins hsp90 and hsp70. *Mol. Endocrinol.*, **10**, 682.

157. Dittmar, K. D. and Pratt, W. B. (1997). Folding of the glucocorticoid receptor by the reconstituted Hsp90-based chaperone machinery. The initial hsp90.p60.hsp70-dependent step is sufficient for creating the steroid binding conformation. *J. Biol. Chem.*, **272**, 13047.

7 | Chaperones in signal transduction

DAVID F. SMITH

1. Hsp90 and signalling proteins

Hsp90, typically the most abundant chaperone in the cytoplasmic compartment of cells, has gained a reputation for its functional interactions with a wide variety of signalling proteins (for recent reviews see 1, 2–4). Examples include the steroid receptors and several transcription factors unrelated to steroid receptors, a variety of tyrosine and serine/threonine kinases, and other regulatory enzymes such as telomerase (5) and nitric oxide synthase (6–8). *In vitro* refolding assays with model substrates have shown that Hsp90 on its own has the general ability to stabilize proteins in a folding competent state (9–11), but there is limited evidence to support a major role for Hsp90 in general nascent chain folding. Still, unstable or alternative folding appears to be a common feature among the signalling proteins that are native targets for Hsp90. Another striking feature of native Hsp90 complexes is the invariable presence of some combination of partner co-chaperones that bind Hsp90.

Alluding to the biological importance of Hsp90, there are several natural compounds that specifically target Hsp90. Geldanamycin (12), herbimycin A, and the structurally distinct compound radicicol (13, 14) have all been identified over the last 5 years as Hsp90-binding agents, suggesting that evolution has found Hsp90 to be an effective toxicological target. These drugs have proven to be valuable research reagents for assessing Hsp90's involvement in intracellular processes, and variations of these compounds are now in clinical trials for therapeutic outcomes. The most biologically consequential action of the Hsp90-binding drugs is not a general deficiency in protein folding processes, but rather the disruption of signal transduction pathways normally mediated by Hsp90 and its partners.

This article will review some of the better characterized interactions of Hsp90 with target proteins, taking into consideration the participation of particular partner proteins and assessing mechanisms by which Hsp90 influences signalling pathways.

2. Hsp90, Hsp70, and their partners

Multiple chaperones and partner proteins are involved in signalling protein complexes. While Hsp90 is the most conspicuous of these, additional components of the

Table 1 Characteristics of chaperone components identified in complexes with signalling proteins

Chaperone component	Partner with	TPR motif	Known target preferences
Hsp90			General
Hsp70			General
DnaJ	Hsp70		General
Hip	Hsp70	√	SR
BAG-1 isoforms	Hsp70		GR, AR, Bcl2
Hop	Hsp70, Hsp90	√	General
p23	Hsp90		General
Cdc37	Hsp90		Kinases
XAP2/AIP/ARA9	Hsp90	√	AhR
PP5	Hsp90	√	GR
Cyp40	Hsp90	√	OR
FKBP52	Hsp90	√	SR
FKBP51	Hsp90	√	PR, GR

Abbreviations: SR—steroid receptors, GR—glucocorticoid receptor, AR—androgen receptor, AhR—arylhydrocarbon receptor, OR—oestrogen receptor, PR—progesterone receptor.

cytoplasmic chaperone machinery play roles—large, small, or not at all—depending on the particular signalling protein that is the target of chaperone interactions. Table 1 lists these components, identifies whether the component interacts directly with Hsp70 and/or Hsp90, and indicates any known qualitative specificity for particular signalling protein complexes. As also shown, many of the partner proteins contain tetratricopeptide repeats (TPR) that mediate binding to Hsp. The TPR-containing Hsp90 partners compete for binding a common site in the C-terminal region of Hsp90 (15, 16). Among the Hsp90 partners, only Hop contains a second, distinctive TPR which gives Hop the ability to bind both Hsp70 and Hsp90 simultaneously (17, 18). Hsp70 has two TPR-binding sites, one in its N-terminal ATPase domain that accommodates Hip binding (19) and a second near its C-terminus where Hop binds (20, 21).

There are many distinct chaperone complexes that can be formed around Hsp70 or Hsp90 with the various partner proteins, and there are multiple capacities for these complexes to interact with target proteins. All of the components listed in Table 1 cannot co-exist in a single target complex; spatial, biochemical, and temporal restrictions apply. For instance, most of the Hsp90 partners compete for a common binding site on Hsp90 and therefore would be mutually excluded from a target complex. Also, several of the partners bind in a nucleotide-specific manner to Hsp70 or Hsp90, so the nucleotide bound status of these ATPases would limit components in a target complex. Finally, the interaction of a target substrate with some chaperones prepares the target for subsequent interactions with other chaperones. These variables are perhaps best appreciated by focusing on a specific set of target-chaperone interactions. Steroid receptors complexes have been the most extensively characterized for chaperone composition, assembly pathways, time-courses, and mechanisms (extensively reviewed in 22). A brief overview of findings from *in vitro* assembly studies illustrates how multiple chaperone components can cooperate in an ordered, dynamic manner to form and maintain target complexes.

As summarized in Fig. 1, a fully synthesized receptor monomer is initially bound by Hsp70 and DnaJ. Subsequently and probably requiring substrate-dependent ATP hydrolysis by Hsp70, the Hsp70 partners Hip and Hop are recruited into the complex. Since Hop is almost quantitatively associated with Hsp90 and independently binds both Hsp70 and Hsp90, Hop effectively functions as an adaptor to localize Hsp90 on the receptor complex (23). Hsp90 in the complex establishes direct contacts with the receptor, and through a mechanism that likely involves ATP hydrolysis, the Hsp90 partner p23 enters the fray and stabilizes Hsp90-receptor interactions. Concomitantly, Hop dissociates from Hsp90 and is replaced by one of the immunophilins or PP5 (24, 25). Along the way, Hsp70, Hip, and DnaJ are also displaced from the maturing complex. Significantly, for either progesterone receptor (PR) or glucocorticoid receptor (GR), the receptor-Hsp90-p23-immunophilin complex is the functionally relevant form for hormone binding (26–29). Earlier forms bind hormone poorly at physiologically relevant concentrations. In a final phase of the receptor assembly pathway, Hsp90 dissociates from the receptor target as would be expected for any chaperone-substrate interaction. In the absence of bound hormone, the assembly pathway is re-initiated as Hsp70 and DnaJ bind the free receptor, leading to a dynamic steady-state cycling of receptor through chaperone interactions. In its hormone-bound conformation, receptor is no longer recognized by Hsp70-DnaJ and the assembly cycle is broken.

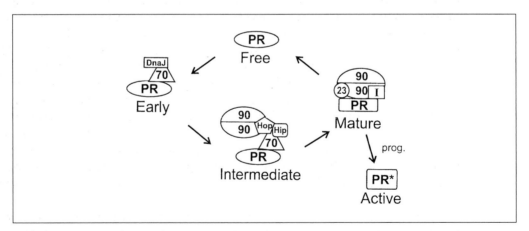

Fig. 1 PR assembly cycle with chaperones. Free, unliganded PR rapidly associates with Hsp70 in a DnaJ-dependent manner to form an early complex. At an intermediate assembly stage, the Hsp70-binding partners Hip and Hop, in association with Hsp90, are incorporated. In a transitional process that remains poorly understood, a mature PR complex is formed in which the receptor is capable of binding hormone. This functionally mature complex contains Hsp90 plus Hsp90-bound partners p23 and one of the immunophilin (I) proteins—FKBP51, FKBP52, Cyp40, or the related protein phosphatase PP5. If progesterone (P) binds while receptor is in a mature complex, the dissociated receptor becomes an active transcription factor. In the absence of progesterone, the dissociated receptor re-enters the assembly pathway.

Assembly studies in a defined system show that Hsp70, DnaJ, Hop, Hsp90, and p23 are minimally required to establish efficient hormone binding by GR or PR (29–33). Additional components observed in complexes assembled in complete reticulocyte lysate may help maintain dynamics in the assembly pathway or may serve biochemical or cellular functions beyond establishing the receptor's hormone binding conformation. For example, Hip and BAG-1 have opposing effects on the activity of Hsp70 in protein refolding assays (34–37), so it may be that their counter-actions become important in a more complex assembly system. Perhaps related to this possibility, certain mutant forms of Hip (38) or excessive amounts of BAG-1 (39) have been shown to inhibit receptor assembly in reticulocyte lysate. On the other hand, the mere presence of a component in a receptor complex does not prove a functional requirement for that component. The immunophilins FKBP51, FKBP52, and Cyp40, which compete for binding to Hsp90, have been shown to exchange in a highly dynamic manner on mature PR complexes (40). Hence, the receptor complex is constantly sampling its environment for different immunophilins; conversely, the immunophilins are sampling Hsp90-target complexes that are potential substrates for immunophilin actions. The sampling process could be mechanistically important in the cellular milieu even while a particular immunophilin in a target complex is functionally insignificant.

3. Target-specific chaperone preferences

Do all signalling proteins that are targets for Hsp90 have chaperone interaction patterns identical to those of steroid receptors? Clearly not, based on compositional differences in various target complexes. Some differences in native complexes could potentially be attributed to tissue-specific expression or restricted sub-cellular local-ization of chaperone components which would thus limit the range of possible target complexes in a natural setting. However, many chaperone components are expressed in a broad range of tissues and exist in both cytoplasmic and nuclear compartments. For unknown reasons, various kinase complexes with Hsp90 commonly contain Cdc37 (41–43), but very little Cdc37 has been observed in steroid receptor or other target complexes. Similarly, the Hsp90-associated TPR protein AIP/XAP2/ARA9 appears to be specific for arylhydrocarbon receptor complexes (44–47). In addition to these conspicuous differences in target complexes, there are likely to be qualitative and kinetic differences in the pathways by which various complexes are formed and stabilized.

Reticulocyte lysate contains a complex mixture of chaperone components and provides a convenient cell-free medium for detailed biochemical analyses and for comparing chaperone interactions between different target proteins. In one study comparing four unrelated Hsp90 targets—oestrogen receptor, heat shock transcription factor 1, the arylhydrocarbon receptor, and the tyrosine kinase Fes—a qualitative similarity in target complex compositions was observed with the most striking difference being the kinase-specific presence of Cdc37 (48). In another more quanti-tative assessment that directly compared the related receptors for progesterone, oestrogen, and glucocorticoids, each receptor had a distinctive pattern of preference for Hsp90-associated immunophilin components (49). From these examples, it seems likely that to fully appreciate the significance of chaperone interactions with various signalling proteins, characterizations are required for each target protein that address the full range of chaperone interactions, the ordering and kinetics of assembly steps, and the functional importance of individual chaperone components in the target complex.

4. Functional significance

Chaperone interactions impact the function of signalling proteins in multiple ways, but the folding status of the target protein is known to be a key factor in many chaperone interactions. Thus, it is tempting to generalize that target proteins simply have inherent folding limitations and this alone explains their interactions with chaperones. In some cases, this generalization is probably accurate, but an argument can be made that, at least with a subset of signalling proteins, the protein target has evolutionarily adapted folding characteristics that are designed to enlist chaperones for regulatory purposes.

4.1 Signalling proteins compared to model misfolded substrates for chaperone interactions

An important issue to address is whether signalling proteins that form complexes with chaperones are any different than typical misfolded or incompletely folded protein substrates. Several laboratories have used partially denatured enzymes as model chaperone substrates for *in vitro* folding assays in a highly purified, simplified medium containing one or a small number of chaperone components. Two readouts in these assays are the recovery of enzymatic activity after denaturing conditions have been halted and the degree of irreversible protein aggregation that occurs over time. These parameters isolate, respectively, two chaperone-like activities: 1) folding —the ability to actively promote refolding of non-aggregated, denatured protein and 2) holding—the ability to maintain a misfolded protein in a folding competent, non-aggregated state. Hsp70 actively promotes folding and renaturation of enzymatic activity in these assays, while Hsp90 has holding activity but no ability to renature the misfolded substrate. Interestingly, several of the Hsp90 partner proteins on their own have a holding activity similar to Hsp90 (50, 51).

Based on these observations and knowing that PR interacts with Hsp70, Hsp90, and various partner chaperones, one might reasonably predict that Hsp90 and its partners first interact with PR to hold it in a folding competent state until Hsp70 can act to promote folding of PR's hormone binding conformation. In fact, as previously illustrated in Fig. 1, something closer to the opposite occurs in which Hsp70's initial binding is obligate for the subsequent binding of Hsp90, and only when Hsp90 is properly situated on PR is hormone binding established. Furthermore, the Hsp90 partners p23 and FKBP52, which have independent holding activity similar to Hsp90 by *in vitro* refolding assays, exhibit no capacity to bind PR except through their association with PR-bound Hsp90. Therefore, we can conclude that the behaviour of chaperones toward PR—and probably most other signalling protein targets—is not reflected by activities of the same chaperones toward model enzyme substrates in minimal refolding assays. One possible explanation for the apparently contradictory results between enzyme refolding and steroid receptor assembly studies is that cytoplasmic chaperones have alternative modes of acting and cooperating that are substrate- and context-dependent. If so, this limits our ability to generalize about chaperone-substrate interactions, but it also enriches the range of biological functions that chaperones can serve.

4.2   Stabilization and folding of signalling proteins by chaperones

There is no consensus sequence shared by targets of Hsp90, but all targets in the absence of Hsp90 or other chaperone activities are characterized by functional defects that most likely relate to folding insufficiencies of the target. There have been no direct observations, for example by X-ray crystallographic approaches, that specify conformational differences in a target associated with Hsp90 versus a free target. However, distinctive conformational states of Hsp90 targets are implicit from frequently observed differences in proteolytic sensitivity, enzymatic activity, or the ability to bind ligands. The following examples illustrate the reliance of targets on Hsp90 for functional maturation and stability.

4.2.1   Steroid receptors

Although free steroid receptors display no particular tendency toward aggregation that would suggest gross misfolding of the polypeptide, there are several indications that chaperones are needed to establish and maintain functionally important receptor conformations. The ability of receptor to bind hormone is chaperone-dependent, but to variable degrees among steroid receptors. All chaperone interactions are localized to the large hormone-binding domain of steroid receptors, so the structure of this domain is worth considering. Three-dimensional crystallographic structures have been solved for the isolated hormone-binding domains from several steroid receptors (52–54) and other nuclear receptor family members (reviewed in 55). In the absence of ligand, all of these domains have a rather open structure containing a large hydrophobic pocket into which ligand will bind. In the presence of ligand, the domain structure becomes more compact as several contacts are made between pocket side chains and ligand. As a result, the ligand-bound conformation is stable, albeit variable in a ligand-specific manner that has important functional implications for transcriptional activity of the receptor. Conversely, the ligand-free conformation is highly labile for some steroid receptors and readily influences the receptor's competence for hormone binding.

GR has the greatest reliance on Hsp90 and other chaperone components for maintaining hormone binding ability (26, 30). PR is similarly dependent on chaper-ones at physiological temperatures, but the conformation that is competent for bind-ing hormone can be maintained in the absence of chaperones if the receptor is kept on ice (27). The receptors for androgens and mineralocorticoids have not been characterized in as much detail, but Hsp90 also influences their hormone binding ability (56–59). On the other hand, the hormone binding conformation of oestrogen receptor (OR) is stable at elevated temperatures in the absence of chaperones; how-ever, it may be that initial chaperone interactions are required to establish hormone binding ability in the nascent OR polypeptide. Since the overall conformation revealed in crystal structures for other nuclear receptors is similar to the steroid re-ceptors, the question is raised as to why the non-steroid nuclear receptors have never been observed in chaperone complexes. This issue will be re-addressed in a later section, but the typical nuclear receptor may have a greater reliance on chaperones than is generally appreciated. Evidence for chaperone interactions with other nuclear receptors comes from yeast models in which Hsp90 levels are genetically restricted to the minimum required for viability. In a minimal Hsp90 background, steroid receptors fail to function in the hormone-dependent expression of a reporter gene. Historically, this finding provided the first evidence *in vivo* for the importance of Hsp90 in steroid receptor function (60) and correlated nicely with the previously observed recovery of Hsp90 in steroid receptor complexes. However, using Hsp90-deficient yeast to test the function of a retinoic acid receptor, for which no chaperone complexes have been observed, it was similarly found that ligand-dependent reporter activity and hormone binding ability were lost (61). Perhaps Hsp90 and other chaperones are generally required in a transient manner by nascent nuclear receptors to assist in folding the ligand binding domain.

Apart from hormone binding ability, the proteolytic stability of steroid receptors *in vivo* is dependent on continuous interactions with the Hsp90 machinery. Geldana-mycin treatment of cells expressing steroid receptors leads to a dramatic decrease in the steady state receptor level (58, 59, 62, 63). For GR and PR, there is also a rapid loss of hormone binding ability (28, 58, 63), reflecting the constant dependence of these receptors on Hsp90 to maintain a hormone binding conformation. The more general property of increased receptor degradation in response to geldanamycin could be attributed either to receptors assuming a more sensitive conformation or to Hsp90 masking a pre-existing protease sensitive site on these receptors. In hormone-treated cells, though, receptors become more susceptible to proteolysis after hormone-binding, chaperone dissociation, and transcriptional activation; this degradation may contribute to the cellular off-signal following hormonal stimulation. Since the hormone-bound receptor has a more compact, stable conformation, one cannot simply equate degradation rates with folding status of the hormone binding domain.

4.2.2 Src-family tyrosine kinases

The association of several viral oncogenic tyrosine kinases with Hsp90 and Cdc37 has long been recognized (41), and a variety of studies have shown that these chaperones are required for the functional maturation and stability of the viral kinases. More recent evidence has pointed to the importance of chaperones for the function of normal cellular counterparts to the viral kinases. In one case, Lindquist and co-workers used a yeast model to compare the folding status of v-Src and c-Src proteins with their requirements for Hsp90 (64). While v-Src was more sensitive to changes in Hsp90 level, c-Src kinase activity and its ability to serve as an autocatalytic substrate were perturbed at lower Hsp90 levels. Analysis of c-Src/v-Src chimeras showed that differences in sensitivity to Hsp90 relate to multiple mutations in v-Src, suggesting that v-Src is less stably folded than c-Src.

Geldanamycin and herbimycin A were at one time purported to be tyrosine kinase inhibitors due to their ability to reverse v-Src transformation of cells, but efforts to show that these drugs directly inhibit kinase activity were unsuccessful (65). Only after Whitesell *et al.* discovered the specific binding of Hsp90 to a geldanamycin affinity matrix was it realized that these drugs function indirectly as selective tyrosine kinase inhibitors by inhibiting Hsp90 function and promoting proteolytic degradation of Hsp90-associated kinases (12).

4.3   Functional repression of signalling proteins by chaperones

To become active transcription factors, steroid receptors must typically dimerize and bind to a specific DNA sequence element in the vicinity of a target gene. Moreover, the receptor must establish protein–protein interactions either with the basal transcriptional machinery or with co-regulatory proteins that alter chromatin structure (recently reviewed in 66, 67). Most members of the nuclear receptor superfamily are competent for these interactions and often have transcriptional influences in the absence of their cognate ligands, with ligand binding resulting in a rearrangement of receptor interactions. Steroid receptors are distinctive in that they are efficiently silenced in the absence of ligand, and chaperone interactions are a major mechanism for repressing receptor's activity (68). Chaperone complexes, which localize to the receptor's hormone binding domain, probably sterically hinder the receptor's ability to bind DNA, dimerize, or interact with cotranscriptional proteins. Hormone binding leads to a conformational change in receptor that interrupts chaperone interactions and frees the receptor to bind DNA and establish new, functionally relevant protein–protein interactions. By virtue of its hormone-sensitive association with 200–300 kDa chaperone complexes, the hormone binding domain of steroid receptors has been used in a variety of experimental constructs to confer hormone-dependent silencing on a chimeric protein in cells (reviewed in 69).

The reversible repression of steroid receptor transcriptional activity may have had adaptive advantages that led to extended chaperone interactions with ancient steroid receptors. In fact, the current reliance of PR and GR on chaperones for establishing and maintaining hormone binding conformations may be a consequence of pre-existing regulatory roles played by chaperones. In this scenario, internal structural elements in the hormone binding domain may have given way over time to compensating external scaffolding provided by Hsp90 and other chaperone interactions. Structural drift subsequent to regulatory interactions could help explain why OR,

PR, and GR exist in similar chaperone complexes, yet OR has a stable hormone binding conformation.

The mammalian heat shock transcription factor 1 (HSF1) offers another example of chaperone interactions that serve repressive regulatory purposes. HSF1 binds a specific DNA sequence in the promoter region of stress-inducible genes and is a key regulator of responses to cellular stress (for a recent review see 70). The activity of HSF1 is primarily regulated at the post-translational level since HSF1 is constitutively present in cells but functionally silent. Active HSF1 is a homotrimer, but inactive HSF1 exists as a monomer in association with cytoplasmic chaperones that inhibit trimer formation. *In vitro* assembly of recombinant HSF1 with chaperones in reticulocyte lysate looks similar in many ways to assembly of steroid receptor complexes (48), but the functional importance of various chaperone constituents is still being determined. Hsp90 is a key player in maintaining a quiescent HSF1. Hsp90 directly associates with HSF1, and selective depletion of Hsp90 from HeLa cell extracts leads to activation of HSF1 in a cell-free model (71). In addition, the Hsp90-binding ansamycins stimulate Hsp expression in cells (72–74); although this activity was originally attributed in some cases to tyrosine kinase inhibition, it now appears due to drug-induced Hsp90 dissociation from HSF1. Hsp70 also binds HSF1 (75, 76), and studies from Morimoto's lab have shown that Hsp70, along with the HSF1-binding protein HBP1, is important in dissociating HSF1 trimers and reversing the transcriptional activity of HSF1 during stress recovery (77–79). Studies in a Xenopus oocyte model suggest that p23 and other partner proteins from the steroid receptor assembly pathway may play important roles, along with Hsp90 and Hsp70, in

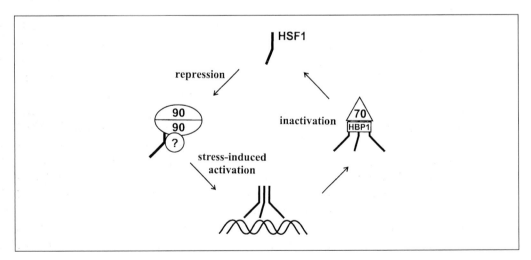

Fig. 2 Model for chaperone-mediated regulation of HSF1. The transcription factor HSF1 is maintained as an inactive monomer through its interaction with Hsp90 and perhaps other chaperone components. When non-native proteins accumulate in response to elevated temperature or other cellular stresses, Hsp90 dissociates and HSF1 subunits trimerize and undergo a series of modifications in becoming an activator of heat shock gene expression. As protein folding problems are resolved, Hsp70 and HBP1 act in concert to inactivate and dissociate HSF1 subunits that then re-associate with Hsp90.

repressing HSF1 activity in this setting (80, 81). The findings relating to chaperone-mediated regulation of HSF1 are summarized in Fig. 2.

4.4 Functional modulation of signalling proteins by chaperones

In addition to target folding and steric repression, chaperones may also modulate the activity of a target signalling protein, as illustrated in the following example. Mature steroid receptor complexes are in a dynamic association with the Hsp90-binding immunophilins, and selective patterns of association are displayed by different steroid receptors. In glucocorticoid receptor complexes assembled in reticulocyte lysate, FKBP51 is recovered preferentially over FKBP52 or Cyp40 (49). Moreover, FKBP51 gene expression is stimulated by glucocorticoids in several GR-containing cell lines (82, 83; D. F. Smith, unpublished observations). The functional significance of FKBP51 association with GR and its inducibility by GR became apparent from studies on cortisol resistance in squirrel monkeys. The New World monkeys in general have markedly higher levels of free, circulating cortisol than do Old World primates, and GR in various squirrel monkey and other New World monkey cells have been shown to have a decreased affinity for cortisol (84, 85). In several human and other mammalian cortisol-resistance syndromes, the defect has been traced to mutations in the hormone binding domain of GR that lower the receptor's affinity for cortisol. Surprisingly, however, the decreased affinity of squirrel monkey GR is not due to amino acid sequence changes in the receptor itself, but instead to a cytoplasmic factor in squirrel monkey cells (86, 87). This factor has recently been identified as FKBP51 (88).

FKBP51 is expressed at constitutively high levels in squirrel monkey cells, and the monkey protein contains sequence differences relative to human FKBP51 that confer a heightened preference for this immunophilin's association with PR and GR (D. F. Smith and J. G. Scammell, unpublished observations). Mouse GR's hormone binding affinity is lowered 11-fold in the presence of squirrel monkey FKBP51, while hormone binding by human GR was reduced over 40% when associated with human FKBP51 and to an even greater extent when assembled with squirrel monkey FKBP51 (88). The evolutionary significance of FKBP51-mediated cortisol-resistance in squirrel monkeys is unknown, but these studies have illuminated a regulatory mechanism whereby FKBP51 modulates GR's hormone responsiveness in a glucocorticoid-inducible manner. In essence, glucocorticoid exposure of naive cells could, through induction of FKBP51, temporarily attenuate the cells' response to subsequent hormone exposures. Moreover, induction or basal expression of FKBP51 in a tissue-specific manner could stratify glucocorticoid response intensities between various GR-containing tissues.

It seems likely that other instances of signalling pathway modulation mediated by Hsp90 partners will be recognized as more careful structural/functional comparisons are made for the different chaperone complexes that assemble around a particular signalling protein target.

5. Concluding comments

Chaperones are demonstrably capable of promoting the functional maturation and proteolytic stabilization of numerous, diverse signalling proteins. This fact implies that these signalling proteins have inherent limitations in their abilities to spontaneously assume and maintain active conformations, a situation that would seemingly be resolved through evolutionary pressures to fold better. However, selection for protein structure no doubt occurs on multiple levels, where maximizing protein folding and stability may compromise the protein's functional capabilities. The presence of a ubiquitous, abundant, broadly selective chaperone machinery could afford some tradeoffs in stability to favour functional purposes, but there must be practical limits to this reliance on the chaperone machinery.

Several of the signalling proteins targeted by chaperones exist in a larger family of related proteins—for example the steroid receptors are only 6 out of well more than 100 members of the nuclear receptor superfamily—in which most members have made the necessary tradeoffs for both efficient, stable folding and adequate activity. This leads one to conclude that chaperone interaction may, with some targets, have become functionally advantageous in its own right, beyond a 'need' to assist folding. The purposes served by chaperone interactions with signalling proteins likely lie along a continuum from simple reliance on chaperones for folding to the exploitation of chaperones as regulatory subunits. It would seem prudent, then, to view cause and effect relationships in chaperone-signalling protein complexes not only in the context of extant mechanisms—such as the stabilizing influence of chaperones—but also in an evolutionary context that considers why a chaperone dependency continues.

Acknowledgements

Research in the author's laboratory is supported by NIH grants DK44923 and DK48218.

References

1. Pratt, W. B. (1998). The hsp90-based chaperone system: involvement in signal transduction from a variety of hormone and growth factor receptors. *Proc. Soc. Exp. Biol. Med.*, **217**, 420.
2. Buchner, J. (1999). Hsp90 & Co.—a holding for folding. *Trends Biochem. Sci.*, **24**, 136.
3. Caplan, A. J. (1999). Hsp90's secrets unfold: new insights from structural and functional studies. *Trends Cell. Biol.*, **9**, 262.
4. Mayer, M. P. and Bukau, B. (1999). Molecular chaperones: the busy life of Hsp90. *Curr. Biol.*, **9**, R322.
5. Holt, S. E., Aisner, D. L., Baur, J., Tesmer, V. M., Dy, M., Ouellette, M., et al. (1999). Functional requirement of p23 and Hsp90 in telomerase complexes. *Genes Dev.*, **13**, 817.
6. Garcia-Cardena, G., Fan, R., Shah, V., Sorrentino, R., Cirino, G., Papapetropoulos, A., and Sessa, W. C. (1998). Dynamic activation of endothelial nitric oxide synthase by Hsp90. *Nature*, **392**, 821.

7. Shah, V., Wiest, R., Garcia-Cardena, G., Cadelina, G., Groszmann, R. J., and Sessa, W. C. (1999). Hsp90 regulation of endothelial nitric oxide synthase contributes to vascular control in portal hypertension. *Am. J. Physiol.*, **277**, G463.

8. Bender, A. T., Silverstein, A. M., Demady, D. R., Kanelakis, K. C., Noguchi, S., Pratt, W. B., and Osawa, Y. (1999). Neuronal nitric-oxide synthase is regulated by the Hsp90-based chaperone system in vivo. *J. Biol. Chem.*, **274**, 1472.

9. Wiech, H., Buchner, J., Zimmermann, R., and Jakob, U. (1992). Hsp90 chaperones protein folding in vitro. *Nature*, **358**, 169.

10. Young, J. C., Schneider, C., and Hartl, F. U. (1997). In vitro evidence that hsp90 contains two independent chaperone sites. *FEBS Lett.*, **418**, 139.

11. Scheibel, T., Weikl, T., and Buchner, J. (1998). Two chaperone sites in Hsp90 differing in substrate specificity and ATP dependence. *Proc. Natl. Acad. Sci. USA*, **95**, 1495.

12. Whitesell, L., Mimnaugh, E., De Costa, B., Myers, C., and Neckers, L. (1994). Inhibition of heat shock protein HSP90-pp60v-src heteroprotein complex formation by benzoquinone ansamycins: essential role for stress proteins in oncogenic transformation. *Proc. Natl. Acad. Sci. USA*, **91**, 8324.

13. Schulte, T. W., Akinaga, S., Soga, S., Sullivan, W., Stensgard, B., Toft, D., and Neckers, L. M. (1998). Antibiotic radicicol binds to the N-terminal domain of Hsp90 and shares import-ant biologic activities with geldanamycin. *Cell Stress Chap.*, **3**, 100.

14. Sharma, S. V., Agatsuma, T., and Nakano, H. (1998). Targeting of the protein chaperone, HSP90, by the transformation suppressing agent, radicicol. *Oncogene*, **16**, 2639.

15. Young, J. C., Obermann, W. M., and Hartl, F. U. (1998). Specific binding of tetratrico-peptide repeat proteins to the C-terminal 12-kDa domain of hsp90. *J. Biol. Chem.*, **273**, 18007.

16. Carrello, A., Ingley, E., Minchin, R. F., Tsai, S., and Ratajczak, T. (1999). The common tetratricopeptide repeat acceptor site for steroid receptor-associated immunophilins and hop is located in the dimerization domain of Hsp90. *J. Biol. Chem.*, **274**, 2682.

17. Chen, S., Prapapanich, V., Rimerman, R. A., Honore, B., and Smith, D. F. (1996). Inter-actions of p60, a mediator of progesterone receptor assembly, with heat shock proteins hsp90 and hsp70. *Mol. Endocrinol.*, **10**, 682.

18. Lassle, M., Blatch, G. L., Kundra, V., Takatori, T., and Zetter, B. R. (1997). Stress-inducible, murine protein mSTI1. Characterization of binding domains for heat shock proteins and in vitro phosphorylation by different kinases. *J. Biol. Chem.*, **272**, 1876.

19. Hohfeld, J., Minami, Y., and Hartl, F. U. (1995). Hip, a novel cochaperone involved in the eukaryotic Hsc70/Hsp40 reaction cycle. *Cell*, **83**, 589.

20. Gebauer, M., Zeiner, M., and Gehring, U. (1998). Interference between proteins Hap46 and Hop/p60, which bind to different domains of the molecular chaperone hsp70/hsc70. *Mol. Cell. Biol.*, **18**, 6238.

21. Demand, J., Luders, J., and Hohfeld, J. (1998). The carboxy-terminal domain of Hsc70 provides binding sites for a distinct set of chaperone cofactors. *Mol. Cell. Biol.*, **18**, 2023.

22. Pratt, W. B., and Toft, D. O. (1997). Steroid receptor interactions with heat shock protein and immunophilin chaperones. *Endocr. Rev.*, **18**, 306.

23. Chen, S. and Smith, D. F. (1998). Hop as an adaptor in the heat shock protein 70 (Hsp70) and hsp90 chaperone machinery. *J. Biol. Chem.*, **273**, 35194.

24. Chen, M.-S., Silverstein, A. M., Pratt, W. B., and Chinkers, M. (1996). The tetratricopeptide repeat domain of protein phosphatase 5 mediates binding to glucocorticoid receptor heterocomplexes and acts as a dominant negative mutant. *J. Biol. Chem.*, **271**, 32315.

25. Silverstein, A. M., Galigniana, M. D., Chen, M.-S., Owens-Grillo, J. K., Chinkers, M., and

Pratt, W. B. (1997). Protein phosphatase 5 is a major component of glucocorticoid re-ceptorHsp90 complexes with properties of an FK506-binding immunophilin. *J. Biol. Chem.*, **272**, 16224.

26. Scherrer, L., Dalman, F., Massa, E., Meshinchi, S., and Pratt, W. (1990). Structural and functional reconstitution of the glucocorticoid receptor-hsp90 complex. *J. Biol. Chem.*, **265**, 21397.

27. Smith, D. F. (1993). Dynamics of heat shock protein 90-progesterone receptor binding and the disactivation loop model for steroid receptor complexes. *Mol. Endocrinol.*, **7**, 1418.

28. Smith, D. F., Whitesell, L., Nair, S. C., Chen, S., Prapapanich, V., and Rimerman, R. A. (1995). Progesterone receptor structure and function altered by geldanamycin, an hsp90-binding agent. *Mol. Cell. Biol.*, **15**, 6804.

29. Dittmar, K. D., Demady, D. R., Stancato, L. F., Krishna, P., and Pratt, W. B. (1997). Folding of the glucocorticoid receptor by the heat shock protein (hsp) 90-based chaperone machinery. The role of p23 is to stabilize receptor.hsp90 heterocomplexes formed by hsp90.p60.hsp70. *J. Biol. Chem.*, **272**, 21213.

30. Dittmar, K. D., Hutchison, K. A., Owens-Grillo, J. K., and Pratt, W. B. (1996). Reconstitu-tion of the steroid receptor.hsp90 heterocomplex assembly system of rabbit reticulocyte lysate. *J. Biol. Chem.*, **271**, 12833.

31. Dittmar, K. D. and Pratt, W. B. (1997). Folding of the glucocorticoid receptor by the re-constituted Hsp90-based chaperone machinery. The initial hsp90.p60.hsp70-dependent step is sufficient for creating the steroid binding conformation. *J. Biol. Chem.*, **272**, 13047.

32. Dittmar, K. D., Banach, M., Galigniana, M. D., and Pratt, W. B. (1998). The role of DnaJ-like proteins in glucocorticoid receptor.hsp90 heterocomplex assembly by the reconstituted hsp90.p60.hsp70 foldosome complex. *J. Biol. Chem.*, **273**, 7358.

33. Kosano, H., Stensgard, B., Charlesworth, M. C., McMahon, N., and Toft, D. (1998). The assembly of progesterone receptor-hsp90 complexes using purified proteins. *J. Biol. Chem.*, **273**, 32973.

34. Hohfeld, J. and Jentsch, S. (1997). GrpE-like regulation of the hsc70 chaperone by the anti-apoptotic protein BAG-1 [published erratum appears in *EMBO J.* 1998 Feb 2; 17(3): 847]. *EMBO J.*, **16**, 6209.

35. Takayama, S., Bimston, D. N., Matsuzawa, S., Freeman, B. C., Aime-Sempe, C., Xie, Z., *et al.* (1997). BAG-1 modulates the chaperone activity of Hsp70/Hsc70. *EMBO J.*, **16**, 4887.

36. Bimston, D., Song, J., Winchester, D., Takayama, S., Reed, J. C., and Morimoto, R. I. (1998). BAG-1, a negative regulator of Hsp70 chaperone activity, uncouples nucleotide hydrolysis from substrate release. *EMBO J.*, **17**, 6871.

37. Hohfeld, J. (1998). Regulation of the heat shock conjugate Hsc70 in the mammalian cell: the characterization of the anti-apoptotic protein BAG-1 provides novel insights. *Biol. Chem.*, **379**, 269.

38. Prapapanich, V., Chen, S., and Smith, D. F. (1998). Mutation of Hip's carboxy-terminal region inhibits a transitional stage of progesterone receptor assembly. *Mol. Cell. Biol.*, **18**, 944.

39. Kanelakis, K. C., Morishima, Y., Dittmar, K. D., Galigniana, M. D., Takayama, S., Reed, J. C., and Pratt, W. B. (1999). Differential effects of the hsp70-binding protein BAG-1 on gluco-corticoid receptor folding by the hsp90-based chaperone machinery. *J. Biol. Chem.*, **274**, 34134.

40. Nair, S. C., Rimerman, R. A., Toran, E. J., Chen, S., Prapapanich, V., Butts, R. N., and Smith, D. F. (1997). Molecular cloning of human FKBP51 and comparisons of immunophilin interactions with Hsp90 and progesterone receptor. *Mol. Cell. Biol.*, **17**, 594.

41. Brugge, J. S. (1986). Interaction of the Rous sarcoma virus protein pp60src with the cellular proteins pp50 and pp90. *Curr. Top. Microbiol. Immunol.*, **123**, 1.

42. Stepanova, L., Leng, X., Parker, S. B., and Harper, J. W. (1996). Mammalian p50Cdc37 is a protein kinase-targeting subunit of Hsp90 that binds and stabilizes Cdk4. *Genes Dev.*, **10**, 1491.

43. Kimura, Y., Rutherford, S. L., Miyata, Y., Yahara, I., Freeman, B. C., Yue, L., Morimoto, R. I., and Lindquist, S. (1997). Cdc37 is a molecular chaperone with specific functions in signal transduction. *Genes Dev.*, **11**, 1775.

44. Ma, Q. and Whitlock, J. P., Jr. (1997). A novel cytoplasmic protein that interacts with the Ah receptor, contains tetratricopeptide repeat motifs, and augments the transcriptional response to 2,3,7,8-tetrachlorodibenzo-p-dioxin. *J. Biol. Chem.*, **272**, 8878.

45. Carver, L. A. and Bradfield, C. A. (1997). Ligand-dependent interaction of the aryl hydrocarbon receptor with a novel immunophilin homolog in vivo. *J. Biol. Chem.*, **272**, 11452.

46. Meyer, B. K., Pray-Grant, M. G., Vanden Heuvel, J. P., and Perdew, G. H. (1998). Hepatitis B virus X-associated protein 2 is a subunit of the unliganded aryl hydrocarbon receptor core complex and exhibits transcriptional enhancer activity. *Mol. Cell. Biol.*, **18**, 978.

47. Carver, L. A., LaPres, J. J., Jain, S., Dunham, E. E., and Bradfield, C. A. (1998). Characterization of the Ah receptor-associated protein, ARA9. *J. Biol. Chem.*, **273**, 33580.

48. Nair, S., Toran, E., Rimerman, R., Hjermstad, S., Smithgall, T., and Smith, D. (1995). A pathway of multi-chaperone interactions common to diverse regulatory proteins: estrogen receptor, Fes tyrosine kinase, heat shock transcription factor Hsf1, and the aryl hydrocarbon receptor. *Cell Stress Chap.*, **1**, 237.

49. Barent, R. L., Nair, S. C., Carr, D. C., Ruan, Y., Rimerman, R. A., Fulton, J., Zhang, Y., and Smith, D. F. (1998). Analysis of FKBP51/FKBP52 chimeras and mutants for Hsp90 binding and association with progesterone receptor complexes. *Mol. Endocrinol.*, **12**, 342.

50. Bose, S., Weikl, T., Bugl, H., and Buchner, J. (1996). Chaperone function of Hsp90-associated proteins. *Science*, **274**, 1715.

51. Freeman, B. C., Toft, D. O., and Morimoto, R. I. (1996). Molecular chaperone machines: chaperone activities of the cyclophilin Cyp-40 and the steroid aporeceptor-associated protein p23. *Science*, **274**, 1718.

52. Brzozowski, A. M., Pike, A. C., Dauter, Z., Hubbard, R. E., Bonn, T., Engstrom, O., *et al.* (1997). Molecular basis of agonism and antagonism in the oestrogen receptor. *Nature*, **389**, 753.

53. Tanenbaum, D. M., Wang, Y., Williams, S. P., and Sigler, P. B. (1998). Crystallographic comparison of the estrogen and progesterone receptor's ligand binding domains. *Proc. Natl. Acad. Sci. USA*, **95**, 5998.

54. Shiau, A. K., Barstad, D., Loria, P. M., Cheng, L., Kushner, P. J., Agard, D. A., and Greene, G. L. (1998). The structural basis of estrogen receptor/coactivator recognition and the antagonism of this interaction by tamoxifen. *Cell*, **95**, 927.

55. Kumar, R. and Thompson, E. B. (1999). The structure of the nuclear hormone receptors. *Steroids*, **64**, 310.

56. Nemoto, T., Ohara-Nemoto, Y., Sato, N., and Ota, M. (1993). Dual roles of 90-kDa heat shock protein in the function of the mineralocorticoid receptor. *J. Biochem.*, **113**, 769.

57. Fang, Y., Fliss, A. E., Robins, D. M., and Caplan, A. J. (1996). Hsp90 regulates androgen receptor hormone binding affinity in vivo. *J. Biol. Chem.*, **271**, 28697.

58. Segnitz, B. and Gehring, U. (1997). The function of steroid hormone receptors is inhibited by the hsp90-specific compound geldanamycin. *J. Biol. Chem.*, **272**, 18694.

59. Bamberger, C. M., Wald, M., Bamberger, A. M., and Schulte, H. M. (1997). Inhibition of

mineralocorticoid and glucocorticoid receptor function by the heat shock protein 90-binding agent geldanamycin. *Mol. Cell. Endocrinol.*, **131**, 233.

60. Picard, D., Khursheed, B., Garabedian, M. J., Fortin, M. G., Lindquist, S., and Yamamoto, K. R. (1990). Reduced levels of hsp90 compromise steroid receptor action in vivo. *Nature*, **348**, 166.

61. Holley, S. J. and Yamamoto, K. R. (1995). A role for Hsp90 in retinoid receptor signal transduction. *Mol. Biol. Cell*, **6**, 1833.

62. Whitesell, L. and Cook, P. (1996). Stable and specific binding of heat shock protein 90 by geldanamycin disrupts glucocorticoid receptor function in intact cells. *Mol. Endocrinol.*, **10**, 705.

63. Czar, M. J., Galigniana, M. D., Silverstein, A. M., and Pratt, W. B. (1997). Geldanamycin, a heat shock protein 90-binding benzoquinone ansamycin, inhibits steroid-dependent translocation of the glucocorticoid receptor from the cytoplasm to the nucleus. *Biochemistry*, **36**, 7776.

64. Xu, Y., Singer, M. A., and Lindquist, S. (1999). Maturation of the tyrosine kinase c-src as a kinase and as a substrate depends on the molecular chaperone Hsp90. *Proc. Natl. Acad. Sci. USA*, **96**, 109.

65. Whitesell, L., Shifrin, S. D., Schwab, G., and Neckers, L. M. (1992). Benzoquinonoid ansamycins possess selective tumoricidal activity unrelated to src kinase inhibition. *Cancer Res.*, **52**, 1721.

66. Lemon, B. D. and Freedman, L. P. (1999). Nuclear receptor cofactors as chromatin re-modelers. *Curr. Opin. Genet. Dev.*, **9**, 499.

67. McKenna, N. J., Xu, J., Nawaz, Z., Tsai, S. Y., Tsai, M. J., and O'Malley, B. W. (1999). Nuclear receptor coactivators: multiple enzymes, multiple complexes, multiple functions. *J. Steroid Biochem. Mol. Biol.*, **69**, 3.

68. Cadepond, F., Schweizer-Groyer, G., Segard-Maurel, I., Jibard, N., Hollenberg, S. M., Giguere, V., *et al.* (1991). Heat shock protein 90 as a critical factor in maintaining gluco-corticosteroid receptor in a nonfunctional state. *J. Biol. Chem.*, **266**, 5834.

69. Picard, D. (1994). Regulation of protein function through expression of chimaeric proteins. *Curr. Opin. Biotechnol.*, **5**, 511.

70. Cotto, J. J. and Morimoto, R. I. (1999). Stress-induced activation of the heat-shock re-sponse: cell and molecular biology of heat-shock factors. *Biochem. Soc. Symp.*, **64**, 105.

71. Zou, J., Guo, Y., Guettouche, T., Smith, D. F., and Voellmy, R. (1998). Repression of heat shock transcription factor HSF1 activation by HSP90 (HSP90 complex) that forms a stress-sensitive complex with HSF1. *Cell*, **94**, 471.

72. Murakami, Y., Uehara, Y., Yamamoto, C., Fukazawa, H., and Mizuno, S. (1991). Induction of hsp 72/73 by herbimycin A, an inhibitor of transformation by tyrosine kinase oncogenes. *Exp. Cell. Res.*, **195**, 338.

73. Hegde, R. S., Zuo, J., Voellmy, R., and Welch, W. J. (1995). Short circuiting stress protein expression via a tyrosine kinase inhibitor, herbimycin A. *J. Cell. Physiol.*, **165**, 186.

74. Morimoto, R. I. and Santoro, M. G. (1998). Stress-inducible responses and heat shock proteins: new pharmacologic targets for cytoprotection. *Nat. Biotechnol.*, **16**, 833.

75. Baler, R., Welch, W. J., and Voellmy, R. (1992). Heat shock gene regulation by nascent polypeptides and denatured proteins: hsp70 as a potential autoregulatory factor. *J. Cell Biol.*, **117**, 1151.

76. Abravaya, K., Myers, M. P., Murphy, S. P., and Morimoto, R. I. (1992). The human heat shock protein hsp70 interacts with HSF, the transcription factor that regulates heat shock gene expression. *Genes Dev.*, **6**, 1153.

77. Morimoto, R. I. (1998). Regulation of the heat shock transcriptional response: cross talk between a family of heat shock factors, molecular chaperones, and negative regulators. *Genes Dev.*, **12**, 3788.

78. Satyal, S. H., Chen, D., Fox, S. G., Kramer, J. M., and Morimoto, R. I. (1998). Negative regulation of the heat shock transcriptional response by HSBP1. *Genes Dev.*, **12**, 1962.

79. Shi, Y., Mosser, D. D., and Morimoto, R. I. (1998). Molecular chaperones as HSF1-specific transcriptional repressors. *Genes Dev.*, **12**, 654.

80. Bharadwaj, S., Ali, A., and Ovsenek, N. (1999). Multiple components of the HSP90 chaperone complex function in regulation of heat shock factor 1 in vivo. *Mol. Cell. Biol.*, **19**, 8033.

81. Ali, A., Bharadwaj, S., O'Carroll, R., and Ovsenek, N. (1998). HSP90 interacts with and regulates the activity of heat shock factor 1 in Xenopus oocytes. *Mol. Cell. Biol.*, **18**, 4949.

82. Baughman, G., Lesley, J., Trotter, J., Hyman, R., and Bourgeois, S. (1992). Tcl-30, a new T cell-specific gene expressed in immature glucocorticoid-sensitive thymocytes. *J. Immunol.*, **149**, 1488.

83. Baughman, G., Wiederrecht, G. J., Campbell, N. F., Martin, M. M., and Bourgeois, S. (1995). FKBP51, a novel T-cell-specific immunophilin capable of calcineurin inhibition. *Mol. Cell. Biol.*, **15**, 4395.

84. Chrousos, G. P., Renquist, D., Brandon, D., Eil, C., Pugeat, M., Vigersky, R., *et al.* (1982). Glucocorticoid hormone resistance during primate evolution: receptor-mediated mechanisms. *Proc. Natl. Acad. Sci. USA*, **79**, 2036.

85. Brandon, D. D., Markwick, A. J., Chrousos, G. P., and Loriaux, D. L. (1989). Glucocorticoid resistance in humans and nonhuman primates. *Cancer Res.*, **49**, 2203s.

86. Brandon, D. D., Kendall, J. W., Alman, K., Tower, P., and Loriaux, D. L. (1995). Inhibition of dexamethasone binding to human glucocorticoid receptor by New World primate cell extracts. *Steroids*, **60**, 463.

87. Reynolds, P. D., Pittler, S. J., and Scammell, J. G. (1997). Cloning and expression of the glucocorticoid receptor from the squirrel monkey (Saimiri boliviensis boliviensis), a glucocorticoid-resistant primate. *J. Clin. Endocrinol. Metab.*, **82**, 465.

88. Reynolds, P. D., Ruan, Y., Smith, D. F., and Scammell, J. G. (1999). Glucocorticoid resistance in the squirrel monkey is associated with overexpression of the immunophilin FKBP51. *J. Clin. Endocrinol. Metab.*, **84**, 663.

8 | Molecular chaperone systems in the endoplasmic reticulum

MARC F. PELLETIER, JOHN J. M. BERGERON, and DAVID Y. THOMAS

1. Introduction

The endoplasmic reticulum (ER) is an organelle defined microscopically as a network of saccular and tubular membranes in continuity with the nuclear envelope. This definition can be further refined into a distinction between the rough endoplasmic reticulum to which ribosomes are attached, and the smooth endoplasmic reticulum which consists of a network of ribosome-free tubules and is the site of exit of secretory cargo vesicles. The ER can also be defined functionally as the site of lipid metabolism, carbohydrate metabolism, drug detoxification, and the entry point for proteins into the secretory pathway. Proteins that are secreted from cells, and proteins which are destined for cellular membranes, are synthesized on the ribosomes of the rough endoplasmic reticulum and translocated into the lumen of the ER, where they are glycosylated, become folded, disulfide bonds are formed, and some multiprotein complexes are assembled. Before proteins are transported out of the ER they are subjected to a 'quality control' process which assesses their state of folding and then, if they are correctly folded, allows them to exit the ER by packaging into transport vesicles. Proteins that are not folded correctly can have several fates. They are usually retained in the ER for further cycles of the quality control process or they are retrotranslocated out of the ER and degraded in the cytosol by the proteasomal system, or they can accumulate in the ER as aggregates.

Many of these processes in the ER are controlled by proteins which are collectively termed 'molecular chaperones'. Christian Anfinsen won the Nobel Prize in 1972 for his demonstration of a decade earlier that the correct folding of ribonuclease A is determined solely by the primary sequence of the protein. That is, the purified enzyme can be denatured and then can refold *in vitro* to restore enzymatically active ribonuclease. However, he also showed subsequently that the rate of folding could be increased closer to physiological rates by the addition of the ER enzyme protein disulfide isomerase (PDI). Hence arose the idea of one protein assisting another to attain its correct folding. This concept was further refined by others and the term 'molecular chaperone' was used by Laskey and co-workers to describe the activity of a protein involved in the assembly of nucleoplasmin (1). The current working definition of molecular chaperones is that they can recognize unfolded proteins and promote protein folding by preventing irreversible aggregation with other proteins (2). They associate with a protein during the folding process but during folding or when folding is complete the chaperones disassociate and participate in the folding of new proteins. Another implication of the recognition of unfolded proteins by molecular chaperones is that they are indiscriminate and can interact with any un-folded protein and therefore any specialization arises from other mechanisms, for example cellular location. The results of studies of molecular chaperone systems in the cytosol show that to accomplish their functions there is a requirement for differ-ent types of molecular chaperone. In the ER, in addition to a variety of molecular chaperones there are also folding enzymes such as PDIs and protein prolyl isomer-ases (PPIs), and the lectin-like chaperones calnexin and calreticulin.

We review here what is known about some selected functions of the ER, the Calnexin Cycle and protein quality control, retrotranslocation of proteins out of the ER, and signalling to the nucleus of the presence of unfolded proteins in the ER. We have not reviewed another ER 'molecular machine', which is known in some detail, that is involved in the translocation of proteins into the ER. Since there have been several comprehensive reviews which catalogue the molecular chaperones and folding enzymes of the ER, we will only discuss in detail those which fit into this functional framework. Fig. 1 provides the reader with a flow chart, depicting the structural and dynamic protein interactions discussed.

2. The Calnexin Cycle

2.1 Discovery of calnexin, and its characterization as an ER lectin-like chaperone

Calnexin provides a link between N-glycosylation and protein folding. The oligo-saccharide $Glc_3Man_9GlcNAc_2$ is synthesized stepwise on the cytosolic surface of the ER on an isoprenoid lipid dolicholpyrophosphate and is transferred into the ER, and attached by oligosaccharyl transferase to nascent polypeptide chains on the asparagine residue in the sequence motif Asn-X-Ser/Thr. Then the glucose residues are sequentially removed by ER glucosidases I and II (Gluc I and Gluc II in Fig. 1) (3). This seemingly redundant process is explained by the recognition of the $Glc_1Man_9GlcNAc_2$ glycoform of glycoproteins by calnexin and also by its ER lumenal homologue calreticulin. Some of the key components of the Calnexin Cycle have been identified and a model of how it functions is emerging. But there remain several questions concerning its mechanism.

Since the cDNA cloning in 1991 the study of calnexin has been a remarkably inten-sive area of endeavour. In a screen for ER membrane proteins which are phosphoryl-ated by endogenous kinases (4), Ikuo Wada isolated a complex of four proteins, one

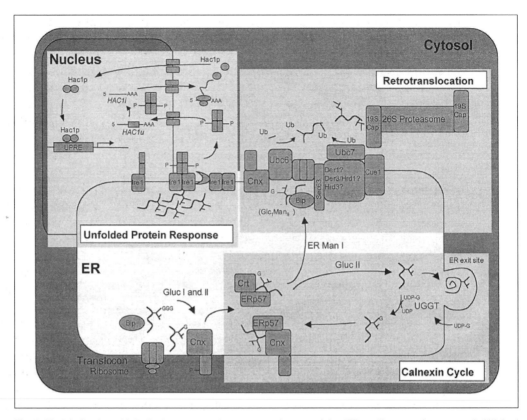

Fig. 1 Models for three biological processes that are constituents of the 'ER quality control apparatus'. N-linked glycosylation of nascent secretory or membrane proteins with a $Glc_3Man_9GlcNAc_2$ oligosaccharide occurs during co-translational translocation into the endoplasmic reticulum through the translocon channel. This subjects the glycoproteins to a recently characterized and highly effective quality control apparatus. This 'molecular machine' prevents the progression of proteins that have not yet attained, or cannot attain, native conformation through the secretory pathway. The interaction with the quality control apparatus is initiated by the trimming of the oligosaccharide to the $Glc_1Man_9GlcNAc_2$ form (through the action of the glucosidases I and II), which permits calnexin/calreticulin binding and thus entry into the Calnexin Cycle (lower-right shaded panel). Prolonged and repeated interactions with calnexin/calreticulin permits the slow acting ER mannosidase I to hydrolyse the glycan to a $Glc_1Man_8GlcNAc_2$ form, preventing its release from calnexin/calreticulin by glucosidase II. Calnexin is then thought to bring the misfolded proteins back to the translocon for retrotranslocation, polyubiquitination by ubiquitin conjugating enzymes (UBCs), and degradation by the 26S proteasome (upper-right shaded panel). If unfolded proteins accumulate in the ER, a signal is transmitted from the ER to the nucleus which initiates the transcription of ER chaperones that are under the transcriptional control of an unfolded protein response element (UPRE) (upper-left shaded panel).

of which had similarity with an ER lumenal protein termed calreticulin. This had been originally identified as a high affinity calcium binding protein of the sarcoplasmic reticulum (5). This membrane protein shared some properties such as calcium binding with calreticulin and the name calnexin was suggested by Professor A. Schachter from the History Department of McGill University. Studies on the kinetics of the assembly of MHC1 β microglobulin in the ER had identified a protein of 88 kDa which associates with unassembled MHC1 heavy chain but is itself retained in the ER (6). Upon analysis, this 88 kDa protein proved to be calnexin and this provided the first indication that it might function as a molecular chaperone (7). This was also tested in mammalian cells after the development of antibodies which could immunoprecipitate calnexin. These experiments (8) used HepG2 cells derived from human embryonic liver which have a well characterized set of secreted proteins. Pulse-chase experiments showed secretory glycoproteins complexed with calnexin, and that during the chase they dissociated and continued through the secretory pathway. However, it was clear that not all proteins interact with calnexin, for example, human serum albumin, an abundant secretory non-glycoprotein in HepG2 cells does not interact with calnexin whereas the related glycoprotein α-fetoprotein, which is also expressed in HepG2 cells, does bind. This suggested that calnexin binds only glycoproteins and this was confirmed with HepG2 cells labelled in the presence of the glycosylation inhibitor tunicamycin where no glycoproteins bind with calnexin. It was also established that calnexin recognizes glycoproteins from studies of the effects of the glucosidase II inhibitor bromoconduritol and the retention of glycoproteins in treated human hepatic cells, and a prescient suggestion was made that there may be a receptor in the ER which recognizes $Glc_1Man_9GlcNAc_2$ and prevents secretion of glycoproteins (9). Further experiments with inhibitors of the glucosidases confirmed that the $Glc_1Man_9GlcNAc_2$ glycoform was the most likely to be retained calnexin and that calnexin is the anticipated receptor. *In vitro*, calnexin was shown to have a preference and weak affinity for free $Glc_1Man_9GlcNAc_2$ oligosaccharides (10).

However, the original experiments of Ou *et al.* (8) also used a standard experimental method of unfolding proteins *in vivo* by treating cells with a proline analogue L-azetidine-2-carboxylic acid (AZC), which is incorporated into proteins and prevents them from folding correctly. Analysis of secretory proteins in cells with AZC showed that they were retained for prolonged periods and thus calnexin was also assumed to recognize unfolded proteins. On the basis of the results of *in vivo* experiments from many laboratories, there has been controversy on whether calnexin acts as a lectin or whether it can recognize unfolded proteins.

2.2 The lectin/true molecular chaperone debate

Studies of the association *in vitro* of calnexin and calreticulin with defined substrates offer a way of answering the question of whether they recognize the oligosaccharide moiety, the protein, or both. Purified and well-characterized components for the *in vitro* systems are available. The form of calnexin used in these studies is the ER lumenal portion that has been shown to have properties similar to full length calnexin in that it interacts identically with HIV-1 gp120 in a reconstructed calnexin interaction system in insect cells, and *in vitro* in terms of calcium binding, protease resistance, and ATP binding it behaves like the intact calnexin molecule (11, 12). In addition the lumenal domain of calnexin recognizes free $Glc_1Man_9GlcNAc_2$ oligosaccharide and not the $Glc_2Man_9GlcNAc_2$ or $Glc_3Man_9GlcNAc_2$ glycoforms (13).

There is also additional support for the assumption that the lumenal domain of cal-nexin contains all the necessary elements for the recognition of glycoproteins from the experiments of Ikuo Wada, who has shown that calreticulin, the closely related lumenal ER protein, can be attached to the transmembrane domain and cytosolic tail of calnexin and now binds some 'calnexin preferred' glycoproteins (14). Thus, by these criteria the lumenal domain of calnexin used in these *in vitro* experiments has all the necessary elements for functionally binding glycoproteins.

The availability of suitable glycoprotein substrates is critical for the *in vitro* ex-periments. Most secreted glycoproteins have lost the high mannose oligosaccharide and the outer chains have been extensively modified by enzymes during transit through the Golgi. There are some well-documented cases where a minor portion of a glycoprotein carries a high mannose oligosaccharide such as thyroglobulin and RNaseB, but recent results from studies on the glycosylation process in yeast and the knowledge that some of the genes responsible are not essential for growth in this organism makes possible the engineering of glycoproteins with specific glycoforms. Thus yeast strains have been constructed which secrete glycoproteins with uniform high mannose oligosaccharides of various forms (15). For the experiments described below, preparations of RNaseB that were enriched in the $Man_9GlcNAc_2$ were used. The *in vitro* experiments took advantage of the considerable amount of information which has accumulated on the folding of this protein and its non-glycosylated form RNaseA since the pioneering work of Anfinsen. To construct the $Glc_1Man_9GlcNAc_2$ glycoform, the ER enzyme UDP-glucose:glycoprotein glucosyltransferase (UGGT) was used with proteins bearing Man_9 N-linked oligosaccharides. This enzyme was first described by Armando Parodi and his colleagues as an activity that can transfer a glucose moiety from UDP-glucose to a $Man_9GlcNAc_2$ acceptor on an unfolded glycoprotein (15, 16). Thus UGGT has a unique role in the cellular glycosylation process in that it is reversible and points to a potential role in the quality control of glycoproteins in that it is specific for unfolded proteins carrying $Man_9GlcNAc_2$ glycans. The experimental design was to glucosylate unfolded RNaseB with UGGT and UDP-glucose which is [^{3}H]-labelled in the glucose moiety and then produce the unfolded, the fully folded, and the partially folded intermediates by using the estab-lished refolding protocols for this enzyme. Measurements of binding to calnexin showed that fully folded, unfolded, and partially folded forms of RNaseB all bind calnexin (17). Thus calnexin (and calreticulin) are defined by these *in vitro* experiments as lectins. The provisos of this seemingly definitive experiment are that although the affinity constants of calnexin for each of the folding forms of $Glc_1Man_9GlcNAc_2$-RNaseB were not individually measured it can be inferred from the experimental design that they must differ less than an order of magnitude. In apparent contra-diction, *in vivo* experiments (8, 11) have shown that the form of proteins, in these cases transferrin and HIV gp120 that are bound to calnexin are unfolded. Confirming these results of Zapun *et al.* (17), Rodan *et al.* (18) showed that in an *in vitro* microsomal system that RNaseB which had been mutated to have two N-linked glycosylation sites bound to calnexin independently of the folding status of the enzyme. Thus the experimental results confirm that for RNaseB, a classical protein folding substrate, the sole determinant of calnexin binding is the correct oligosaccharide moiety Glc_1Man_9. However, there are some aspects of this result that are worth considering. One of the advantages of using RNaseB in protein folding studies is that it does not have extensive hydrophobic regions and does not aggregate when denatured, and perhaps studies with proteins with more extensive hydrophobic regions will reveal other functions of calnexin. Recently, support for this view has been provided by the demonstration that calnexin can function as a molecular chaperone with classical substrates used in protein folding studies which are not glycoproteins. This surprising result comes from a series of experiments performed by David Williams and colleagues (19).

In these experiments calnexin, when added in stoichiometric amounts, protected the glycoprotein soybean agglutinin (SBA), and the nonglycosylated proteins citrate synthase (CS) and malate dehydrogenase (MDH) against thermal aggregation as measured by changes in light scattering. Interestingly, they examined the role of the oligosaccharide by assaying various glycoforms of SBA and found that the monoglucosylated form ($Glc_1Man_9GlcNAc_2$) was best protected from thermal aggregation by calnexin while it required eightfold more calnexin to equivalently protect the $Man_9GlcNAc_2$ form, and 20-fold more for SBA with only one GlcNAc residue. However, definitive evidence for interaction with the polypeptide portions of the unfolded substrates came from experiments with non-glycosylated proteins CS and MDH. Not only did calnexin protect these unfolded proteins from aggregation, calnexin maintained CS in a folding competent state which allowed for the reactivation of its enzymatic activity. The property of maintaining a protein in a folding competent state had previously been shown for the chaperones Hsp90, Hsp70, and the small heat shock protein chaperones (20–22, and see Chapters 1, 5, and 7), and this result suggests that calnexin may also act as a 'bona fide' molecular chaperone. Moreover, calnexin associated only with the unfolded form of MDH to produce a complex that was hypersensitive to trypsin, which is most effective on unfolded proteins, while no association could be detected with enzymatically active fully folded MDH. Thus calnexin can discriminate between folded and non-folded substrates, satisfying a criterion of a molecular chaperone. The conclusion from these *in vitro* experiments, which was also found for calreticulin by Saito *et al.* (23), suggests that calnexin may interact with the protein portion of its substrate in addition to the glycan. There is some caution needed in interpreting these results as *in vivo* the proteins that have been found complexed with calnexin are always glycoproteins. Based on these results, the model predicts that the oligosaccharide is the ligand that initiates and stabilizes the subsequent polypeptide interactions with calnexin. Thus calnexin can exhibit both the properties of a lectin and may also act as a molecular chaperone.

2.3   The calnexin/ERp57 protein folding complex

What happens to glycoproteins while they are bound to calnexin is not clear but there are some clues to possible mechanisms. Glycoproteins bound to calnexin and calreticulin interact with a member of the protein disulfide isomerase family termed ERp57 (24). It was found using *in vitro* cross-linking, that as well as calnexin and calreticulin, ERp57 can be cross-linked in dog pancreatic microsomes to a glycoprotein precursor *N*-glycosylated with the $Glc_1Man_9GlcNAc_2$ form of oligosaccharide (24). An interpretation of this data is that ERp57 is itself either a lectin or that it associates with calnexin and calreticulin which are themselves bound to glycoproteins. *In vitro* studies established that there was no enhancement of the rate of folding of monoglucosylated ($Glc_1Man_9GlcNAc_2$) RNaseB by ERp57 alone (25). Based on the crosslinking results and the similarity of ERp57 with PDI, we speculated that it might be a disulfide isomerase for protein substrates bound to calnexin. Using an *in vitro* system that was developed to determine the functions of calnexin and calreticulin, using recombinant UGGT (15) and [^{3}H]UDP-glucose to produce labelled monoglucosylated RNaseB, it was shown that ERp57 does not recognize this oligosaccharide directly. However, the protein disulfide isomerase activity of ERp57 on the refolding of monoglucosylated RNaseB is greatly enhanced in the presence of calnexin, in contrast to the result with PDI whose activity on monoglucosylated RNaseB is decreased by the presence of calnexin (25). Calnexin had no effect on the refolding of non-monoglucosylated RNaseB catalysed by either PDI or ERp57. Thus, it was concluded that ERp57 is a protein disulfide isomerase dedicated to glycoproteins that are bound to calnexin and calreticulin. This result provides an interesting new insight into calnexin function and we can speculate that other PDI family members such as ERp72 and PDI itself may each have their ER molecular chaperone partners.

2.4 Quality control of protein folding for *N*-linked glycoproteins in the ER

Quality control within Calnexin Cycle takes place at the release of the glycoprotein substrates from calnexin following the action of glucosidase II, which removes the remaining glucose residue from the $Glc_1Man_9GlcNAc_2$ oligosaccharide. The process in itself is defined by the ability of the quality control apparatus to recognize and selectively bind proteins that are misfolded or have not yet attained their native conformation, and prevent their premature progression though the secretory pathway. Attaining native conformation or proper subunit assembly for protein complexes within the lumen of the ER has been shown in a number of cases to require cyclical interactions with both calnexin and calreticulin.

The protein folding sensor of the Calnexin Cycle is the ER enzyme UDP-glucose: glycoprotein glucosyltransferase (UGGT), which acts only on unfolded glycoproteins, reglucosylating them, allowing for further binding to calnexin and calreticulin. Demonstrating this cyclical process under physiological conditions was achieved by using novel and innovative experimental approaches by both Wada *et al.* (26) and Cannon and Helenius (27).

2.4.1 The role of de/reglucosylation in quality control and protein folding.

The importance of deglucosylation and reglucosylation in the quality control of protein folding for a glycoprotein has been demonstrated in a series of ingenious experiments (26). As a glycoprotein substrate, the iron binding protein transferrin, which has two homologous domains, 19 pairs of cysteines that form disulfide bridges, and two N-linked glycans was selected. HepG2 cells were pulse labelled with [^{35}S]methionine, and microsomes were prepared on a discontinuous sucrose gradient. The process of transferrin folding was followed by non-reducing SDS-PAGE. In this gel system, reduced unfolded transferrin migrates slowly as a diffuse band, while fully folded transferrin moves more rapidly as a sharper band. Raising the temperature of the cell preparations to 43 °C resulted in an increase in high molecular weight disulfide-linked transferrin aggregates at the top of the gel, and they could control transferrin misfolding. After labelling microsomes with [^{3}H]UDP-glucose at 37 °C, followed by an unlabelled UDP-glucose chase, the association of transferrin with both calnexin and calreticulin was determined by co-immunoprecipitation. The half lives of glucose removal from the calnexin and calreticulin complexes were similar (5.8 minutes and 4.7 minutes, respectively). Alternatively, by examining the kinetics of substrate dissociation from calnexin and calreticulin using [^{35}S]methionine labelled microsomes (which labels the polypeptide moiety of the glycoprotein), it was shown that the addition of UDP-glucose prolonged their association with transferrin. This result demonstrated that transferrin undergoes cyclical interactions with these lectin-like chaperones. It was also noticed that at increased temperature the addition of UDP-glucose suppressed the formation of disulfide-linked aggregates and promoted transferrin folding. Thus at 43°C folding intermediates and folded transferrin were only faintly detectable in the absence of UDP-glucose. The significant contribution to folding efficiencies by the addition of UDP-glucose at 43 °C demonstrated that not only are cyclical interaction occurring under these physiological conditions, but also that the actions of de- and reglucosylation of the ER enzymes glucosidase II and UGGT promote multiple interactions with calnexin and calreticulin, a process that both prevents protein misfolding and facilitates the attainment of native protein conformation.

While work by Wada *et al.* (26) and Zapun *et al.* (17) also gave insight on the role of glucose trimming of N-linked glycans, and its effect on mediating cyclical interactions with the lectin-like chaperones calnexin and calreticulin, they used either a reconstituted system involving microsomal preparations or a completely *in vitro* system. Cannon and Helenius (27) developed a system that used a mutant form of glycoprotein from the ts045 vesicular stomatitis virus. The mutant VSV G protein has a thermoreversible folding defect that can be manipulated *in vivo* by shifting the temperature from permissive (30 °C) to nonpermissive temperatures (40 °C). VSV G protein is a type I transmembrane glycoprotein with two N-linked glycans, and while the mutant form folds rapidly and moves on the Golgi apparatus at 30 °C, at 40 °C it fails to acquire a complete set of disulfide bonds and is retained in the ER. In these experiments, VSV G folding was monitored by immunoprecipitation with conformation specific antisera.

To examine the effects of reglucosylation on quality control, cells were infected with ts045 VSV, pulsed labelled, and then chased at permissive temperature to allow G protein to fold correctly and be released from calnexin. Upon a shift to non-permissive temperature, VSV G protein unfolded and rebound to calnexin. The rates of unfolding, reglucosylation, and rebinding to calnexin at nonpermissive temperatures were examined. Similar kinetics were observed for VSV G unfolding and reglucosylation by UGGT, while, quite surprisingly, the rebinding to calnexin was bimodal, with a rapid interaction phase being followed by a slower second phase. This could be explained by the intracellular distribution of VSV G. Under permissive conditions, VSV G exhibited a faintly reticular distribution with punctate spots that co-localized with UGGT at putative ER exit sites. Alternatively, when under conditions of shift to nonpermissive temperature, the ER staining was more prominent with a diminished spotty pattern. Hence, as VSV G protein changed from a folded to unfolded state, it returned from the ER exit sites. Cannon and Helenius (27) showed *in vivo* that glycoproteins are redirected back for further binding to the lectin-like chaperones under conditions that induce protein unfolding. It was also noted that the putative ER exit sites were devoid of calnexin, and thus they ascribed the quality control function of the Calnexin Cycle to a late ER-exit site location. Moreover, both Wada *et al.* (26) and Cannon and Helenius (27) demonstrated the importance of the $Glc_1Man_9GlcNAc_2$ form of the oligosaccharide in mediating calnexin–substrate interactions, suggesting that calnexin acts primarily as a lectin.

3. Retrotranslocation of proteins from the ER

3.1 ER-associated degradation and genetic disease

There are components of the Calnexin Cycle that can discriminate between folded and unfolded proteins, and there is also an associated quality control system that allows correctly folded proteins to be exported to subsequent cellular compartments. Misfolded secretory proteins, however, which are usually the consequence of mutations or their over-expression, can have several fates. The α1PiZ allele of α1-antitrypsin is the underlying genetic defect of the most common genetic liver disease in children and results in a protein that can accumulates together with other ER resident proteins within distinct regions of the ER termed 'Russell Bodies'. The accumulation of protein aggregates is hepatotoxic, causing liver cirrhosis, while the ensuing serum depletion of α1-antitrypsin can lead to emphysema, in which alveolar destruction occurs as the result of the inability to block neutrophil elastase activity in the lungs (28). In the case of cystic fibrosis, the most common mutation (CFTRΔF508) in the gene that encodes the cystic fibrosis transmembrane conductance regulator, which is a plasma membrane chloride channel, leads to a prolonged association with calnexin and subsequent degradation of almost all the mutant CFTR protein. The maturation of wild-type CFTR is, also, relatively inefficient, with 75% of the wild-type form being degraded, versus 99% of the mutant form. However, the mutant protein has been shown to form an active channel both *in vitro* and in cells cultured under conditions of reduced temperatures or over-expression, suggesting that the pathology of the disease may be based largely on trafficking (29–32). Many other protein trafficking diseases such as familial hypercholesterolemia, Tay-Sachs, and congenital sucrase-isomaltase deficiency have similar pathologies resulting in the quality control of N-linked glycoproteins, where the corresponding proteins are unable to fold or assemble correctly and become degraded by the ER-associated degradation pathway (ERAD).

3.2 Where does ER-associated degradation occur, in the ER or cytosol?

The degradation pathway used to eliminate folding incompetent proteins and alleviate the cellular stresses incurred by their accumulation was originally thought to reside in the ER (33). But as suggested by Ron Kopito in a later review: 'The abundance of unfolded and partially folded polypeptide chains that are highly susceptible to proteolysis is difficult to reconcile with the presence in the same compartment of an aggressive proteolytic apparatus' (34). Experimental evidence of a cytosolically located proteolytic apparatus became apparent with the work of Sommer and Jentsch (35) who quite serendipitously made the first link between the degradation of ER membrane proteins and the ubiquitin–proteasome degradation pathway.

Sommer and Jentsch characterized a novel integral membrane ubiquitin-conjugating enzyme (UBC6p) from *Saccharomyces cerevisiae* that was localized on the ER membrane, with its catalytic domain in a cytosolic orientation. Ubiquitin-conjugating enzymes (UBCs) target proteins for degradation by the 26S proteasome by covalently attaching a 76 residue ubiquitin protein to lysine residues on substrate proteins (36). The initial ubiquitin molecule itself becomes ubiquitinated and this processes continues, leading to multiple ubiquitin chains that form a degradation signal. The 26S proteasome is composed of the 20S core catalytic complex which forms a multi-subunit cylindrical structure with multiple proteolytic activities that include tryptic-, chymotryptic-, and post-glutamyl peptidyl hydrolytic-like specificities. The cylinder is capped by two 19S regulatory complexes at each end that are thought to mediate substrate recognition through the interaction of specific subunits with the poly-ubiquitin chain.

Based on the localization of the UBC6p protein, Sommer and Jentsch sought synthetic lethal phenotypes of UBC mutants with mutants of the secretory pathway. Surprisingly, they found that the *ubc6Δ* mutant suppressed the *sec61* mutant phenotype. *SEC61* encodes a multi-spanning ER membrane subunit of heterotrimeric complex that also includes Sbh1p and Sss1p contributing to the translocon channel (37). In the *sec61 ubc6Δ* strain both translocation and glycosylation levels of the secretory proteins carboxypeptidase Y (CPY) and α-factor were restored to near wild-type levels. The *sec61* defect was thus shown to require active UBC6p. Since the mutant sec61p was found to be stable in *sec61 ubc6Δ* cells, it was concluded that the substrates for UBC6p-mediated degradation were membrane proteins, including proteins that interacted directly with *sec61*. These experiments provided the first conclusive evidence of a cytosolic degradation pathway associated with ER, at least for membrane proteins.

3.3 ERAD of membrane and soluble ER proteins

3.3.1 The proteasome degrades ER membrane proteins

A large number of reports from 1995 though 1996 suggested that ER protein degradation used the ubiquitin–proteasome pathway in the cytosol (38–42). Two illuminating papers published together in the same edition of *Cell* (39, 41) made significant contributions to our understanding of the pathology of cystic fibrosis by determining the protein degradation pathway that affects CFTR.

Jensen *et al.* (39) characterized the rapid turnover of the CFTR precursor and its maturation using peptide aldehyde protease inhibitors N-acetyl-L-leucinyl-L-leucinyl-L-norleucinal (ALLN), N-acetyl-L-leucinyl-L-leucinyl-L-methionyl (ALLM), and N-carbobenzoxyl-L-leucinyl-L-leucinyl-L-leucinyl (MG-132). ALLN was more effective at blocking the rapid degradation than ALLM, which agreed with their relative abilities to block protein degradation by purified proteasome preparations (43). Moreover, MG-132 completely blocked the ATP-dependent conversion of the precursor CFTR to the native folded form capable of transport from the ER, a process that was not affected by the other proteasome inhibitors suggesting that there is a second proteolytic pathway. Lactacystin, which is highly specific for the multiple activities of the proteasome (44) was the most effective in blocking turnover, and demonstrated that the proteasome is responsible for the degradation of the CFTR precursor.

Ward *et al.* (41) demonstrated the inhibitory effects of ALLN and lactacystin; they also noted the accumulation of polyubiquitinated CFTR. Upon examination of immunoblots of insoluble CFTR, they noticed a ladder of bands spaced at about 6.5 kDa, which is diagnostic of polyubiquitination. Several methods were used to characterize the polyubiquitinated CFTR. Firstly, they immunoprecipitated ΔF508 CFTR, ran a gel with the immunoprecipitate, and then immunoblotted with anti-ubiquitin. They only found ubiquitin immunoreactivity in the precipitations from cells expressing CFTR, and with pronounced accumulation of polyubiquitinated species in the presence of lactacystin. They purified polyubiquitinated CFTR by Ni^{2+} chelation chromatography of cells expressing a His_6 tagged form of ubiquitin. Ubiquitinated CFTR was observed in the eluate of columns loaded with either soluble or insoluble material from cells expressing both ΔF508 CFTR and His_6M-Ub. They also showed that the expression of a dominant negative ubiquitin (Ub_{K48R}), which prevents polyubiquitin chain formation on the Lys-48, attenuated the ubiquitination of CFTR. Together, these experiments demonstrated the polyubiquitination of CFTR *in vivo*, and are in agreement with the result of Jensen *et al.* (39) on its degradation by the ubiquitin-linked pathway.

3.3.2 Soluble lumenal proteins are also degraded by a cytosolic degradation pathway

Many membrane proteins of different cellular compartments share the same topology and their degradation appears to follow a common pathway. Non-membrane proteins of the ER (lumenal proteins) are compartmentalized such that they are apparently inaccessible to the 26S proteasomal system. The demonstration that lumenal and soluble secretory proteins follow the same degradation pathway as membrane proteins comes from results from experiments *in vitro* and *in vivo* using both yeast and mammalian cells (38, 40, 42).

Werner *et al.* (42) examined the ER associated degradation of mutant form of α1-antitryspin (α1-ATZ) heterologously expressed in *S. cerevisiae* mutants. When expressed in strains with a proteasomal mutant genetic background (*pre1-1 pre2-2*) deficient in chymotrypsin-like activity, the intracellular half-life of α1-ATZ increased threefold from 56 to 163 minutes compared with the wild-type. McCracken and Brodsky (45) developed an *in vitro* assay with purified ER (microsomes) and the mating pheromone precursor of *S. cerevisiae* (prepro-α factor, ppαF). When ppαF was added to these microsomal preparations, it was post-translationally translocated into the ER lumen and its location was verified by resistance to proteinase K digestion. A mutant form that had its *N*-linked glycosylation sites mutated (ΔGppαF) and was known to be subjected to ERAD was also translocated into the ER lumen, however, it was subsequently retrotranslocated out of the microsomes. It should be noted that in that same study, McCracken and Brodsky (45) found that degradation required the addition of cytosol, ATP, and the presence of calnexin. Using a similar system, Werner *et al.* (42) found that when cytosol of proteasomal mutant *pre1-1 pre2-2* was added, the degradation of ΔGppαF was significantly decreased. The role of the proteasome in degradation was confirmed with the proteasome specific inhibitors, decreasing degradation at least twofold.

Genetic screens using endogenous yeast misfolded proteins have been used to identify other components of the ERAD pathway (38). Hiller *et al.* (38) used a mutant form of soluble vacuolar protein carboxypeptidase Y (CPY*) that fails to reach the vacuole, is retained in the ER, and then degraded. In a screen for mutants defective in the ERAD of intracellular CPY*, they isolated the *der2-1* mutant which was found to be identical to a known gene *UBC7*. *UBC7* encodes a ubiquitin-conjugating enzyme responsible for the ATP-dependent addition of ubiquitin from an intermediate to a target protein. This prompted an investigation of other genes in the ubiquitination pathway and they demonstrated the role of Ubc6p, Ubc7p, polyubiquitination, and a functional proteasome in the degradation of CPY*.

3.4 Membrane and soluble secretory proteins must cross the ER membrane barrier, but how?

Although the role of the ubiquitin–proteasome degradation pathway in ERAD has been established, there is a question of how misfolded proteins selected by the ER quality control apparatus traverse the ER membrane? One possibility was that the Sec61 translocon channel may work in both translocation into the ER for unfolded glycoproteins and that they can be retrotranslocated through the same channel. This was demonstrated with surprising results from the study of two human cytomegalovirus (HCMV) proteins that participated in an intriguing host immunity response evasion strategy (46, 47). Expression of the HCMV genes *US2* or *US11* in cells lead to the rapid degradation ($t_{1/2} < 1$ minute) of newly synthesized class I heavy chains (MHCs) in a lactacystin dependent manner. That is, the MHCs were dislocated to the cytosol where deglucosylation occurred by a cytosolic N-glucanase, followed by proteasomal degradation (46). Breakdown intermediates found associated with the Sec61 complex suggested that retrograde transport occurred through the same protein conducting channel that allowed the original insertion of the MHCs. This provided the first evidence that the Sec61 translocon channel participates in both translocation of proteins into the ER and their retrotranslocation into the cytosol.

Genetic studies in yeast with CPY* confirm the role of the Sec61 translocon as the channel for retrotranslocation of proteins from the ER and also a role of BiP and its co-chaperone Sec63p (48). CPY* was stabilized in the three mutants of *SEC61*, *BiP*, and *SEC63*, and different mutations were found that blocked either the import function of the Sec61 translocon channel or the export of misfolded proteins to the cytosol. For example, the allele *sec61-2* had a twofold stabilization effect on ERAD substrates which was the result of blocked export from the ER, while a mutation in *Sec62* (which encodes a subunit of the tetrameric complex including Sec62p, Sec63p, Sec71p, and Sec72p associated with the Sec61 complex) led to an import defect. The *sec61-2* allele, which encodes Sec61p, only expressed a mutant phenotype with respect to retrotranslocation. Another example is the *kar2-113* allele (BiP), which is fully functional in CPY* import at 25 °C. Expression of this mutation lead to a twofold increase in CPY* half-life under these same conditions, again suggesting that the processes of import and export from the ER defined by BiP activity can be functionally independent. It was also noted by Plemper *et al.* (48) that with the pore size of the translocon channel of approximately 20Å, CPY* unfolding would be required, be it partially or fully, prior to retrotranslocation, also suggesting a role for BiP in retrotranslocation.

3.5   Some possible mechanisms for retrotranslocation through the Sec61 translocon channel

While the experiments by Plemper *et al.* (48) provide some interesting possibilities for the mechanisms of retrotranslocation of ERAD substrates involving the Sec61 translocon channel and BiP, a recent paper (49) reported that the import and export from the ER are mechanistically distinct. Pulse-chase experiments were performed *in vivo* with two alleles of *KAR2* (*kar2-113* and *kar2-133*) showing that they are functional for import but incapable in the *in vitro* retrotranslocation assay (Section 3.3.2) of export from the ER, using ΔGppαF and α1-ATZ. Conversely, export could be demonstrated in the presence of cytosol from *ssa1* mutants which are defective for import into the

ER and ΔGppαF as a substrate. Thus, there is additional support for the mechanistically distinct processes using common components.

BiP has been shown in yeast to act as a ratchet or motor for protein import into the ER (50), but the driving force mediating export from the ER has not yet been elucidated. An interesting development emerged when an ER integral membrane protein termed Cue1p (factor for coupling of ubiquitin conjugation to ER degradation) was isolated (51). Cue1p protein was found to recruit Ubc7p to the ER membrane, and to be essential for its function. Ubiquitination by both Cue1p-assembled Ubc7p and Ubc6p is necessary for retrograde transport, demonstrating that ubiquitination and export are mechanistically coupled.

Proteasomal activity has been implicated as a driving force for retrotranslocation (52). The evidence came from experiments which used fusion proteins consisting of an N-terminal degradation signal with the doubly membrane spanning Sec62p. In cells with proteasomal mutations, it was demonstrated that the rapid proteolytic activity of the proteasome on the N-terminal cytosolic domain was followed by slowed proteolysis at the point when membrane extraction is necessary. Retrotranslocation of Sec62p from the membrane, and proteasomal activity on transmembrane ERAD substrates appeared to be coupled.

3.6 Proteins with unknown function

Other proteins involved in selective retrotranslocation have been found primarily though genetic screens in *S. cerevisiae*, which used the accumulation of the ERAD substrates such as 3-hydroxy-3-methylglutaryl-CoA reductase (HGM-R), CPY*, and α1-ATZ. All three screens looked for mutants that are inefficient in degrading these substrates. HMG-R is integral membrane protein of the ER and an essential enzyme involved in sterol synthesis. Hampton *et al.* (53) expressed a Myc-tagged version of HMG-R (6myc-Hmg2p), and selected for cells with increased steady state levels of the protein on lovastatin, an inhibitor of HMG-R that induces its more rapid degradation. They isolated three mutant alleles designated *hrd1* to *hrd3*. The *HRD1* (also known as *DER3*, see below) and *HRD3* genes encode proteins, predicted to be membrane bound, which are also involved in ERAD of CPY* (54). While the function of these proteins is still unclear, *HRD2* is homologous to the p97 (TRAP-2) and encodes a component of the mature 26S proteasome.

In similarly designed experiments (55), three mutant alleles were found that result in the accumulation of CPY*, termed *der1* to *der3*. *DER1* encodes a membrane protein of still unknown function, (55), *DER2* was found to be identical to *UBC7* (38), and *DER3* is identical to *HRD1*.

3.7 The mannosidase clock: a model for the selective process associated with ERAD

It is still largely unclear how proteins are selected for ERAD. The Calnexin Cycle through de/reglucosylation of immature glycoproteins ensures that unfolded proteins do not progress along the secretory pathway. The selective mechanism by which misfolded proteins are brought to the translocon still has not been determined. Experiments in *S. cerevisiae* and mammalian cells have examined how *N*-linked oligosaccharide structure can influence ERAD. Slight differences exist between the two systems; however, a common denominator does arise. Both Knop *et al.* (55) and Jakob *et al.* (56) examined the turnover of CPY* in yeast. Knop showed that cells devoid of the ER-processing α-1,2-mannosidase displayed reduced degradation of CPY*, while Jakob demonstrated that the glycoproteins with *N*-linked $Man_8GlcNAc_2$ oligosaccharides were most effectively degraded when compared to glycoproteins with $Man_9GlcNAc_2$, $Man_7GlcNAc_2$, $Man_6GlcNAc_2$ glycan structures.

It was also shown recently that ER mannosidase I activity was a requirement for ERAD in mammalian cells using truncated α1-antitrypsin (57). ER associated degradation of misfolded α1-antitrypsin was found to be dependent on calnexin binding. Moreover, ER mannosidase trimming resulted in prolonged association with calnexin, a prerequisite for ERAD to occur. This was to be expected since glucosidase II, which mediates release from calnexin by trimming the last glucose from the $Glc_1Man_9GlcNAc_2$ structure, exhibits much slower kinetics on the $Glc_1Man_8GlcNAc_2$ (58). Therefore, the model suggests that the ER mannosidase induces prolonged association of misfolded glycoproteins with calnexin, which then directs them to the translocon and subsequently to the proteasome for degradation. The model itself is stochastic, whereby the rates of glycan trimming on the *N*-linked oligosaccharides of misfolded glycoproteins, combined with the residency time within the Calnexin Cycle, influence their interactions with the ER chaperones, determining the selective process underlying ERAD.

Results from the mammalian cells model contrast with those from the yeast model with respect to the role of calnexin. Knop *et al.* (55) demonstrated in strains deleted for *CNE1* (the *S. cerevisiae* calnexin homologue (59)) had little effect on CPY* degradation, while Jakob *et al.* (56) also reported the optimal glycan structure for ERAD as being $Man_8GlcNAc_2$. Jakob proposed the existence of an as yet unknown Man_8 lectin that selectively targets misfolded proteins for retrotranslocation and degradation.

4. Signalling from the ER

The accumulation of unfolded proteins in the ER leads to a stress response termed the unfolded protein response, UPR. The UPR is defined by the signalling of the presence of unfolded proteins in the ER to the nucleus and leads to the increased transcription of some ER proteins and folding enzymes. Experimentally, there are several ways in which this response can be generated. For example, compounds such as DTT, which inhibit disulfide bond formation, or tunicamycin which inhibits *N*-glycosylation, are potent inducers of the transcription of ER proteins such as BiP. This upregulation of the synthesis of molecular chaperones and folding enzymes leads to an increased capability of the ER to process unfolded proteins. The UPR is found in both mammalian cells and yeast, and recent results show that the components responsible for the response are remarkably conserved between species.

4.1 The unfolded protein response in yeast

In *S. cerevisiae* the *IRE1* gene was identified in a screen for genes involved in the UPR (60). The screen was to use conditions which lead to unfolded proteins in the ER and to use fusion of the *KAR2* promoter (the yeast BiP) with a *LacZ* gene acting as a reporter. The screen was for mutants which no longer upregulated the *KAR2* promoter. Several mutants were isolated and the genes responsible were cloned by complementation of the mutants using a yeast plasmid library. The remarkable finding was that the gene responsible, *IRE1*, had been previously identified as an inositol auxotroph (61). The *IRE1* gene codes for a predicted type 1 transmembrane protein, Ire1p, with an N-terminal ER lumenal sequence, a transmembrane domain, and a C-terminal cytosolically oriented domain which has both a serine/threonine kinase domain and an endoribonuclease domain. Genetic analysis has shown that the Ire1p protein dimerizes due to the accumulation of unfolded proteins in the ER, but the ligand which is responsible for this dimerization remains unknown (62).

Oligomerization of Ire1p has been shown to lead to *trans*-autophosphorylation of the kinase domain and apparently activation of the ribonuclease activity (63). The ribonuclease domain of Ire1p has sequence similarity with mammalian RNase L that is activated upon treatment of cells with interferon. In the case of RNaseL the ligand for dimerization is a 2´-5´-linked oligoadenylate and the consequence of this binding is activation of the ribonuclease activity (64). In the case of Ire1p deletion of the predicted ribonuclease domain leads to a defective UPR but the kinase activity is not affected. The involvement of a ribonuclease activity in the UPR was rationalized by the isolation in a genetic over-expression screen of another member of the UPR pathway, the gene for *HAC1* (65). The protein Hac1p is a bZIP transcription factor and deletion of this gene also abolishes the UPR. The surprising result is that *HAC1* mRNA is a substrate for the site specific endoribonuclease activity of Ire1p. *HAC1* is not itself a gene under the control of the transcriptional consequences of the UPR but its mRNA is constitutively transcribed but poorly translated. This defect was found to be due to an internal 252 base pair sequence originally thought to be a region which affected the stability of the Hac1p protein, but subsequently it was shown that both forms of mRNA give rise to Hac1p proteins that are unstable. Another gene identified in the original screen as required for the UPR was *RLG1*, which codes for the ligase responsible for processing the introns in tRNAs (66). Although this was initially difficult to reconcile with the initial discovery of *IRE1*, with the later realization that it has site-specific endoribonuclease activity a rational and novel mechanism of control of the UPR was proposed (66).

Thus the current model for the UPR in yeast is that the Ire1p oligomerizes due to an unknown ligand in the ER, and the oligomerization induces autophosphorylation by the kinase domains which leads to activation of the intrinsic ribonuclease activity. *HAC1* mRNA is spliced and ligated, exported to the cytoplasm, and translated; the

Hac1p protein is imported into the nucleus and combines with other transcription factors to increase the rate of transcription of ER molecular chaperones and folding enzymes (Fig. 1).

One requirement of this model is that the *HAC1* mRNA is accessible to the processing activity of Ire1p and the Rlg1p. The latter was known to be restricted to the nucleus and recent results with a mammalian hIrep homologue have shown that it is located on the inner nuclear membrane, which is contiguous with the ER and needs an intact kinase domain to signal the UPR (67). Other members of the UPR signalling pathway that have been identified by yeast two hybrid as interacting with both Ire1p and with Hac1p are the members of the yeast transcriptional coactivator complex Gcn5p/Ada, containing Gcn5p, Ada2p, Ada3p, and Ada5p (68). In support of their functional role, the disruption of the *GCN5*, *ADA2*, and *ADA3* genes leads to a reduced UPR and deletion of the *ADA5* gene abolishes it (69). Lastly, a serine/threonine phosphatase, Ptc2p, was identified as a direct interactor with Ire1p and can dephosphorylate it (70).

4.2   The mammalian UPR

Two isoforms of Ire1 have been cloned from human and mouse, and termed Ire1α and Ire1β (67, 71). A remarkable set of experiments (67, 72) characterizing the human homologues was based on the result that demonstrated the endoribonuclease activity of the yeast Ire1p on *HAC1* mRNA (63). Tirasophon and colleagues set out to test the endoribonuclease activity of human Ire1 (Ire1αp); however, neither a mammalian homologue for *HAC1*, nor a putative substrate for the enzyme were known. Therefore they tested its activity on the yeast *HAC1* mRNA prepared *in vitro*. They expressed hIre1αp in COS-1 cells, and using immunoprecipitation demonstrated the cleavage of *HAC1* mRNA at the 5′ splice site junction into two species, a 224-nucleotide 5′ exon and a 326-nucleotide intron/3′ exon. The precise location of the splice site was determined by primer extension analysis and confirmed to be identical to the 5′ splice site junction processed by a GST-yeast Ire1p fusion protein (which itself, however, could cleave both 5′ and 3′ junctions of the intron). Tirasophon and co-workers suggested that Ire1α and Ire1β may each have evolved with specificities for 5′ and 3′ splice site junctions, respectively. They also tested the endoribonuclease activity of a kinase defective form of hIre1αp in which a conserved lysine residue in the ATP binding site had been mutated (K599A). The mutant form failed to cleave the *HAC1* mRNA substrate, suggesting that the kinase activity is a prerequisite for endoribonuclease activity. Moreover, while over-expression of wild-type Ire1p constitutively activated a *LacZ* reporter gene under the control of the rat BiP promoter, a result also observed in *S. cerevisiae* (63), they found that the mutant hIre1αp-K599A effectively blocked the unfolded protein response, thus acting in a *trans*-dominant manner. Alternatively, Wang *et al.* (71) confirmed the functional homology of the murine Ire1p through its over-expression in 293T cells, leading to an increase in the levels of GRP78/BiP, but not to the same extent as with tunicamycin.

Recently, Niwa *et al.* (72) examined the capacity of mammalian cells to *in vivo* splice yeast *HAC1* mRNA. They transfected HeLa cells with a plasmid containing the yeast *HAC1* gene, and the UPR was induced with tunicamycin. PCR was performed on cDNAs generated from both control and induced cells. The PCR product of the cDNA from control cells was identical in size to the uncleaved *HAC1*, while a 370 bp product was retrieved from cells induced for the UPR. Sequencing the PCR product confirmed the accurate splicing of the *HAC1* mRNA. This result confirms the functional conservation of a highly specific mRNA splicing apparatus from yeast to mammalian cells. The question was now raised as to how the yeast mRNA was spliced, testing the model in which the hIre1αp and hIre1βp homologues each cleave specific splice site junctions of the *HAC1* mRNA. Both hIre1p isoforms were made in a baculovirus/insect cell system, and [α-^{32}P] UTP labelled and *in vitro* transcribed yeast HAC1 mRNA was prepared and used as a substrate to test the endoribonuclease specificities for each isoform. Both were highly specific and exhibited identical ribonuclease activities to the yeast Ire1p.

Niwa *et al.* (72) found that both hIre1αp and hIre1βp were localized to the ER membrane, but were translocated to the nucleus upon stimulation of the UPR. By indirect immunofluorescence with anti-peptide antibodies that were isoform specific, they found a localization pattern identical to the β subunit of the signal recognition particle and mp30, which are both ER-resident membrane proteins. It was noted, however, that 24% of hIre1αp cells and 54% of hIre1βp cells exhibited nuclear staining and upon the addition of tunicamycin, the number of cells with nuclear localization patterns increased to 80% and 100%, respectively. Moreover, the pattern was diffuse throughout the nucleus, with no nuclear rim staining, which is inconsistent with them being integral membrane proteins. This anomaly was resolved by Western blot analysis, which revealed that the 140 kDa predominant form of Ire1p underwent proteolysis upon UPR induction, which led to an increase in a 60 kDa band. Based on the size of the fragment and localization of the epitope, it was suggested that proteolysis occurred cytosolically just C-terminal of the transmembrane domain. Thus, a highly complex and dynamic model has been developed which can be summarized as follows: unfolded proteins promote the oligomerization of Ire1p, then *trans*-autophosphorylation occurs on the cytoplasmic domain activating an endoribonuclease domain, which is proteolytically cleaved from the cytosolic surface of the ER and is then targeted to the nucleus, where it processes an as yet unknown substrate similar to to *HAC1*. This eliminates the transcriptional attenuation caused by the presence of the intron, allowing for the translation, and subsequent binding, of the transcriptional activator to the UPR upstream *cis*-element (UPRE), transcriptionally activating the UPR induced genes (Fig. 1).

One known protease that cytosolically cleaves integral membrane proteins is γ-secretase. PS1, a gene thought to encode a protein that either exhibits γ-secretase activity or is involved in activating γ-secretase, was a potential candidate for the proteolysis of hIrep. It has been shown that γ-secretase cleaves amyloid precursor protein (APP) generating Aβ, the major component of amyloid plaques seen in Alzheimer's patients (73). To test the involvement of PS1, Niwa's group examined the localization of both that hIre1αp and hIre1βp in fibroblast cells derived from homozygous knockout mice (PS1$^{-/-}$). Localization of both forms of Ire1p upon UPR induction remained unchanged in PS1$^{-/-}$ cells, while 100% of Ire1αp and 95% of Ireβp displayed nuclear distribution patterns of in PS1$^{+/+}$ cells after treatment with tunicamycin. Moreover, they examined UPR induction in PS1$^{-/-}$ cells in a 7-hour time course monitoring changes in BiP mRNA levels. BiP was induced to threefold that of basal levels, while PS1$^{+/+}$ cells exhibited a fivefold induction. It was clear from these experiments that PS1 was involved in UPR-dependent nuclear localization of both hIre1αp and hIre1βp, and also an important mediating factor in the mammalian UPR (72).

4.3 Ire1p and PERK in the regulation of other cellular responses to ER stress

While stress signalling from the lumen of the ER is comprehensively reviewed elsewhere (74), recent and exciting work has been performed by the Ron lab at the New York University Medical School, on Ire1p and the related PERK kinase, and their role in mediating the activity of stress-activated protein kinases (SAPKs) and translational attenuation in response to ER stress. SAPKs, also known as JNKs for cJUN NH$_s$-terminal kinases, are activated in response to ER stress induced by perturbations that promote the UPR (75, 76). These kinases then activate transcription through the phosphorylation of transcriptional activators such as cJUN and ATF2 (77, 78). The Ron group disrupted m*IRE1*α in mouse embryonic stem (ES) cells (79). Fibroblasts derived from wild-type embryos exhibited a twofold increase of JNK activity upon the induction of ER stress, whereas *IRE1*α$^{-/-}$ embryos not only failed to induce JNK activity, but displayed reduced levels of activity. Thus mIre1αp was found to mediate JNK activation in embryonic cells in response to ER stress.

They then searched for the proteins that couple the UPR receptor to SAPK pathway activation using the yeast two hybrid screen with mIre1αp as a bait and isolated TRAF2. TRAFs are adapter proteins that are recruited to the cytosolic domains of ligated receptors, activating downstream JNKs. This has been best demonstrated for the tumour necrosis factor (TNF) receptor in the activation of the stress-activated pathway (80, 81). The interaction was confirmed *in vivo* by co-immunoprecipitation experiments. The model was then proposed that Ire1p mediated activation of the JNK stress-activated pathway directly through the recruitment of TRAF2, in a mechanism that is similar to JNK activation by cell surface receptors such as the TNF receptor.

Coincidentally with the upregulation of ER resident proteins by upstream *cis*-acting elements (UPRE) in response to accumulation of unfolded proteins, cells also down regulate transcription through the phosphorylation of eIF2α. The phosphorylated form of eIF2α interferes with the assembly of the 43S translation-initiation complex (82). One of the kinases that phosphorylates eIF2α, the interferon-inducible RNA-dependent protein kinase (PKR), has been shown to be activated in response to treatments that lead to Ca^{2+} depletion in the ER (83). However, the Ron group found that the inhibition of protein synthesis in *PKR*$^{-/-}$ cells was close to that of wild-type cells in response to ER stress. Using a sequence homology based approach, they searched for kinases related to PKR and HRI (heme-regulated eIF2α kinase) that could be involved in the response to ER stress. They identified a *Caenorhabditis elegans* cosmid clone (CEF46C3) that encodes a predicted type I transmembrane protein with a kinase domain highly similar to PKR and HRI, and a lumenal domain homologous to Ire1p. Using a human expressed sequence tag (EST) clone that encoded a peptide fragment similar to the predicted sequence to the *C. elegans* protein, they isolated the mouse PERK cDNA, which encoded a protein with 20% homology to the lumenal domain of Ire1p and 40% homology of the cytosolic domain to PKR. *In vitro* experiments demonstrated that bacterially expressed GST-PERK fusion (for PKR-like ER kinase) could phosphorylate eIF2α, while translation-competent reticulocyte lysates were profoundly inhibited by the addition of the fusion proteins. Lastly, they expressed the lumenal domain of PERK. This had a dominant-negative effect on ER stress induction of the UPR marker CHOP, suggesting a similar dominant negative effect on Ire1p as seen in yeast and mammalian cells with the expression of the lumenal domain of Ire1p (62, 71). This suggested that the lumenal domains of Ire1p and PERK are functionally conserved, and share similar ligands or upstream transducers. Nevertheless, they mediate different signalling pathways, one upregulating the expression of ER resident chaperones and folding enzymes (Ire1p), and the second attenuating translation (PERK). Remarkably, this model is supported with the recent results of Bertolotti *et al.* (84). They found that the lumenal domains of Ire1p and PERK were interchangeable, and that both form stable complexes with BiP in unstressed cells. It was also found that BiP dissociates from the lumenal domains of both Ire1p and PERK upon ER stress, leading to the activation and oligomerization of both Ire1p and PERK. Thus, BiP is likely to be the ligand that regulates the activation of both ER stress receptor kinases.

5. Concluding remarks

The three ER molecular systems that we have described comprise only a few of its known functions. It is clear that for the processes of the Calnexin Cycle, ERAD, and the unfolded protein response, most of the key molecules are known but the details of their mechanisms remain to be elucidated. Similarly, for some of the ER processes such as protein translocation into the ER, glycosylation, and vesicular transport, some of the components are known and mechanisms are being elaborated. The role of the ER in a variety of diseases has been known for some time, and we are becoming increasingly aware of the underlying mechanisms.

Acknowledgements

The results from the authors' (JJMB and DYT) laboratories were supported by grants from the Medical Research Council of Canada.

References

1. Ellis, R. J. and Hemmingsen, S. M. (1989). Molecular chaperones: proteins essential for the biogenesis of some macromolecular structures. *Trends Biochem. Sci.*, **14**, 339–42.
2. Hendrick, J. P. and Hartl, F. U. (1993). Molecular chaperone functions of heat-shock proteins. *Annu. Rev. Biochem.* **62**, 349–84.
3. Hubbard, S. C. and Ivatt, R. J. (1981). Synthesis and processing of asparagine-linked oligosaccharides. *Annu. Rev. Biochem.*, **50**, 555–84.
4. Wada, I., Rindress, D., Cameron, P. H., Ou, W. J., Doherty, J. J. D., Louvard, D., *et al.* (1991). SSR alpha and associated calnexin are major calcium binding proteins of the endoplasmic reticulum membrane. *J. Biol. Chem.*, **266**, 19599–610.
5. Fliegel L., Burns K., MacLennan D. H., Reithmeier R. A., Michalak, M. (1989). Molecular cloning of the high affinity calcium-binding protein (calreticulin) of skeletal muscle sarcoplasmic reticulum. *J. Biol. Chem.*, **264**, 21522–8.
6. Degen, E. and Williams, D. B. (1991). Participation of a novel 88 kDa protein in the biogenesis of murine class I histocompatability. *J. Cell. Biol.*, **112**, 1099–115.
7. Ahluwalia, N., Bergeron, J. J., Wada, I., Degen, E., and Williams, D. B. (1992). The p88 molecular chaperone is identical to the endoplasmic reticulum membrane protein, calnexin. *J. Biol. Chem.*, **267**, 10914–18.
8. Ou, W. J., Cameron, P. H., Thomas, D. Y., and Bergeron, J. J. (1993). Association of folding intermediates of glycoproteins with calnexin during protein maturation. *Nature*, **364**, 771–6.
9. Yeo, T. K., Yeo, K. T., and Olden, K. (1989). Accumulation of unglycosylated liver secretory glycoproteins in the rough endoplasmic reticulum. *Biochem. Biophys. Res. Commun.*, **160**, 1421–8.
10. Ware, F. E., Vassilakos, A., Peterson, P. A., Jackson, M. R., Lehrman, M. A., and Williams, D. B. (1995). The molecular chaperone calnexin binds Glc1Man9GlcNAc2 oligosaccharide as an initial step in recognizing unfolded glycoproteins. *J. Biol. Chem.*, **270**, 4697–704.
11. Li, Y., Bergeron, J. J., Luo, L., Ou, W. J., Thomas, D. Y., and Kang, C. Y. (1996). Effects of inefficient cleavage of the signal sequence of HIV-1 gp 120 on its association with calnexin, folding, and intracellular transport. *Proc. Natl. Acad. Sci. USA*, **93**, 9606–11.
12. Ou, W. J., Bergeron, J. J., Li, Y., Kang, C. Y., and Thomas, D. Y. (1995). Conformational changes induced in the endoplasmic reticulum luminal domain of calnexin by Mg-ATP and Ca2+. *J. Biol. Chem.*, **270**, 18051–9.
13. Vassilakos, A., Michalak, M., Lehrman, M. A., and Williams, D. B. (1998). Oligosaccharide binding characteristics of the molecular chaperones calnexin and calreticulin. *Biochemistry*, **37**, 3480–90.
14. Wada, I., Imai, S., Kai, M., Sakane, F., and Kanoh, H. (1995). Chaperone function of calreticulin when expressed in the endoplasmic reticulum as the membrane-anchored and soluble forms. *J. Biol. Chem.*, **270**, 20298–304.
15. Tessier, D. C., Dignard, D., Zapun, A., Radominska, A., Parodi, A. J., Bergeron, J. J. M., and Thomas, D. Y. (2000). Cloning and characterization of mammalian UDP-glucose glycoprotein:glucosyltransferase and the development of a specific substrate for this enzyme. *Glycobiology*, **10**, 403–12.
16. Trombetta, S. E., Bosch, M., and Parodi, A. J. (1989). Glucosylation of glycoproteins by mammalian, plant, fungal, and trypanosomatid protozoa microsomal membranes. *Biochemistry*, **28**, 8108–16.

17. Zapun, A., Petrescu, S. M., Rudd, P. M., Dwek, R. A., Thomas, D. Y., and Bergeron, J. J. (1997). Conformation-independent binding of monoglucosylated ribonuclease B to calnexin. *Cell*, **88**, 29–38.

18. Rodan A. R., Simons J. F., Trombetta E. S., and Helenius A. (1996). N-linked oligosaccharides are necessary and sufficient for association of glycosylated forms of bovine RNase with calnexin and calreticulin. *EMBO J.*, **15**, 6921–30.

19. Ihara, Y., Cohen-Doyle, M. F., Saito, Y., and Williams, D. B. (1999). Calnexin discriminates between protein conformational states and functions as a molecular chaperone in vitro [In Process Citation]. *Mol. Cell*, **4**, 331–41.

20. Freeman, B. C. and Morimoto, R. I. (1996). The human cytosolic molecular chaperones hsp90, hsp70 (hsc70) and hdj-1 have distinct roles in recognition of a non-native protein and protein refolding. *EMBO J.*, **15**, 2969–79.

21. Ehrnsperger, M., Graber, S., Gaestel, M., and Buchner, J. (1997). Binding of non-native protein to Hsp25 during heat shock creates a reservoir of folding intermediates for reactivation. *EMBO J.*, **16**, 221–9.

22. Lee, G. J., Roseman, A., Saibil, H. R., and Vierling, E. (1997). A small heat shock protein stably binds heat-denatured model substrates and can maintain a substrate in a folding-competent state. *EMBO J.*, **16**, 659–71.

23. Saito, Y., Ihara, Y., Leach, M. R., Cohen-Doyle, M. F., and Williams, D. B. (1999). Calreticulin functions *in vitro* as a molecular chaperone for both glycosylated and non-glycosylated proteins. *EMBO J.*, **18**, 6718–29.

24. Oliver, J. D., van der Wal, F. J., Bulleid, N. J., and High, S. (1997). Interaction of the thiol-dependent reductase ERp57 with nascent glycoproteins. *Science*, **275**, 86–8.

25. Zapun, A., Darby, N. J., Tessier, D. C., Michalak, M., Bergeron, J. J., and Thomas, D. Y. (1998). Enhanced catalysis of ribonuclease B folding by the interaction of calnexin or calreticulin with ERp57. *J. Biol. Chem.*, **273**, 6009–12.

26. Wada, I., Kai, M., Imai, S., Sakane, F., and Kanoh, H. (1997). Promotion of transferrin folding by cyclic interactions with calnexin and calreticulin. *EMBO J.*, **16**, 5420–542.

27. Cannon, K. S. and Helenius, A. (1999). Trimming and readdition of glucose to N-linked oligosaccharides determines calnexin association of a substrate glycoprotein in living cells. *J. Biol. Chem.*, **274**, 7537–44.

28. Perlmutter, D. H. (1993). Liver disease associated with alpha-1-antitrypsin deficiency. *Prog. Liver Dis.*, **11**, 139–65.

29. Cheng, S. H., Gregory, R. J., Marshall, J., Paul, S., Souza, D. W., White, G. A., *et al.* (1990). Defective intracellular transport and processing of CFTR is the molecular basis of most cystic fibrosis. *Cell*, **63**, 827–34.

30. Dalemans, W., Barbry, P., Champigny, G., Jallat, S., Dott, K., Dreyer, D., *et al.* (1991). Altered chloride ion channel kinetics associated with the ΔF508 cystic fibrosis mutation. *Nature*, **354**, 526–8.

31. Denning, G. M., Anderson, M. P., Amara, J. F., Marshall, J., Smith, A. E., and Welsh, M. J. (1992). Processing of mutant cystic fibrosis transmembrane conductance regulator is temperature-sensitive. *Nature*, **358**, 761–4.

32. Drumm, M. L., Wilkinson, D. J., Smit, L. S., Worrell, R. T., Strong, T. V., Frizzell, R. A., *et al.* (1991). Chloride conductance expressed by ΔF508 and other mutant CFTRs in Xenopus oocytes. *Science*, **254**, 1797–19.

33. Klausner, R. D. and Sitia, R. (1990). Protein degradation in the endoplasmic reticulum. *Cell*, **62**, 611–14.

34. Kopito, R. R. (1997). ER quality control: the cytoplasmic connection. *Cell*, **88**, 427–30.

35. Sommer, T. and Jentsch, S. (1993). A protein translocation defect linked to ubiquitin conjugation at the endoplasmic reticulum. *Nature*, **365**, 176–9.

36. Ciechanover, A. (1998). The ubiquitin-proteasome pathway: on protein death and cell life. *EMBO J.*, **17**, 7151–60.

37. Rapoport, T. A., Rolls, M. M., and Jungnickel, B. (1996). Approaching the mechanism of protein transport across the ER membrane. *Curr. Opin. Cell Biol.*, **8**, 499–504.

38. Hiller, M. M., Finger, A., Schweiger, M., and Wolf, D. H. (1996). ER degradation of a misfolded luminal protein by the cytosolic ubiquitin-proteasome pathway. *Science*, **273**, 1725–8.

39. Jensen, T. J., Loo, M. A., Pind, S., Williams, D. B., Goldberg, A. L., and Riordan, J. R. (1995). Multiple proteolytic systems, including the proteasome, contribute to CFTR processing. *Cell*, **83**, 129–35.

40. Qu, D., Teckman, J. H., Omura, S., and Perlmutter, D. H. (1996). Degradation of a mutant secretory protein, alpha1-antitrypsin Z, in the endoplasmic reticulum requires proteasome activity. *J. Biol. Chem.*, **271**, 22791–5.

41. Ward, C. L., Omura, S., and Kopito, R. R. (1995). Degradation of CFTR by the ubiquitin-proteasome pathway. *Cell*, **83**, 121–7.

42. Werner, E. D., Brodsky, J. L., and McCracken, A. A. (1996). Proteasome-dependent endoplasmic reticulum-associated protein degradation: an unconventional route to a familiar fate. *Proc. Natl. Acad. Sci. USA*, **93**, 13797–801.

43. Rock, K. L., Gramm, C., Rothstein, L., Clark, K., Stein, R., Dick, L., *et al*. (1994). Inhibitors of the proteasome block the degradation of most cell proteins and the generation of peptides presented on MHC class I molecules. *Cell*, **78**, 761–71.

44. Fenteany, G., Standaert, R. F., Lane, W. S., Choi, S., Corey, E. J., and Schreiber, S. L. (1995). Inhibition of proteasome activities and subunit-specific amino-terminal threonine modification by lactacystin. *Science*, **268**, 726–31.

45. McCracken, A. A. and Brodsky, J. L. (1996). Assembly of ER-associated protein degradation in vitro: dependence on cytosol, calnexin, and ATP. *J. Cell Biol.*, **132**, 291–8.

46. Wiertz, E. J., Jones, T. R., Sun, L., Bogyo, M., Geuze, H. J., and Ploegh, H. L. (1996). The human cytomegalovirus US11 gene product dislocates MHC class I heavy chains from the endoplasmic reticulum to the cytosol. *Cell*, **84**, 769–79.

47. Wiertz, E. J., Tortorella, D., Bogyo, M., Yu, J., Mothes, W., Jones, T. R., *et al*. (1996). Sec61-mediated transfer of a membrane protein from the endoplasmic reticulum to the proteasome for destruction. *Nature*, **384**, 432–8.

48. Plemper, R. K., Bohmler, S., Bordallo, J., Sommer, T., and Wolf, D. H. (1997). Mutant analysis links the translocon and BiP to retrograde protein transport for ER degradation. *Nature*, **388**, 891–5.

49. Brodsky, J. L., Werner, E. D., Dubas, M. E., Goeckeler, J. L., Kruse, K. B., and McCracken, A. A. (1999). The requirement for molecular chaperones during endoplasmic reticulum-associated protein degradation demonstrates that protein export and import are mechanistically distinct. *J. Biol. Chem.*, **274**, 3453–60.

50. Brodsky, J. L. (1998). Translocation of proteins across the endoplasmic reticulum membrane. *Int. Rev. Cytol.*, **178**, 277–328.

51. Biederer, T., Volkwein, C., and Sommer, T. (1997). Role of Cue1p in ubiquitination and degradation at the ER surface. *Science*, **278**, 1806–9.

52. Mayer, T. U., Braun, T., and Jentsch, S. (1998). Role of the proteasome in membrane extraction of a short-lived ER-transmembrane protein. *EMBO J.*, **17**, 3251–7.

53. Hampton, R. Y., Gardner, R. G., and Rine, J. (1996). Role of 26S proteasome and HRD genes in the degradation of 3-hydroxy-3-methylglutaryl-CoA reductase, an integral endoplasmic reticulum membrane protein. *Mol. Biol. Cell*, **7**, 2029–44.

54. Plemper, R. K., Bordallo, J., Deak, P. M., Taxis, C., Hitt, R., and Wolf, D. H. (1999). Genetic interactions of Hrd3p and Der3p/Hrd1p with Sec61p suggest a retro-translocation complex mediating protein transport for ER degradation. *J. Cell Sci.*, **112**, 4123–34.

55. Knop, M., Hauser, N., and Wolf, D. H. (1996). N-Glycosylation affects endoplasmic reticulum degradation of a mutated derivative of carboxypeptidase yscY in yeast. *Yeast*, **12**, 1229–38.

56. Jakob, C. A., Burda, P., Roth, J., and Aebi, M. (1998). Degradation of misfolded endoplasmic reticulum glycoproteins in Saccharomyces cerevisiae is determined by a specific oligosaccharide structure. *J. Cell Biol.*, **142**, 1223–33.

57. Liu, Y., Choudhury, P., Cabral, C. M., and Sifers, R. N. (1999). Oligosaccharide modification in the early secretory pathway directs the selection of a misfolded glycoprotein for degradation by the proteasome. *J. Biol. Chem.*, **274**, 5861–7.

58. Hubbard, S. C. and Robbins, P. W. (1979). Synthesis and processing of protein-linked oligosaccharides in vivo. *J. Biol. Chem.*, **254**, 4568–76.

59. Parlati, F., Dominguez, M., Bergeron, J. J., and Thomas, D. Y. (1995). Saccharomyces cerevisiae CNE1 encodes an endoplasmic reticulum (ER) membrane protein with sequence similarity to calnexin and calreticulin and functions as a constituent of the ER quality control apparatus. *J. Biol. Chem.*, **270**, 244–53.

60. Cox, J. S., Shamu, C. E., and Walter, P. (1993). Transcriptional induction of genes encoding endoplasmic reticulum resident proteins requires a transmembrane protein kinase. *Cell*, **73**, 1197–206.

61. Nikawa, J. I. and Yashamita, S. (1992). *IRE1* encodes a putative protein kinase containing a membrane-spanning domain and is required for inositol photrophy in *Sacchaomyces cerevisiae. Mol. Microbiol.*, **6**, 1441–6.

62. Shamu, C. E. and Walter, P. (1996). Oligomerization and phosphorylation of the Ire1p kinase during intracellular signaling from the endoplasmic reticulum to the nucleus. *EMBO J.*, **15**, 3028–39.

63. Sidrauski, C. and Walter, P. (1997). The transmembrane kinase Ire1p is a site-specific endonuclease that initiates mRNA splicing in the unfolded protein response. *Cell*, **90**, 1031–9.

64. Zhou, A., Hassel, B. A., and Silverman, R. H. (1993). Expression cloning of 2-5A-dependent RNAase: a uniquely regulated mediator of interferon action. *Cell*, **72**, 7537–65.

65. Cox, J. S. and Walter P. (1996). A novel mechanism for regulating activity of a transcription factor that controls the unfolded protein response. *Cell*, **87**, 391–404.

66. Sidrauski, C., Cox, J. S., and Walter, P (1996). tRNA ligase is required for regulated mRNA splicing in the unfolded protein response. *Cell*, **87**, 405–13.

67. Tirasophon, W., Welihinda, A. A., and Kaufman, R. J. (1998). A stress response pathway from the endoplasmic reticulum to the nucleus requires a novel bifunctional protein kinase/endoribonuclease (Ire1p) in mammalian cells. *Genes Dev.*, **12**, 1812–24.

68. Welihinda, A. A., Tirasophon, W., Green, S. R., and Kaufman, R. J. (1997). Gene induction in response to unfolded protein in the endoplasmic reticulum is mediated through Ire1p kinase interaction with a transcriptional coactivator complex containing Ada5p. *Proc. Natl. Acad. Sci. USA*, **94**, 4289–94.

69. Welihinda, A. A., Tirasophon, W., and Kaufman, R. J. (2000). The transcriptional co-activator *ADA5* is required for *HAC1* mRNA processing in vivo. *J. Biol. Chem.*, **275**, 3377–81.

70. Welihinda, A. A., Tirasophon, W., Green, S. R., and Kaufman, R. J. (1998). Protein serine/threonine phosphatase Ptc2p negatively regulates the unfolded-protein response by dephosphorylating Ire1p kinase. *Mol. Cell Biol.*, **18**, 1967–77.

71. Wang, X. Z., Harding, H. P., Zhang, Y., Jolicoeur, E. M., Kuroda, M., and Ron, D. (1998). Cloning of mammalian Ire1 reveals diversity in the ER stress responses. *EMBO J.*, **17**, 5708–17.

72. Niwa, M., Sidrauski, C., Kaufman, R. J., and Walter, P. (1999). A role for presenilin-1 in nuclear accumulation of Ire1 fragments and induction of the mammalian unfolded protein response. *Cell*, **99**, 691–702.

73. Selkoe, D. J. (1998) The cell biology of beta-amyloid precursor protein and presinilin in Alzheimer's disease. *Trends Cell Bio.*, **8**, 447–53.

74. Kaufman, R. J. (1999). Stress signaling from the lumen of the endoplasmic reticulum: coordination of gene transcriptional and translational controls. *Genes Dev.*, **13**, 1211–33.

75. Kyriakis, J. M., Banerjee, P., Nikolakaki, E., Dai, T., Rubie, E. A., Ahmad, M. F., *et al.* (1994). The stress-activated protein kinase subfamily of c-Jun kinases. *Nature*, **369**, 156–60.

76. Srivastava, R. K., Sollott, S. J., Khan, L., Hansford, R., Lakatta, E. G., and Longo, D. L. (1999). Bcl-2 and Bcl-X(L) block thapsigargin-induced nitric oxide generation, c-Jun NH(2)-terminal kinase activity, and apoptosis. *Mol. Cell Biol.*, **19**, 5659–74.

77. Sanchez, I., Hughes, R. T., Mayer, B. J., Yee, K., Woodgett, J. R., Avruch, J., *et al.* (1994). Role of SAPK/ERK kinase-1 in the stress-activated pathway regulating transcription factor c-Jun. *Nature*, **372**, 794–8.

78. Gupta, S., Campbell, D., Derijard, B., and Davis, R. J. (1995). Transcription factor ATF2 regulation by the JNK signal transduction pathway. *Science*, **267**, 389–93.

79. Urano, F., Wang, X., Bertolotti, A., Zhang, Y., Chung, P., Harding, H. P., and Ron, D. (2000). Coupling the stress in the Er to activation of JNK protein kinases by transmembrane protein kinase IRE1. *Science*, **287**, 664–6.

80. Lee, S. Y., Reichlin, A., Santana, A., Sokol, K. A., Nussenzweig, M. C., and Choi Y. (1997). TRAF2 is essential for JNK but not NF-kappaB activation and regulates lymphocyte proliferation and survival. *Immunity*, **7**, 703–13.

81. Yeh, W. C., Shahinian, A., Speiser, D., Kraunus, J., Billia, F., Wakeham, A., *et al.* (1997). Early lethality, functional NF-kappaB activation, and increased sensitivity to TNF-induced cell death in TRAF2-deficient mice. *Immunity*, **7**, 715–25.

82. Hinnebusch, A. G. (1994). The eIF-2 alpha kinases: regulators of protein synthesis in starvation and stress. *Semin. Cell Biol.*, **5**, 417–26.

83. Prostko, C. R., Dholakia, J. N., Brostrom, M. A., and Brostrom, C. O. (1995). Activation of the double-stranded RNA-regulated protein kinase by depletion of endoplasmic reticular calcium stores. *J. Biol. Chem.*, **270**, 6211–15.

84. Bertolotti, A., Zhang, Y., Hendershot, L. M., Harding, H. P., and Ron, D. (2000). Dynamic interaction of BiP and ER stress transducers in the unfolded-protein response. *Nature Cell Biol.*, **2**, 326–32.

9 | The function of chaperones and proteases in protein quality control and intracellular protein degradation

MICHAEL R. MAURIZI

1. General introduction

During the last few years we have witnessed a major revolution in our understanding of the importance and the mechanism of intracellular protein degradation. We have come to appreciate that highly selective protein degradation occurs in all organisms and is a component of many essential regulatory pathways. We have also learned that protein degradation plays an important part in maintaining protein quality control within the cell. Most excitingly, we now have a grasp of the basic mechanism of ATP-dependent protein degradation and have begun to develop an understanding of how cells protect their proteins from inadvertent destruction while assuring that appropriate proteins are efficiently targeted for degradation. Studies of the structure and mechanism of the ATP-dependent proteases responsible for the majority of intracellular protein degradation have revealed unanticipated parallels between the proteolytic machinery of cells and the molecular chaperone systems responsible for the folding of cellular proteins and the assembly or disassembly of macromolecular complexes. This progress has led to a unified model of protein degradation and protein quality control, in which protein surveillance and structural remodelling mediated by chaperone systems and the chaperone components of ATP-dependent proteases serve as the keystone.

In this chapter I will describe current models of the structure and activities of the ATP-dependent proteases and the relationship between these proteases and molecular chaperones. I will also highlight those areas in which there is still much to be learned and propose working hypotheses to answer four important questions.

- First, on substrate recognition, how do proteases recognize both very specific substrates as well as any unfolded protein?
- Second, on self-compartmentalization, how does the structure of ATP-dependent proteases allow complete degradation of select substrates and avoid damage to other proteins?
- Third, on the energy-dependence of degradation, how does ATP binding and hydrolysis contribute to protein unfolding and translocation to the protease?
- Fourth, what is the functional and biochemical interaction between autonomous molecular chaperones and proteases?

Throughout this review, I will refer to molecular chaperone systems, such as GroEL/GroES, Hsp70/Hsp40, Hsp90, whose function is primarily to solubilize and to promote folding and assembly of proteins, as autonomous chaperones. The chaperone components of ATP-dependent proteases, which unfold proteins to facilitate their degradation, will be referred to as charonins (1).

2. Intracellular protein degradation

Protein degradation serves as a vital endpoint for many important processes in the cell. These processes fall into two general categories, dependent on whether they serve in protein quality control or they contribute to specific regulatory pathways. In protein quality control, proteolysis serves to rid the cell of aberrant proteins (2–6). In wild-type cells, between 5 and 20% of newly synthesized polypeptides are rapidly degraded, implying that proteolysis is especially important during biosynthesis and assembly of proteins (7). Proteolysis also increases in response to stress, such as heat shock or nutritional deprivation, and cells with mutations in specific proteases show increased sensitivity to agents or conditions that produce misfolded or incomplete proteins (8).

In regulatory pathways, selective proteolysis is used to remove specific functional proteins rapidly from the cell and thereby imposes a requirement for *de novo* biosynthesis in order to restore their activities (9, 10). Such regulatory proteolysis plays a crucial role in many essential pathways affecting cell cycle control, stress responses, and developmental and environmental signalling. Regulation of the proteolytic systems that carry out these degradative activities is essential to maintain an adequate level of proteolytic activity within the cell while protecting functional proteins from unwarranted damage.

2.1 Proteolysis as part of protein quality control

Post-translational quality control is required for proper folding of a significant fraction of newly synthesized proteins, and to repair or remove potentially harmful damaged and misfolded proteins from the cell (2, 6, 11). Protein quality control is dependent on autonomous chaperones, which assist proteins in folding and assembling, and ATP-dependent proteases, which eliminate proteins that cannot attain a functional native

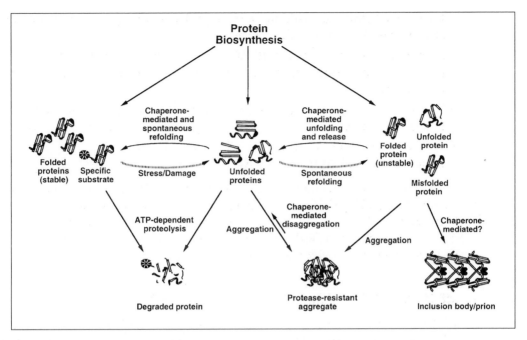

Fig. 1 Folding, unfolding, and degradation pathways for proteins. Most proteins fold without assistance from chaperones. Chaperones bind proteins that have a tendency to aggregate and help them refold or keep them soluble until proteases can degrade them. Chaperone components of ATP-dependent proteases can help proteins fold but usually unfold proteins to facilitate their degradation. Proteins that cannot be refolded or degraded will aggregate. In some cases, unfolded or misfolded proteins undergo specific aggregation or polymerization to form highly structured inclusion bodies or self-propagating prions.

state. Other aspects of protein quality control, such as protein disulfide and proline *cis-trans* isomerization (12–14), repair of oxidative damage (15), glycosylation and deglycosylation (16), N-terminal processing and modification (17), and many others, are also very important, but will not be discussed here. Integration of chaperone and protease activities is needed to assure that proteins are not degraded before they can reach a stable native state, while proteins that are too defective to become functional are efficiently removed before they can aggregate.

A summary of the protein quality control pathways in which chaperones and proteases participate is presented in Fig. 1. The chaperone-mediated unfolding/ refolding and the protease-mediated degradation pathways overlap in a number of interesting ways. For one, the study of ATP-dependent proteases has led to the discovery of whole new classes of chaperones (charonins), such as the Clp/Hsp100 family (18) and the chaperone domains of Lon (19) and FtsH (20). Autonomous chaperones and charonins can interact with unfolded or misfolded proteins, preventing their aggregation. They can also interact with exposed motifs on specific substrates and unfold or rearrange the proteins in preparation for a subsequent reaction (e. g., assembly or degradation). Separation of the unfolding and degradation steps in ATP-dependent proteases allows a second phase of substrate discrimination in which proteins that can easily refold or that lose a recognition signal prior to degradation may be released rather than degraded. For some substrates, repeated cycles of binding and unfolding may be necessary before the protein is processed in a way that either commits it to degradation or enables it to refold. Thus, kinetic partitioning of proteins between the refolding and degradation pathways may be carried out both by chaperones and proteases working together and by the ATP-dependent proteases themselves.

2.2 Recognition of appropriate substrates for degradation

The question of how ATP-dependent proteases recognize appropriate substrates and target them for degradation is fundamental to the biological regulation and the biochemical mechanism of protein degradation (8, 21). As a safeguard to avoid degradation of the improper substrates the recognition and degradation processes are performed by separate components of the proteases, although different strategies have been adopted for this purpose by proteases such as Clp and Lon and by the eukaryotic 26S proteasome. The proteasome degrades only proteins that have been modified by polyubiquitin chains or are complexed with specific factors called anti-zymes (22, 23). The binding sites for polyubiquitin or antizyme on the proteasome are distinct from the proteolytic component. Initial targeting for ubiquitination is carried out independently of the protease by complex interactions involving a myriad of ubiquitin conjugating enzymes and ubiquitin ligases that are responsive to various signalling mechanisms and regulatory factors (24). Lon and Clp can interact directly with their unmodified substrates (8). The binding sites for motifs displayed by their respective substrates are located on components or domains that constitute integral parts of the proteases. Regulation of Clp and Lon dependent degradation is also effected by factors that interact with the substrate proteins and affect the accessibility of the recognition motifs (see section 4.5).

Another interesting aspect of substrate recognition is the ability of proteases to specifically degrade unfolded or misfolded proteins, implying that all proteins can be targeted to one or another protease *in vivo*. In eukaryotes, this problem is partially solved by having a subset of E2 enzymes responsible for tagging abnormal proteins for degradation, although the nature or mechanism of recognition is only beginning to be defined (22, 24). The proteases that directly interact with substrates vary in their ability to degrade denatured proteins *in vivo*, and that ability can differ between organisms. For example, in *E. coli*, Lon protease plays the major role in degradation of canavanine-containing proteins and puromycyl fragments, while ClpAP and ClpXP make relatively little contribution (25); on the other hand, in Lactococcus, ClpXP appears to play a major role in degrading puromycyl fragments (26). Thus, proteases such as Clp and Lon can target highly specific proteins as well as a broad range of unfolded proteins.

Several possibilities exist to explain the diversity of substrates recognized by ATP-dependent proteases *in vivo*. First, molecular chaperones could bind unfolded proteins and present them to the proteases; this would require a receptor site on proteases for a chaperone/substrate complex. No data showing direct chaperone/ protease interaction has been obtained, although recent examples of coordinated action of two chaperones, in which the first prevents aggregation of an unfolded protein and the second promotes its refolding, could provide a general model for chaperone–protease cooperation (27, 28). Second, there could be an enzymatic system that tags proteins with a motif recognized by particular proteases. Such a tagging system has been described for truncated polypeptides made from damaged mRNA or on stalled ribosomes and results in targeting them for degradation by ClpXP and ClpAP (29, 30). Another tagging system adds a single hydrophobic amino acid to the N-terminus of proteins with basic N-terminal amino acids targeting them to ClpAP (31). Although no tagging systems have been described for other types of misfolded or damaged proteins, comprehensive genetic screens for loci affecting abnormal protein degradation have not been conducted. Third, all proteins may have evolved to contain specific degradation motifs which are in buried positions and only become exposed when the protein is unfolded. A corollary would be that motifs recognized by proteases should mimic sites that are common to internal regions of most proteins. There is some experimental data (discussed in the next section) that tends to support this last model.

2.3 Motifs used to target proteins for degradation

Studies with peptides that bind to molecular chaperones and ATP-dependent proteases suggest peptides with clusters of 3–5 hydrophobic residues or peptides that form amphipathic helices with one hydrophobic surface tend to bind favourably (32–34). Peptides with such hydrophobic character would be expected to be found in the interior of proteins or within buried regions of intersubunit bonding domains. Thus, a simple model for substrate recognition by chaperones and ATP-dependent proteases is that they bind to hydrophobic regions of polypeptides exposed upon unfolding or dissociation of subunits.

Identification of targeting sequences in highly unstable proteins subject to regulatory degradation suggests some overlap between specific and general recognition motifs. Many motifs found to target proteins for degradation fall into two general categories: relatively short, hydrophobic amino acids sequences, and somewhat larger structured regions with a hydrophobic surface character (8, 21). Motifs that target specific eukaryotic proteins for degradation are recognized by complexes containing E2 and E3 components of the ubiquitination system (22, 24). Among the motifs recognized by Lon and Clp proteases, the most clearly defined is an 11 amino acid tag, AANDENYALAA, sometimes referred to as the SsrA tag, added co-translationally to the C-terminus of incomplete polypeptides by the transfer-mRNA system (29). While this tag can target otherwise stable proteins for degradation, it is recognized by different proteins, such as ClpX and ClpA (30), as well as FtsH under some circumstances (35). Other naturally occurring targeting motifs are found at the C-terminus of MuA (36), MuC (37), and SulA (38), as well as at the N-terminus of UmuD (39, 39a), HemA (40), and RepA (40a). Those motifs specifically target the proteins for

Table 1 Degradation motifs in unstable proteins. Amino acids sequences in the N-terminus or C-terminus of substrates of *E. coli* Lon and Clp proteases. Only proteins that have been shown by mutation or protein fusion analysis to have degradation motifs are listed. Regions identified as important for recognition by mutation are underlined. Sequences marked with an asterisk have been shown to confer susceptibility to proteolysis when fused to heterologous proteins. Sequences marked with a bullet have not been shown to constitute recognition motifs on their own but are part of larger N-terminal domains that contain a motif.

Protease	Substrate	Sequence
N-terminal motifs		
Lon	UmuD[39]	MLFIKPADLREIVT<u>FPLF</u>SDLVQCG<u>FPS</u>PA ....
Lon/ClpA	HemA[40]	MT<u>LL</u>ALGINHKTAPVSLRERVSFSPDKLDQ ....
ClpA	RepA[40a]	MNQSFISDILYADIESKAKELTVNSNNTVQ ....
	ClpA[41]	MLNQELELSLNMAFARAREJRJEFMTVEHL ....
ClpX	Lambda O[42]	TNTAKILNFGRGNFAGQERNVADLDDGYA ....
	RpoS[43]	MSQNTLKVHDLNEDAEFDENGVEVFDEKAL ....
	UmuD[39a]	MLFIKPAD<u>LREI</u>VTFPLFSDLVQCGFPSPA ....
N-terminal motifs		
Lon	SulA[44]	....SSHATRQLSGLKIHSNLYH
	MinC[45]	....GKAARLQLVENALTVQPLN
ClpX	SssrA-tagged proteins[29]	....AANDENYALAA
	Mu_{vir} 3060[37]	....SVVVPFR<u>N</u>HRR
	MuA[36]	....ILEQN<u>RRKK</u>AI
	Mu_{vir} 3061[37]	....QSMGFMNRKVL

degradation and, importantly, can act as targeting motifs when transferred to heterologous proteins. The motifs recognized by ClpX and Lon tend to have a cluster of hydrophobic amino acids preceded by a region with charged amino acids. Such motifs resemble the sequences comprising internal hydrophobic regions that are expected to become exposed upon unfolding of a protein. Thus, targeting motifs may have evolved to mimic regions that signal protein structural perturbations.

Another kind of targeting motif recognized by *E. coli* ClpA and by specialized components of the ubiquitin system is located at the amino terminus of some proteins (46). Bulky hydrophobic amino acids at the amino terminus tend to destabilize a protein *in vivo*. In *E. coli*, such proteins are directly targeted by the ClpAP protease (47), and an enzymatic system exists for specifically adding phenylalanine or leucine to the amino terminus of a subset of proteins bearing acidic or basic amino terminal residues, resulting in their being targeted for degradation also (31). Since bulky hydrophobic, acidic, and basic amino acids do not occur at the amino terminus of most *E. coli* proteins (17), this system appears to be designed to recognize proteins that have been improperly processed or have been partially degraded by other proteases.

The location of specific recognition motifs near the ends of polypeptides has several consequences that contribute to degradation. First, the site would be located in what is usually an accessible region of a protein structure or one that can be made accessible by relatively slight perturbations. Second, tethering of the substrate protein by one end would be expected to facilitate its unfolding as additional hydrophobic regions become exposed and interact with other subsites on the protease. Third, if substrate translocation in protease complexes is vectorial, for example, proceeds via the end bearing the motif, general targeting motifs exposed in unfolded proteins would be used only if located near the end favoured by a particular protease. Discrimination of targeting motifs on the basis of both sequence and location in the protein would help explain why proteases make unequal contributions toward degradation of misfolded and other damaged proteins.

2.4 A bipartite binding model for protein degradation

The ability of proteases to recognize both specific and general motifs suggests a fundamental principle that may apply to all protein degradation, namely, bipartite or even multipartite recognition as a basis for regulating protein degradation. I have suggested a model (48–50) in which 'targeting' for degradation occurs in two or more binding steps involving both specific motifs and exposed hydrophobic regions of a substrate (see Fig. 2). These interactions would participate simultaneously or sequentially to maintain the substrate–protease complex until translocation and degradation

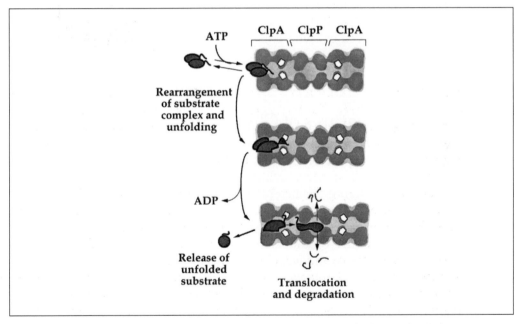

Fig. 2 Kinetic partitioning of substrates by ATP-dependent proteases. Substrates encountering an ATP-dependent protease bind to the ATPase/chaperone and are unfolded. The unfolded protein can be released and allowed to refold or translocated to the proteolytic component and degraded. What determines the partitioning of the substrate is not known. In the illustration, a possible scenario showing the fate of a dimer is shown. Unfolding of a subunit would disrupt the bonding domains holding the dimer together. Depending on the size and surface properties of the protein, the unfolded dimer could be translocated to the protease while the remaining partially unfolded dimer would dissociate and refold.

could proceed. The importance of specific versus general motifs in targeting a particular protein would depend on their number and location in the protein, as well as on the degree to which motifs conform to high-affinity consensus sites. All chaperone components of ATP-dependent proteases are multimeric, and it may be that the presence of six or seven binding sites is crucial to facilitating multipartite binding of substrate proteins.

3. The major ATP-dependent proteolytic systems

Cellular proteases can be grouped into two classes: ATP-dependent and ATP-independent. ATP-dependent proteases are responsible for the major portion of protein degradation within the soluble compartments of the cell and have been implicated both genetically and biochemically in the degradation of many important regulatory proteins (9, 10, 51). ATP-dependent proteases include the 26S proteasome (52), the major cytosolic protease in eukaryotic cells, and the Clp (53), Lon (54), and FtsH proteases (55), which are found abundantly in prokaryotes and within the mitochondria or chloroplasts of eukaryotes (Fig. 3). In addition to their regulatory functions, the ATP-dependent proteases play a major role in the protein quality control system (5, 6, 11). Proteases that are not ATP-dependent, such as calpains, caspases, cathepsins, and others, have important specific biological roles as well and have been reviewed elsewhere (56–58). Here, we will focus on ATP-dependent proteases, and describe how these proteases have functional, structural, and mechanistic parallels to molecular chaperones.

3.1 Structural features of ClpAP and ClpXP

ClpAP and ClpXP from *E. coli* are two component, ATP-dependent proteases consisting of a common proteolytic core, ClpP, associated with a homologous regulatory component, ClpA or ClpX, which have ATPase and charonin activity (53). The architectural design is a barrel-like structure with two heptameric rings of ClpP flanked on either side by hexameric rings of ClpA or ClpX (59). The interior of the barrel has a series of aqueous chambers connected by axial channels. The aqueous chamber within ClpP contains the active sites, which are therefore isolated from the surrounding medium and inaccessible in a direct manner to cellular proteins (60, 61). The isolation of the proteolytic sites is observed also with the proteasome (62) and ClpQ (63), and has led to their being referred to as self-compartmentalizing proteases (52, 64).

Another aqueous chamber is formed at the interface between ClpP and the ATPase (59). In ClpA, which appears to be a fusion of two non-homologous though distantly related ATPases (65), there is a third aqueous chamber between the tiers formed by the tandem ATPase domains. The chambers within the ATPase domains may function in an analogous way to the unfolding chambers found in the GroEL family of chaperonins (66). This structural feature allows the charonins and chaperones to

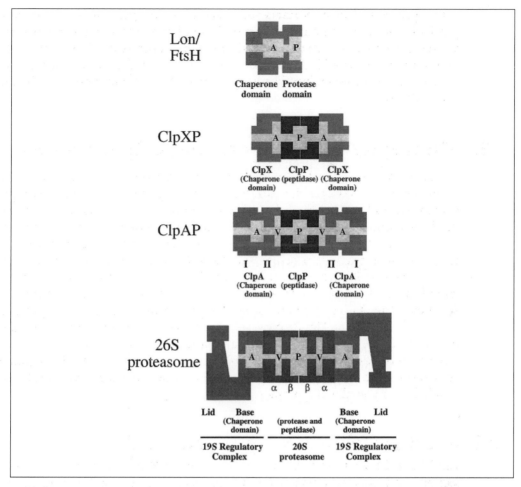

Fig. 3 Schematic drawing of cross-sections showing similar architectures of ATP-dependent proteases. The drawings are based on electron micrographs in work cited in the text. ATP-dependent proteases are high molecular weight multimers. Assembly of the subunits form hexagonal or heptagonal rings which enclose aqueous chambers containing the sites for both chaperone and proteolytic activities. ATPase/chaperone domains are directly associated with the protease domains or, in the case of Lon and FtsH, are fused together in a single polypeptide. Access to the proteolytic active sites, in the centre of the complex, is via axial channels through the rings and requires an extended polypeptide chain. Regulatory components and substrate binding sites are associated with the distal surface of the ATPase/chaperone rings. The means by which substrates gain access to the chaperone cavity has not been shown, but, by analogy with the chaperonins, probably involves major conformational changes that open an axial passage through the distal ring into the interior.

interact with exposed hydrophobic regions of damaged or unfolded proteins in an isolated environment, thus avoiding non-specific interactions with other cellular macromolecules (67).

The placement of the catalytic sites of ATP-dependent protease on the interior limits their availability to cellular proteins and imposes two requirements in order that proteolysis occur. First there must be a gating mechanism consisting of a recognition step, allowing appropriate substrates to be captured, and a processing step to remodel or unfold the substrate, so the substrate can fit through the connecting channels to the next chamber. Second, there must be a driving mechanism to facilitate the passage of the unfolded substrate into the interior proteolytic chamber. Biological control of proteolysis is exercised by factors that influence binding of appropriate substrates at recognition sites on the exterior surface or within internal chambers of the proteolytic complex. A second biological control point occurs following substrate binding and unfolding, because at that stage the substrate can either be translocated to the proteolytic chamber or released from the complex, dependent on factors that either affect the protease itself or influence the refolding of the substrate.

3.2 Functional domains of ClpA and ClpX

Clp ATPases belong to a subgroup of the AAA protein family (65). Sequence and structural alignments suggest that ClpA and ClpX consist of three domains: a variable N-terminal region, a conserved ATPase domain, and a C-terminal region analogous to the sensor-2 domain of AAA proteins (68). The function of the N-terminal domain is not known, but given its divergent sequence and structure in different Clp ATPases, it is likely to play a role reflecting a unique function for each protein, such as favouring different sub-cellular localizations, interacting with auxiliary factors, or binding of specific substrates. The ATPase domains bind and hydrolyze ATP. ATP binding stabilizes association of subunits in hexameric rings, which is required for expression of ATPase activity and association with ClpP (69).

Domains comparable to the C-terminal regions of Clp ATPases are also found internally in FtsH, in Lon protease, and within the N-terminal half of ClpA itself. When expressed and purified, the C-terminal domains from ClpA, ClpX, and others are able to specifically bind to their respective substrates by virtue of unique motifs that target those proteins for degradation (70, 71). These domains have therefore been referred to as SSD (for sensor and substrate discrimination). domains (71). In the presence of stabilizing nucleotides, both ClpA and ClpX are cleaved by proteases between the ATPase and SSD domains (72). Cleavage at this site is blocked by association with ClpP, suggesting that ClpP might interact with the ATPases through the C-terminal domain. If SSD domains are part of the interface with ClpP, the presence of substrate binding sites within them presents somewhat of a dilemma, since substrate sites might be expected to be on the surface of ATPase ring away from ClpP. In fact, electron micrographs of complexes between ClpAP and RepA in the presence of ATPγS show increased density on the distal ring surface (73). These data can be reconciled if the binding sites within the SSD domains are used after the substrate has initially interacted with surface binding sites on ClpAP or ClpXP and has entered one of the aqueous cavities. The SSD domain sites, which by analogy to NSF (74, 75) might form part of the wall of the aqueous cavity, could then serve to anchor or tether the substrate by specific interaction with the amino acids in the motif.

3.3. Comparison of structures of ClpAP and the proteasome

The structure of the 26S proteasome has a number of similarities to ClpAP. The proteolytic core of the proteasome is composed of two types of subunits, alpha and beta (76). The inner beta rings are composed of seven subunits of either identical gene products in prokaryotes or homologous products of separate genes in eukaryotes (77). In yeast and mammals, only three of the beta subunits have catalytic activity (78, 79). The active sites of the beta subunits are located within an aqueous cavity enclosed by the rings. The alpha subunits are related evolutionarily to the beta subunits, but they have no catalytic activity and their function is not known. The aqueous chamber between alpha and beta rings may serve as an intermediate binding region prior to substrates being translocated into the proteolytic chamber (76). Electron microscopy of isolated proteasome core particles with nanogold-conjugated peptide substrate showed nanogold situated on the apical surface of the beta rings over the axial channel leading to the internal chambers (80), suggesting that substrates enter the core particle through the axial channel.

The 19S regulatory component of the proteasome has two main parts—a base complex containing six ATPase subunits in addition to two other proteins, and a lid complex, which contains a number of other regulatory factors, including ubiquitin binding components (81). The base complex directly contacts the core particle, thus forming a minimal particle that resembles ClpAP or ClpXP. The complex of the base and the core also has proteolytic activity against unfolded proteins, which do not need to be ubiquitinated to be degraded (82). The base–core complex binds unfolded proteins and has a chaperone-like ability to prevent their aggregation and promote their refolding (82, 83). The lid complex acts as a regulatory component, controlling access to the base and presumably serving a substrate discrimination function by allowing only ubiquitinated proteins to be degraded (84). Given the complex composition of the lid, it is likely that it has additional roles, for example, modulation of the ubiquitination state of bound substrates, and is responsive to a variety of other metabolic and regulatory signals.

3.4 Lon and FtsH, multiple functional domains in a single polypeptide

Many of the activities of Lon (85, 86) resemble those seen with Clp proteases and the proteasome, raising the expectation that there would be underlying similarities in functional organization. Recent studies have suggested that there are indeed parallels in their structures. The molecular weights of Lon protease indicate that they are multimeric (87). Yeast Lon was reported to be a heptamer based on its size and on analysis of electron microscopic images (88). *E. coli* Lon, despite earlier reports, is most likely a hexamer (89). In the presence of ATP, *E. coli* Lon can associate to form a species with two hexamers, which in electron micrographs has an elongated shape (100 × 140 Å) (89), but it is too early to know whether it has the same barrel-like arrangement and internal chambers found in Clp proteases and the proteasome. *In vivo* data suggest that proteolytically inactive Lon can protect substrates from other proteases, implying that there are binding sites or chambers in which proteins are sequestered from the surrounding medium (90). FtsH is a membrane protease held in place by two N-terminal membrane spanning regions (91). The major domains, ATPase and C-terminal metalloprotease, are cytoplasmic. FtsH is also oligomeric, but the number of subunits making up the protomer is not known.

4. Enzymatic properties of ATP-dependent proteases

Dissection of the pathway of ATP-dependent protein degradation has been facilitated by the ability to separate the ATPase and protease components of the Clp proteases and the proteasome and demonstrate their independent functions. The ATPases alone have ATP-dependent chaperone-like activities, which they retain even as part of the holoenzyme complexes (92–94). This activity enables them to promote global unfolding of stable folded proteins (95). The proteolytic components have the ability to catalyse peptide bond cleavage, but that ability is severely restricted in the absence of the ATPase (96). Mutational studies suggest that the domains of proteases such as Lon and FtsH similarly express separate chaperone and protease activities (19).

4.1 Protein and peptide degradation activities

ClpP, ClpQ, and the 20S core of the proteasome can degrade unstructured peptides, but folded polypeptides cannot penetrate into the degradation chamber (80). Interaction with the ATPase components is promoted by nucleotide binding without ATP hydrolysis, and this interaction activates peptidase activity (34). With Lon and FtsH, nucleotide binding also alters interactions between the ATPase and protease domains, leading to activation of peptidase activity (97). In most cases, degradation of folded proteins and polypeptides requires the ATPase component and ATP hydrolysis (98). Protein degradation by the holoenzyme complexes is highly processive, in that the entire protein is usually degraded without the release of partially degraded intermediates (99, 100). The products of degradation are peptides of 5–15 amino acids (101). The size of the products is probably a stochastic function of the frequency of preferred cleavage sites (2–4 hydrophobic residues preceded by a charged or hydrophilic residue), and the rate of escape of peptides from the chamber (through exit sites not yet identified).

4.2 Chaperone activities of ClpA and ClpX

The first direct evidence that the ATPase components of proteases might promote protein unfolding came from studies showing that, independently of ClpP, ClpA and ClpX express molecular chaperone-like activity (92–94). For ClpA, chaperone activity was first shown with the P1 replication protein, RepA, which is purified as an inactive dimer and can be converted by ClpA to stable monomer capable of specifically binding DNA (92, 102). RepA monomers produced by ClpA cannot readily dimerize without structural perturbation, indicating that they have undergone structural remodelling by ClpA. RepA dimers bind to ClpA in the presence of ATPγS, but monomerization requires ATP hydrolysis (103). Chaperone activity requires a functional N-terminal ATPase domain, because mutations in the N-terminal, but not the C-terminal, ATP-binding consensus cause severe defects in RepA activation (104). ClpX chaperone activity *in vivo* and *in vitro* has been shown with the phage Mu transposase, MuA (93). ClpX promotes disassembly of a very stable MuA/DNA strand transfer complex during phage Mu transposition (36). Its ability to disrupt such a strong complex implies that ClpX has the ability to promote at least partial unfolding of a stably folded protein. ClpX also prevents aggregation of phage λ O protein and can promote dissolution of O protein aggregates (94).

The unfoldase activity of ClpA was shown with a stable folded form of green fluorescent protein carrying an SsrA tag at its C-terminus. Unfolding by ClpA requires ATP hydrolysis and globally disrupts the structure of GFP, as shown by exchange of amide deuterium buried in regions inaccessible to solvent in the native protein (95). ClpX has a similar ability to unfold stably folded proteins (50, 105). ClpA and ClpX interact with folded forms of their specific substrates, but unfolded forms of the same proteins bind to ClpA and ClpX with higher affinity and are released more slowly (49, 50). Thus, ClpA and ClpX recognize both specific motifs in native proteins and motifs exposed in unfolded proteins. ClpA binds a broader range of unfolded proteins than does ClpX, and can effectively trap an unfolded protein when that protein has no specific recognition motif. The unfoldase activity of ClpA and ClpX is needed so that the extended substrates may pass through the narrow access channels into the proteolytic sites.

4.3 Chaperone activities of ATP-dependent protease holoenzyme complexes

Demonstration of the chaperone activity of the holoenzyme complexes was important for two reasons. First, it established that this activity could in fact be part of the reaction cycle of ATP-dependent proteolysis. Second, it raised the possibility that the proteases may be bifunctional, both restoring activity to proteins and degrading them (see section 5.2). When an inactive mutant or a chemically inactivated form of ClpP was used, pre-assembled ClpAP complexes activated RepA as efficiently as ClpA alone (104). Moreover, even with wild-type ClpP, a significant fraction of bound RepA was activated at the same time that some RepA was being degraded (104). ClpP binding may actually stimulate the chaperone activity of ClpA; allosteric effects of ClpP binding are most evident with ClpA mutants. ATPase activity of ClpA proteins carrying mutations in the N-terminal ATP-binding site is stimulated by ClpP (69). Also, the chaperone activity of an N-terminal domain mutant, ClpA-K220V, which by itself is unable to activate RepA, is completely restored when ClpP is added (104). ClpP also activates ClpX chaperone activity, as indicated by its ability to stimulate disruption of the MuA transpososome by ClpX without degradation of MuA (106).

Evidence that other ATP-dependent proteases have chaperone-like activities is beginning to accumulate. Complexes of the proteasome that include the ATPases can prevent protein aggregation and promote refolding of unfolded proteins *in vitro* (82, 83). In yeast, a complex of the base (which contains the ATPases) with the 20S core promotes renaturation of heat denatured citrate synthase (82). The mammalian 19S regulatory particle can prevent aggregation of unfolded proteins, and catalytic amounts can also promote refolding of citrate synthase (83). Whether the ATPases are completely responsible for this activity and whether all the activities described are ATP-dependent is still being investigated. Recognition of unfolded proteins by the proteasome appears to be direct, because ubiquitination of the substrate is not required in the refolding reactions. *In vivo*, mutants of Lon and FtsH that lack proteolytic activity can promote assembly of macromolecular complexes, suggesting that the ATPase domains of these proteases also have chaperone-like activity (see below) (107, 108). Extrapolating from the example of ClpA, it is probably safe to conclude that the chaperone activity of all ATP-dependent proteases probably corresponds to an unfoldase activity.

4.4 The reaction pathway for ATP-dependent protein degradation

ATP-dependent proteolysis can be broken down into four steps: binding, unfolding, translocation, and cleavage. Substrates initially interact through the ATPase/ regulatory component. Electron micrographs show that substrates are bound to the distal ring surface in complexes of ClpAP (73). A portion of the substrate overlaps the centre of the ring, suggesting that substrates may pass into the complex through the axial channel. If the primary binding sites were located near the central channel, binding one molecule of most substrates would sterically hinder binding of others, limiting the stoichiometry of substrate binding. With RepA (a dimer of 70 kDa), for example, biochemical studies indicated that only one dimer binds per hexamer of ClpA (103).

In the case of Clp or the proteasome, the question arises whether the substrate interacts directly with the assembled holoenzyme or binds to the isolated regulatory component and is then brought to the protease. *In vitro* studies with ClpAP indicate that both routes are possible. Substrate tightly bound to ClpA in the presence of ATPγS was preferentially degraded upon addition of ATP and ClpP, even when excess competitive substrate was present (109). Similarly, substrate could be bound to pre-assembled ClpAP and preferentially degraded without exchange with competitive substrate (109). Under assay conditions, the ClpAP holoenzyme complex has a half-time for disassembly that is much longer than the turnover time for protein degradation and can therefore degrade a large stoichiometric excess of substrate without dissociation (110). Since the holoenzyme complex is the favoured state ClpAP at physiological concentrations of ATP, whether substrates encounter free

ClpA or ClpAP *in vivo* would appear to depend mostly on the relative intracellular levels of ClpA to ClpP. Association of the 20S proteasome core with the 19S regulatory complex is also very tight in the presence of nucleotide (22), suggesting that the holoenzyme form of the proteasome would also be favoured. It is always possible that regulatory factors *in vivo* could affect these associations or that certain physiological conditions, such as heat stress, could alter them.

The bound protein is probably not unfolded until at least one cycle of ATP hydrolysis occurs. GFP-SsrA bound to ClpA or ClpX in the presence of ATPγS is fluorescent (49, 50), and RepA bound to ClpA and released by disruption of the complex without ATP hydrolysis is not activated (103). How fast unfolding occurs once ATP is added is not known and may vary with different proteins. RepA bound to ClpA can be unfolded without dissociation, as shown by an experiment in which RepA complexed with ClpA was activated after ATP addition in the presence of excess competitive substrate to prevent rebinding. Since ATP turnover occurs at 10–20 per second per hexamer of ClpA (111), multiple cycles of ATP hydrolysis may nonetheless be required for structural remodelling and release.

Of course, degradation requires not release of the unfolded protein but its translocation into the proteolytic chamber. Translocation occurs rapidly. With ClpAP, between 10 and 30% of the bound RepA can be degraded in a single cycle of binding (109), and it would seem likely that this efficiency could be higher under optimized conditions or with specific substrates. At present it is not possible to kinetically distinguish translocation steps from further unfolding steps. Since the protein must be in an extended conformation to pass through the channel into the protease (61, 62), it would seem plausible that the two processes are coupled. Electron micrographs of ClpAP complexes reveal a large aqueous chamber, analogous to an Anfinsen chamber as found in GroEL, in the interior of ClpA (73). ClpP is positioned over the mouth of the chamber such that, once the protein is unfolded, terminal regions or extended polypeptide loops could enter the access channel into the degradative chamber. The distal channel out of the ClpA chamber is narrow, apparently to hinder release of unfolded proteins. Micrographs made after ATP addition to ClpAP in the presence of RepA show additional density attributable to RepA mostly within the degradation chamber of ClpP, suggesting that substrate translocation is concerted (73).

Once the substrate is in the proteolytic chamber, cleavage is very rapid. The turnover for peptide bond cleavage can be as high as 10 000 per minute or more (34). Since, cleavage occurs about every 10 peptide bonds, a protein the size of a RepA subunit (35 kDa) would be completely degraded (ca. 32 cuts) in much less than a second once it is in the chamber. In fact, protein turnover numbers are lower than this. For example, casein degradation turnover number is about 10 per minute, about 40 times slower than the maximum (111). Very few if any partially degraded intermediates arise, even though the ClpP chamber is large enough for about 22 kDa of protein at most. These data suggest that once translocation begins, it continues very rapidly, keeping pace with peptide bond cleavage. Thus, either unfolding or the initiating stages of translocation must be the rate limiting steps in the overall pathway.

4.5 Kinetic partitioning of substrates between release and translocation

ClpAP can simultaneously activate and degrade RepA (104), suggesting that unfolded substrate go through a checkpoint where they can be partitioned between two pathways: release or translocation (Fig. 2). Such a model is attractive because it implies an additional level of proof-reading to ensure that degradation of the protein is appropriate. Whether such partitioning occurs for all proteins is not known. RepA is a dimer, and one possibility is that partitioning occurs because one monomer is unfolded, breaking subunit contacts with the other monomer which is released. Since many substrates for ATP-dependent proteases are either dimers or parts of oligomeric complexes, such a mechanism would allow disruption of the complex and selective degradation of one protein in a complex.

The partitioning mechanism could also underlie trans-targeting phenomena, in which a substrate for degradation is inefficiently recognized by itself but interacts with the protease machinery as part of a complex with another protein (112). The carrier protein might bind to a site on the protease, positioning the target protein to interact with lower affinity sites that serve as substrate sites for the unfolding activity. Unfolding of the target protein would then release the carrier protein. Two possible examples of trans-targeting are the degradation of E. coli RpoS out of complexes of RpoS and RssB without degradation of RssB (113), and degradation of B. subtilis ComK from complexes with MecA. In both instances, the target protein is degraded very slowly or not at all in the absence of its binding partner, which interacts with the protease but is itself not degraded. A different trans-targeting mechanism is exemplified by the degradation of ubiquitin-conjugated proteins by the proteasome (22). Binding of ubiquitin to sites on the regulatory complex possibly positions a substrate protein to interact with the chaperone sites on the base ATPase-complex. In this case, unfolding of the target protein is linked to removal of the modifying ubiquitin chain, although the kinetics of these steps is not known. One possibility is that unfolding must start before ubiquitin is removed, ensuring that the substrate will be trapped and degraded. Premature removal of ubiquitin may result in dissociation of the protein from the chaperone site before it is unfolded.

5. The action and interaction of chaperones and proteases *in vivo*

Mutations in molecular chaperones, such as DnaK or GroEL in E. coli, can cause defects in folding and assembly of specific proteins, resulting in well-defined phenotypes (114). Chaperone mutations are also highly pleiotropic, and generally lead to accumulation of proteins in insoluble aggregates and to decreased intracellular proteolysis (115). These last data establish a connection between the activity of chaperones and the activity of proteases *in vivo*. There has been further speculation that chaperones might directly present unfolded proteins to proteases *in vivo*. Such a model is plausible, since there is evidence that both chaperones and proteases can interact with the same unfolded proteins (116). However, there is as yet no definitive evidence for a direct interaction between any of the known chaperones and any components of ATP-dependent proteases.

5.1 Cooperation between chaperones and proteases

In *E. coli*, abnormal proteins, as well as certain mutant proteins, are degraded more slowly in cells with decreased activity of molecular chaperones such as DnaK, DnaJ, GrpE, or GroEL (117–121). Most of these proteins are targeted by ATP-dependent proteases, particularly Lon (117, 121). Degradation of the *E. coli* Lon substrate, RcsA, is reduced in a *dnaJ* mutant (117). RcsA, which is found mostly in the soluble fraction in *dnaJ*$^+$ cells, is found in an insoluble form in *dnaJ* mutant cells. T4 lysozyme which is degraded when expressed in wild-type *E. coli*, is stabilized and accumulates in inclusion bodies in a *clpB* mutant (122). *clpB* mutants also accumulate aggregated proteins at high temperature, and these proteins may be less susceptible to degradation in extracts (123, and see Chapter 1). These data all suggest that chaperones promote degradation of a protein by maintaining proteins in a non-aggregated state.

In eukaryotic cells, degradation in mitochondria and in the cytosol is affected by molecular chaperones. Degradation of abnormal proteins by the yeast PIM1 protease (Lon) is reduced in the absence of Mdj1 or mt-Hsp70 (124, 125). In the presence of mitochondrial chaperones, proteins that accumulate in PIM1 mutants are found in the soluble fraction, but are found in insoluble aggregates in chaperone mutants (125). In yeast, cytosolic chaperones are required for some ubiquitin-dependent degradation. In the absence of Ydj1 function (126, 127), degradation of abnormal proteins as well as specific proteins, such as Cln3a and a GCN4-β-galactosidase fusion protein, is reduced. Since degradation of cyclins involves a complex process requiring multiple factors, the effect of molecular chaperones could be far removed from the proteolytic step. In the case of Cln3, phosphorylation, which is required for its degradation, is blocked in the Ydj1 mutant (126). Also, not all ubiquitin-dependent degradation (e.g., N-end rule substrates and cyclin Clb5) requires Ydj1, indicating that the chaperones act on specific substrates or on specific components of the degradative pathway. An interesting recent report of a novel activity for Vcp (valosin containing protein) suggests that specific carrier proteins could be involved in presenting ubiquitinated proteins to the proteasome. Since Vcp is an AAA protein family member, one possibility is that it has a chaperone-like activity, either protecting the ubiquitinated protein from deubiquitinating enzymes or unfolding the ubiquitinated protein to facilitate its degradation by the proteasome (C. C. Li, pers. comm.).

Cooperation between chaperones and proteases is highly orchestrated in quality control in the endoplasmic reticulum, as discussed in Chapter 8. Protein import into and folding within the ER is dependent on a number of chaperones resident in the ER, particularly calnexin and BiP/Kar2p and related proteins (4, 11, 128). Proteins that remain unfolded in the ER or which are incompletely translocated into the ER are degraded in a process that requires the cytosolic ubiquitin/proteasome system.

Misfolded proteins are transported back out of the ER by a retrograde transport apparatus that involves some components of the import apparatus, such as Sec61p. ER chaperones are needed to maintain the misfolded proteins in a soluble state for retrograde transport, and because most secreted proteins undergo post-translational glycosylation, glycosidases must remove at least some of the sugars which otherwise retard export. In yeast, the cytosolic chaperones are also required for degradation, but their roles have not been defined (11).

5.2 Is autonomous chaperone and protease interaction direct or indirect?

ATP-dependent proteases and molecular chaperones overlap in the initial recognition of unfolded proteins and other exposed hydrophobic regions. At one level, overlap in specificity should result in competition between chaperones and proteases. However, the net effect of chaperone activity on proteolysis is also influenced by the tendency of proteins with exposed hydrophobic regions to aggregate. Since chaperones are very abundant in cells and generally more abundant than proteases, most unfolded proteins should bind chaperones before interacting with proteases. By keeping the concentration of unfolded proteins free in solution low, chaperones prevent aggregation and in the long run keep them more accessible to proteases.

Is there a more direct functional interaction between chaperones and proteases? Molecular chaperones promote local or global unfolding of proteins and release them in a state that can spontaneously refold to a native or near-native structure. However, not all interactions between chaperones and proteins result in complete refolding. Recently, it has been found that different classes of chaperones can cooperate, and that a partially unfolded protein can bind to one chaperone, which prevents its aggregation, and then be refolded by another chaperone (27, 28, 129, 130). Whether chaperones can cooperate with proteases in the same manner has been a subject of speculation, but little definitive data supporting such a process exist. There is no evidence for direct binding between autonomous chaperones and ATP-dependent proteases, but interaction could occur through the bound protein. Chaperones of the Hsp70 family probably bind limited regions of protein substrates, which could leave portions of the protein accessible to proteases. If a recognition motif for a protease is present on the exposed region, the protease could bind sufficiently tightly to the protein to wrest it from the chaperone and degrade it. The validity of such a model awaits demonstration of ternary complexes between a chaperone, a protease, and a protein substrate.

5.3 Independent chaperone activity of components of ATP-dependent proteases

Evidence exists that the ATPase components of ATP-dependent proteases or homologues of these components have autonomous chaperone activity *in vivo*. Most of these activities are defined by studies showing that some degree of physiological function is retained with mutants that lack proteolytic activity but have retained an active ATPase domain. A clear example of this phenomenon is the growth of phage Mu on *E. coli* hosts lacking ClpP but not on hosts lacking ClpX (93, 131). Replicative transposition of Mu requires disassembly of strand transfer complexes in which MuA is tightly bound to DNA. The transposase, MuA, is a substrate for ClpXP, but apparently the disruption of the MuA/DNA complex by ClpX without degradation is sufficient to allow transposition to proceed (36, 93, 106, 132).

Over-expression of *E. coli* or *Mycobacterium smegmatis* Lon carrying mutations in the proteolytic active site can complement null mutations in the chromosomal *lon* gene (90, 133). In *E. coli*, the over-expressed proteolytically inactive Lon was shown to protect specific proteins from degradation by Lon or other proteases, while Lon lacking ATPase activity did not (90). Since the protected proteins did not aggregate into inclusion bodies, these data suggest that the Lon protease mutant could sequester proteins by unfolding and translocating them in the absence of degradative activity. An alternative explanation for these results is that, by interaction with the Lon chaperone domain, the proteins are converted to a protease resistant form.

In yeast, multicopy *LON* complements mutations that cause defects in assembly of mitochondrial inner membrane complexes (108). Complementation was observed with a mutant Lon lacking proteolytic activity but not with a mutant lacking ATPase activity. Thus, yeast Lon may prevent assembly intermediates from aggregating and eliminate defective components. Similar observations were made with the FtsH homologues, YTA-10 (AFG3) and YTA-12 (RCA1) in yeast. Unassembled subunits of membrane complexes are degraded in wild-type cells, but are stable in YTA-10 or YTA-12 mutants (107). Over-expression of YTA-10 and YTA-12 proteins lacking proteolytic activity allowed ATP synthase assembly and restored respiration. Thus, YTA-10–12 mediates assembly of inner membrane complexes and degradation of unassembled subunits (20).

5.4 Homologues of ATP-dependent proteases with autonomous chaperone activity

A striking example of the overlap in activity between autonomous chaperones and ATP-dependent proteases is the occurrence of homologues of the ATPases that appear to function exclusively as molecular chaperones. *E. coli* ClpB and *S. cerevisiae* Hsp104, close homologues of ClpA ($\sim$ 85% similarity) have no proteolytic activity (134, 135). ClpB/Hsp104 mutants tend to accumulate protein inclusion bodies and are sensitive to heat and other stress (136–138). These data suggest that members of the ClpB subfamily catalyse the disaggregation of proteins (18, and see Chapter 1). Hsp104 also appears to be required for the propagation of aberrant structures of a prion-like particle in yeast cells (139, 140), and *in vitro* Hsp104 can promote the conversion of a prion protein to a protease-resistant form (141). HSP78, a lower molecular weight homologue of ClpB/Hsp104, appears to have chaperone activity in yeast mitochondria. Over-production of HSP78 restores mitochondrial protein import activity in mt-Hsp70 mutants, and, when expressed in the cytosol, can substitute for Hsp104 in conferring thermotolerance (142, 143).

In *B. subtilis* the role of ClpC (a ClpA homologue) in the MecA-mediated degradation of the competence factor, ComK, is complex (144). ClpC/MecA appears to function either for sequestration or for structural remodelling of ComK (145). ComK is degraded by different Clp protease, possibly a complex of ClpE or ClpC and ClpP, which may interact with the ternary MecA/ClpC/ComK complex (144, 146). Thus, one ClpA homologue functions to bind a substrate, while another ClpA homologue forms a complex with ClpP and promotes the expected proteolytic activity.

HtrA, while not an ATP-dependent protease, undergoes a transition from a proteolytic form to a chaperone form that provides an intriguing insight into the complexity of defining chaperone activities for proteases. HtrA (also called DegP) is an essential periplasmic protein in *E. coli* involved in quality control of secreted proteins (147); it is highly conserved and found in human cells (148). HtrA has both protease and chaperone activity (149). At elevated temperatures, HtrA has protease activity and can degrade proteins with C-terminal hydrophobic sequences (150). At lower temperatures (e.g., 20 °C), HtrA undergoes a conformation switch and no longer expresses protease activity but continues to express chaperone-like activity on a variety of unfolded proteins *in vitro* (149). HtrA is predicted to have two PDZ-like domains in its C-terminal region which are required for protease but not chaperone activity. By analogy with other proteins, these domains could bind to exposed C-terminal hydrophobic tails of proteins, but they are also needed for oligomerization of HtrA (151). Chaperone-like activity can be assayed *in vivo* and does not require the proteolytic active site serine residue. The example of HtrA suggests that the interactions of protease or protease domains can have chaperone-like effects on unfolded proteins and could result in refolding or prevention of aggregation if the protease activity is blocked.

The examples of ClpB/Hsp104 and MecA illustrate the diversity of effects possible from the interactions between Clp proteins and folded or unfolded proteins. As more members of the Clp family are discovered, it seems likely that additional functions will be discovered, which will reflect unique modes of interaction between the Clp proteins and their substrates. What is not yet known is whether the diversity of functions will extend only to unique Clp proteins in different organisms, or will be seen for a given Clp protein acting on different substrates in the same cell.

5.5 Regulation by proteases: applying the 'coup de grace'

Do the chaperone components of ATP-dependent proteases have autonomous activity *in vivo*? The activity of Clp ATPases or Lon mutants lacking proteolytic activity suggests that the chaperone components of ATP-dependent proteases may function independently *in vivo*. Certainly, the occurrence of Clp ATPase homologues without cognate proteolytic components lends support to this idea. However, it may not be appropriate to extrapolate from results obtained with protease mutants, because, under normal physiological conditions with wild-type proteins, the chaperone components exist in complexes with the proteolytic components. The dissociation constant for ClpAP is in the nanomolar range and the half-time for dissociation is > 20 min when ATP is saturating (110). As long as ClpP is present in stoichiometric amounts with ClpA or ClpX, the ATPase should remain in stable complexes with it *in vivo*. To date, ClpP appears to be present in quantities to saturate ClpA and ClpX under the conditions tested.

Why is proteolysis necessary? For protein quality control, most of the proteins that interact strongly with the ATP-dependent protease are probably non-functional or cannot adopt a native conformation easily. Efficient transfer of the unfolded protein to the protease thus guarantees that it will not find its way back into the pool of proteins and cause interference with normal functional proteins. For regulatory degradation, the unfolding or disassembling activity of the complex could well be sufficient to accomplish the biological goal of disrupting a complex or eliminating an activity. In these cases, one can think of the protease as delivering the 'coup de grace'. Once the decision to inactivate some function has been made, the protease assures that the activity is not restored without integration of the global regulatory factors governing biosynthesis of the targeted protein. Turnover of the non-functional protein also serves to immediately recycle its constituent amino acids. Thus, the biological advantage of the proteolysis step in some regulatory processes may be subtle and difficult to evaluate under standard growth conditions. For many other systems, proteolysis is required to eliminate the targeted activity, and mutations in either the ATPase or the protease cause a defect in the biological function of the ATP-dependent protease.

The versatility of the ATP-dependent proteases is that they can function in several modes: restoring a proper fold to a protein, unfolding a protein to temporarily eliminate its activity, or rapidly unfolding and degrading a protein to reset its regulatory circuit to the ground state. Many of the proteins subject to rapid degradation by ATP-dependent proteases are proteins involved in global regulatory networks, often as part of the initial response to stress, environmental changes, developmental signals, and intracellular regulatory signals. Such global regulatory proteins are present in low amounts in the cells, and their synthesis is subject to strict transcriptional or translational control. The energy cost to regulate protein amounts in this way is relatively low. Many viruses and phage have also taken advantage of the degradative machinery to regulate their own decisions and reactions pertaining to lysis/lysogeny, integration/excision, and replication or to inactivate host factors detrimental to their survival. Partial activities (e.g., chaperone only) observed with phage proteins may not be sufficient to produce the full biological effect in the endogenous cellular pathways regulated by the ATP-dependent proteases.

6. Challenges for the future

Current research on ATP-dependent proteases is directed at further identifying specific and general recognition motifs in cellular proteins, defining the conditions that affect their exposure and accessibility of motifs; describing the mechanism of protein unfolding and translocation by the chaperone components and the holo-enzyme complexes of ATP-dependent proteases; determining the high resolution structure of these amazing multi-component machines; and visualizing the complexes of protein substrates with the proteases and their components. An additional challenge will be to determine the distribution of the components of ATP-dependent proteases between the free state and the proteolytically active complexed state *in vivo*. For the holoenzymes, it will be necessary to define what properties of the enzymes favor the refolding activity and which favor the degradative, and for individual substrates, what structural parameters affect their kinetic partitioning between release and degradation. Finally, with respect to recognition of general motifs in damaged proteins, it will be important to critically examine the possible functional and substrate-mediated physical interactions between autonomous chaperones and ATP-dependent proteases to define the interplay between these basic components of the protein quality control system.

References

1. Van Melderen, L., Thi, M. H. D., Lecchi, P., Gottesman, S., Couturier, M., and Maurizi, M. R. (1996). ATP-dependent degradation of CcdA by Lon protease. Effects of secondary structure and heterologous subunit interactions. *J. Biol. Chem.*, **271**, 27730.
2. Wickner, S., Maurizi, M. R., and Gottesman, S. (1999). Post-translational quality control: folding, refolding, and degradation of proteins. *Science*, **286**, 1888.
3. Bross, P., Corydon, T. J., Andresen, B. S., Jorgensen, M. M., Bolund, L., and Gregersen, N. (1999). Protein misfolding and degradation in genetic diseases. *Hum. Mutat.*, **14**, 186.
4. Plemper, R. K. and Wolf, D. H. (1999). Endoplasmic reticulum degradation. Reverse protein transport and its end in the proteasome. *Mol. Biol. Rep.*, **26**, 125.
5. Langer, T. and Neupert, W. (1996). Regulated protein degradation in mitochondria. *Experientia*, **52**, 1069.
6. Gottesman, S., Wickner, S., and Maurizi, M. R. (1997). Protein quality control: triage by chaperones and proteases. *Genes Dev.*, **11**, 815.
7. Goldberg, A. L. and John, A. C. S. (1976). Intracellular protein degradation in mammalian and bacterial cells. *Annu. Rev. Biochem.*, **45**, 747.
8. Gottesman, S. (1996). Proteases and their targets in *Escherichia coli*. *Annu. Rev. Genet.*, **30**, 465.
9. Gottesman, S. and Maurizi, M. R. (1992). Regulation by proteolysis: energy-dependent proteases and their targets. *Microbiol. Rev.*, **56**, 592.
10. Gottesman, S. (1999). Regulation by proteolysis: developmental switches. *Curr. Opin. Microbiol.*, **2**, 142.
11. Brodsky, J. L. and McCracken, A. A. (1999). ER protein quality control and proteasome-mediated protein degradation. *Semin. Cell Dev. Biol.*, **10**, 507.
12. Schmid, F. X. (1995). Protein folding. Prolyl isomerases join the fold. *Curr. Biol.*, **5**, 993.
13. Gilbert, H. F. (1997). Protein disulfide isomerase and assisted protein folding. *J. Biol. Chem.*, **272**, 29399.
14. Raina, S. and Missiakas, D. (1997). Making and breaking disulfide bonds. *Annu. Rev. Microbiol.*, **51**, 179.
15. Berlett, B. S. and Stadtman, E. R. (1997). Protein oxidation in aging, disease, and oxidative stress. *J. Biol. Chem.*, **272**, 20313.

16. Trombetta, E. S. and Helenius, A. (1998). Lectins as chaperones in glycoprotein folding. *Curr. Opin. Struct. Biol.*, **8**, 587.

17. Bradshaw, R. A., Brickey, W. W., and Walker, K. W. (1998). N-terminal processing: the methionine aminopeptidase and N alpha-acetyltransferase families. *Trends Biochem. Sci.*, **23**, 263.

18. Schirmer, E. C., Glover, J. R., Singer, M. A., and Lindquist, S. (1996). HSP100/Clp proteins: a common mechanism explains diverse functions. *Trends Biochem. Sci.*, **21**, 289.

19. Suzuki, C. K., Rep, M., van Dijl, J. M., Suda, K., Grivell, L. A., and Schatz, G. (1997). ATP-dependent proteases that also chaperone protein biogenesis. *Trends Biochem. Sci.*, **22**, 118.

20. Leonhard, K., Stiegler, A., Neupert, W., and Langer, T. (1999). Chaperone-like activity of the AAA domain of the yeast Yme1 AAA protease. *Nature*, **398**, 348.

21. Laney, J. D. and Hochstrasser, M. (1999). Substrate targeting in the ubiquitin system. *Cell*, **97**, 427.

22. Hershko, A. and Ciechanover, A. (1992). The ubiquitin system for protein degradation. *Annu. Rev. Biochem.*, **61**, 761.

23. Pickart, C. M. (1997). Targeting of substrates to the 26S proteasome. *FASEB J.*, **11**, 1055.

24. Haas, A. L. and Siepmann, T. J. (1997). Pathways of ubiquitin conjugation. *FASEB J.*, **11**, 1257.

25. Gottesman, S., Clark, W. P., de Crecy-Lagard, V., and Maurizi, M. R. (1993). ClpX, an alternative subunit for the ATP-dependent Clp protease of *Escherichia coli*. Sequence and in vivo activities. *J. Biol. Chem.*, **268**, 22618.

26. Frees, D. and Ingmer, H. (1999). ClpP participates in the degradation of misfolded protein in Lactococcus lactis. *Mol. Microbiol.*, **31**, 79.

27. Zolkiewski, M. (1999). ClpB cooperates with DnaK, DnaJ, and GrpE in suppressing protein aggregation. A novel multi-chaperone system from *Escherichia coli*. *J. Biol. Chem.*, **274**, 28083.

28. Goloubinoff, P., Mogk, A., Zvi, A. P., Tomoyasu, T., and Bukau, B. (1999). Sequential mechanism of solubilization and refolding of stable protein aggregates by a bichaperone network. *Proc. Natl. Acad. Sci. USA*, **96**, 13732.

29. Keiler, K. C., Waller, P. R., and Sauer, R. T. (1996). Role of a peptide tagging system in degradation of proteins synthesized from damaged messenger RNA. *Science*, **271**, 990.

30. Gottesman, S., Roche, E., Zhou, Y., and Sauer, R. T. (1998). The ClpXP and ClpAP proteases degrade proteins with carboxy-terminal peptide tails added by the SsrA-tagging system. *Genes Dev.*, **12**, 1338.

31. Shrader, T. E., Tobias, J. W., and Varshavsky, A. (1993). The N-end rule in *Escherichia coli*: cloning and analysis of the leucyl, phenylalanyl-tRNA-protein transferase gene aat. *J. Bacteriol.*, **175**, 4364.

32. Gierasch, L. M. (1994). Molecular chaperones. Panning for chaperone-binding peptides. *Curr. Biol*, **4**, 173.

33. Rudiger, S., Germeroth, L., Schneider-Mergener, J., and Bukau, B. (1997). Substrate specificity of the DnaK chaperone determined by screening cellulose-bound peptide libraries. *EMBO J.*, **16**, 1501.

34. Thompson, M. W. and Maurizi, M. R. (1994). Activity and specificity of *Escherichia coli* ClpAP protease in cleaving model peptide substrates. *J. Biol. Chem.*, **269**, 18201.

35. Herman, C., Thevenet, D., Bouloc, P., Walker, G. C., and D'Ari, R. (1998). Degradation of carboxy-terminal-tagged cytoplasmic proteins by the *Escherichia coli* protease HflB (FtsH). *Genes Dev.*, **12**, 1348.

36. Levchenko, I., Yamauchi, M., and Baker, T. A. (1997). ClpX and MuB interact with overlapping regions of Mu transposase: implications for control of the transposition pathway. *Genes Dev.*, **11**, 1561.

37. Laachouch, J. E., Desmet, L., Geuskens, V., Grimaud, R., and Toussaint, A. (1996). Bacteriophage Mu repressor as a target for the *Escherichia coli* ATP-dependent Clp protease. *EMBO J.*, **15**, 437.

38. Sonezaki, S., Ishii, Y., Okita, K., Sugino, T., Kondo, A., and Kato, Y. (1995). Overproduction and purification of SulA fusion protein in *Escherichia coli* and its degradation by Lon protease in vitro. *Appl. Microbiol. Biotechnol.*, **43**, 304.

39. Gonzalez, M., Frank, E. G., Levine, A. S., and Woodgate, R. (1998). Lon-mediated proteolysis of the Escherichia coli UmuD mutagenesis protein: in vitro degradation and identification of residues required for proteolysis. *Genes Dev.*, **12**, 3889.

39a. Gonzalez, M., Maurizi, M. R., Rasulova, F., and Woodgate, R. (2000). Trans-targeting of UmuD' in a heterodimeric UmuD/D' complex by ClpXP protease. *EMBO Jou.*, **19**, 5251.

40. Wang, L., Wilson, S., and Elliott, T. (1999). A mutant HemA protein with positive charge close to the N terminus is stabilized against heme-regulated proteolysis in Salmonella typhimurium. *J. Bacteriol.*, **181**, 6033.

40a. Hoskins, J. R., Kim, S. Y., and Wickner, S. (2000). Substrate recognition by the ClpA chaperone component of ClpAP protease. *J. Biol. Chem.*, **275**, 35361.

41. Gottesman, S., Clark, W. P., and Maurizi, M. R. (1990). The ATP-dependent Clp protease of Escherichia coli. Sequence of clpA and identification of a Clp-specific substrate. *J. Biol. Chem.*, **265**, 7886.

42. Gonciarz-Swiatek, M., Wawrzynow, A., Um, S. J., Learn, B. A., McMacken, R., Kelley, W. L., *et al.* (1999). Recognition, targeting, and hydrolysis of the lambda O replication protein by the ClpP/ClpX protease. *J. Biol. Chem.*, **274**, 13999.

43. Hengge-Aronis, R. (1993). Survival of hunger and stress: the role of *rpoS* in early stationary phase gene regulation in *Escherichia coli. Cell*, **72**, 165.

44. Higashitani, A., Ishii, Y., Kato, Y., and Koriuchi, K. (1997). Functional dissection of a cell-division inhibitor, SulA, of Escherichia coli and its negative regulation by Lon. *Mol. Gen. Genet.*, **254**, 351.

45. Sen, M. and Rothfield, L. I. (1998). Stability of the Escherichia coli division inhibitor protein MinC requires determinants in the carboxy-terminal region of the protein. *J. Bacteriol.*, **180**, 175.

46. Varshavsky, A. (1992). The N-end rule. *Cell*, **69**, 725.

47. Tobias, J. W., Shrader, T. E., Rocap, G., and Varshavsky, A. (1991). The N-end rule in bacteria. *Science*, **254**, 1374.

48. Wickner, S. and Maurizi, M. R. (1999). Here's the hook: similar substrate binding sites in the chaperone domains of Clp and Lon. *Proc. Natl. Acad. Sci. USA*, **96**, 8318.

49. Hoskins, J. R., Singh, S. K., Maurizi, M. R., and Wickner, S. (2000). Binding of unfolded proteins by ClpA and degradation by ClpAP. *Proc. Natl. Acad. Sci. USA*, **97**, 8892.

50. Singh, S. K., Grimaud, R., Hoskins, J. R., Wickner, S., and Maurizi, M. R. (2000). Unfolding and internalization of proteins by ClpXP and ClpAP. *Proc. Natl. Acad. Sci. USA*, **97**, 8898.

51. Kirschner, M. (1999). Intracellular proteolysis. *Trends Cell. Biol.*, **9**, M42.

52. Lupas, A., Flanagan, J. M., Tamura, T., and Baumeister, W. (1997). Self-compartmentalizing proteases. *Trends Biochem. Sci.*, **22**, 399.

53. Maurizi, M. R. (1992). Proteases and protein degradation in *Escherichia coli. Experientia*, **48**, 178.

54. Goldberg, A. L. (1992). The mechanism and functions of ATP-dependent proteases in bacterial and animal cells. *Eur. J. Biochem.*, **203**, 9.

55. Schumann, W. (1999). FtsH – a single-chain charonin? *FEMS Microbiol. Rev.*, **23**, 1.

56. Sorimachi, H., Ishiura, S., and Suzuki, K. (1997). Structure and physiological function of calpains. *Biochem. J.*, **328**, 721.

57. Skidgel, R. A. and Erdos, E. G. (1998). Cellular carboxypeptidases. *Immunol. Rev.*, **161**, 129.

58. Stroh, C. and Schulze-Osthoff, K. (1998). Death by a thousand cuts: an ever increasing list of caspase substrates. *Cell Death Differ.*, **5**, 997.

59. Beuron, F., Maurizi, M. R., Belnap, D. M., Kocsis, E., Booy, F. P., Kessel, M., and Steven, A. C. (1998). At sixes and sevens: characterization of the symmetry mismatch of the ClpAP chaperone-assisted protease. *J. Struct. Biol.*, **123**, 248.

60. Kessel, M., Maurizi, M. R., Kim, B., Kocsis, E., Trus, B. L., Singh, S. K., and Steven, A. C. (1995). Homology in structural organization between E. coli ClpAP protease and the eukaryotic 26 S proteasome. *J. Mol. Biol.*, **250**, 587.

61. Wang, J., Hartling, J. A., and Flanagan, J. M. (1997). The structure of ClpP at 2.3 A resolution suggests a model for ATP-dependent proteolysis. *Cell*, **91**, 447.

62. Lowe, J., Stock, D., Jap, B., Zwickl, P., Baumeister, W., and Huber, R. (1995). Crystal structure of the 20S proteasome from the archaeon T. acidophilum at 3.4 A resolution [see comments]. *Science*, **268**, 533.

63. Bochtler, M., Ditzel, L., Groll, M., and Huber, R. (1997). Crystal structure of heat shock locus V (HslV) from *Escherichia coli*. *Proc. Natl. Acad. Sci. USA*, **94**, 6070.

64. De Mot, R., Nagy, I., Walz, J., and Baumeister, W. (1999). Proteasomes and other self-compartmentalizing proteases in prokaryotes. *Trends Microbiol.*, **7**, 88.

65. Neuwald, A. F., Aravind, L., Spouge, J. L., and Koonin, E. V. (1999). AAA+: A class of chaperone-like ATPases associated with the assembly, operation, and disassembly of protein complexes. *Genome Res.*, **9**, 27.

66. Roseman, A. M., Chen, S., White, H., Braig, K., and Saibil, H. R. (1996). The chaperonin ATPase cycle: mechanism of allosteric switching and movements of substrate-binding domains in GroEL. *Cell*, **18**, 241.

67. Bukau, B. and Horwich, A. L. (1998). The Hsp70 and Hsp60 chaperone machines. *Cell*, **92**, 351.

68. Gottesman, S., Wickner, S., Jubete, Y., Singh, S. K., Kessel, M., and Maurizi, M. R. (1995). Selective energy-dependent proteolysis in *Escherichia coli*. *Cold Spring Harbor Sympos. Quant. Biol.*, **Volume LX**, 533.

69. Singh, S. K. and Maurizi, M. R. (1994). Mutational analysis demonstrates different functional roles for the two ATP-binding sites in ClpAP protease from Escherichia coli. *J. Biol. Chem.*, **269**, 29537.

70. Levchenko, I., Smith, C. K., Walsh, N. P., Sauer, R. T., and Baker, T. A. (1997). PDZ-like domains mediate binding specificity in the Clp/Hsp100 family of chaperones and protease regulatory subunits. *Cell*, **91**, 939.

71. Smith, C. K., Baker, T. A., and Sauer, R. T. (1999). Lon and Clp family proteases and chaperones share homologous substrate-recognition domains. *Proc. Natl. Acad. Sci. USA*, **96**, 6678.

72. Singh, S. K., Rozycki, J., and Maurizi, M. R. (2001). Domain organization and structural changes upon nucleotide binding in E. *coli* ClpX and ClpA. (Submitted.)

73. Ishikawa, T., Beuron, F., Kessel, M., Wickner, S., Maurizi, M. R., and Steven, A. C. (2001). ATP-dependent conformational transitions and translocation of bound substrates in ClpAP protease. (Submitted.)

74. Lenzen, C. U., Steinmann, D., Whiteheart, S. W., and Weis, W. I. (1998). Crystal structure of the hexamerization domain of N-ethylmaleimide-sensitive fusion protein. *Cell*, **94**, 525.

75. Yu, R. C., Hanson, P. I., Jahn, R., and Brunger, A. T. (1998). Structure of the ATP-dependent oligomerization domain of N-ethylmaleimide sensitive factor complexed with ATP. *Nat. Struct. Biol.*, **5**, 803.

76. Baumeister, W. and Lupas, A. (1997). The proteasome. *Curr. Opin. Struct. Bio.l*, **7**, 273.

77. Seemuller, E., Lupas, A., Stock, D., Lowe, J., Huber, R., and Baumeister, W. (1995). Proteasome from Thermoplasma acidophilum: a threonine protease [see comments]. *Science*, **268**, 579.

78. Groll, M., Ditzel, L., Lowe, J., Stock, D., Bochtler, M., Bartunik, H. D., and Huber, R. (1997). Structure of 20S proteasome from yeast at 2.4 A resolution [see comments]. *Nature*, **386**, 463.

79. Arendt, C. S. and Hochstrasser, M. (1997). Identification of the yeast 20S proteasome catalytic centers and subunit interactions required for active-site formation. *Proc. Natl. Acad. Sci. USA*, **94**, 7156.

80. Wenzel, T. and Baumeister, W. (1995). Conformational constraints in protein degradation by the 20S proteasome. *Nat. Struct. Biol.*, **2**, 199.

81. Glickman, M. H., Rubin, D. M., Coux, O., Wefes, I., Pfeifer, G., Cjeka, Z., *et al.* (1998). A subcomplex of the proteasome regulatory particle required for ubiquitin-conjugate degradation and related to the COP9-signalosome and eIF3. *Cell*, **94**, 615.

82. Braun, B. C., Glickman, M., Kraft, R., Dahlmann, B., Kloetzel, P. M., Finley, D., and Schmidt, M. (1999). The base of the proteasome regulatory particle exhibits chaperone-like activity. *Nat. Cell Biol.*, **1**, 221.

83. Strickland, E., Hakala, K., Thomas, P. J., and DeMartino, G. N. (2000). Recognition of misfolding proteins by PA700, the regulatory subcomplex of the 26S proteasome. *J. Biol. Chem.*, **275**, 5565.

84. Fu, H., Girod, P. A., Doelling, J. H., van Nocker, S., Hochstrasser, M., Finley, D., and Vierstra, R. D. (1999). Structure and functional analysis of the 26S proteasome subunits from plants. *Mol. Biol. Rep.*, **26**, 137.

85. Chung, C. H. and Goldberg, A. L. (1981). The product of the lon (capR) gene in Escherichia coli is the ATP-dependent protease, protease La. *Proc. Natl. Acad. Sci. USA*, **78**, 4931.

86. Charette, M. F., Henderson, G. W., and Markovitz, A. (1981). ATP hydrolysis-dependent protease activity of the lon (capR) protein of Escherichia coli K-12. *Proc. Natl. Acad. Sci. USA*, **78**, 4728.

87. Goldberg, A. L., Moerschell, R. P., Chung, C. H., and Maurizi, M. R. (1994). ATP-dependent protease La (lon) from Escherichia coli. *Methods Enzymol.*, **244**, 350.

88. Stahlberg, H., Kutejova, E., Suda, K., Wolpensinger, B., Lustig, A., Schatz, G., *et al.* (1999). Mitochondrial Lon of Saccharomyces cerevisiae is a ring-shaped protease with seven flexible subunits. *Proc. Natl. Acad. Sci. USA*, **96**, 6787.

89. Leffers, G. (2000). Oligomeric state of E. coli Lon. (Unpublished.)

90. Van Melderen, L. and Gottesman, S. (1999). Substrate sequestration by a proteolytically inactive Lon mutant. *Proc. Natl. Acad. Sci. USA*, **96**, 6064.

91. Akiyama, Y., Yoshihisa, T., and Ito, K. (1995). FtsH, a membrane-bound ATPase, forms a complex in the cytoplasmic membrane of Escherichia coli. *J. Biol. Chem.*, **270**, 23485.

92. Wickner, S., Gottesman, S., Skowyra, D., Hoskins, J., McKenney, K., and Maurizi, M. R. (1994). A molecular chaperone, ClpA, functions like DnaK and DnaJ. *Proc. Natl. Acad. Sci. USA*, **91**, 12218.

93. Levchenko, I., Luo, L., and Baker, T. A. (1995). Disassembly of the Mu transposase tetramer by the ClpX chaperone. *Genes & Dev.*, **9**, 2399.

94. Wawrzynow, A., Wojtkowiak, D., Marszalek, J., Banecki, B., Jonsen, M., Graves, B., *et al.* (1995). The ClpX heat-shock protein of Escherichia coli, the ATP-dependent substrate specificity component of the ClpP-ClpX protease, is a novel molecular chaperone. *EMBO J.*, **14**, 1867.

95. Weber-Ban, E. U., Reid, B. G., Miranker, A. D., and Horwich, A. L. (1999). Global unfolding of a substrate protein by the Hsp100 chaperone ClpA [see comments]. *Nature*, **401**, 90.

96. Woo, K. M., Chung, W. J., Ha, D. B., Goldberg, A. L., and Chung, C. H. (1989). Protease Ti from *Escherichia coli* requires ATP hydrolysis for protein breakdown but not for hydrolysis of small peptides. *J. Biol. Chem.*, **264**, 2088.

97. Waxman, L. and Goldberg, A. L. (1986). Selectivity of intracellular proteolysis: protein substrates activate the ATP-dependent protease (La). *Science*, **232**, 500.

98. Goldberg, A. L. and Waxman, L. (1985). The role of ATP hydrolysis in the breakdown of proteins and peptides by protease La from *Escherichia coli*. *J. Biol. Chem.*, **260**, 12029.

99. Thompson, M. W., Singh, S. K., and Maurizi, M. R. (1994). Processive degradation of proteins by the ATP-dependent Clp protease from Escherichia coli. Requirement for the multiple array of active sites in ClpP but not ATP hydrolysis. *J. Biol. Chem.*, **269**, 18209.

100. Akopian, T. N., Kisselev, A. F., and Goldberg, A. L. (1997). Processive degradation of proteins and other catalytic properties of the proteasome from *Thermoplasma acidophilum*. *J. Biol. Chem.*, **272**, 1791.

101. Kisselev, A. F., Akopian, T. N., Woo, K. M., and Goldberg, A. L. (1999). The sizes of peptides generated from protein by mammalian 26 and 20 S proteasomes. Implications for understanding the degradative mechanism and antigen presentation. *J. Biol. Chem.*, **274**, 3363.

102. Wickner, S., Hoskins, J., and McKenney, K. (1991). Monomerization of RepA dimers by heat shock proteins activates binding to DNA replication origin. *Proc. Natl. Acad. Sci. USA*, **88**, 7903.

103. Pak, M. and Wickner, S. (1997). Mechanism of protein remodeling by ClpA chaperone [published erratum appears in *Proc. Natl. Acad. Sci. USA* 1997 Sep 16; 94(19): 10485]. *Proc. Natl. Acad. Sci. USA*, **94**, 4901.

104. Pak, M., Hoskins, J. R., Singh, S. K., Maurizi, M. R., and Wickner, S. (1999). ClpAP chaperone and protease activities function concurrently and the chaperone activity requires the N-terminal ClpA ATP-binding site. *J. Biol. Chem.*, **274**, 19316.

105. Kim, Y. I., Burton, R. E., Burton, B. M., Sauer, R. T., and Baker, T. A. (2000). Dynamics of substrate denaturation and translocation by the ClpXP degradation machine. *Mol. Cell*, **5**, 639.

106. Jones, J. M., Welty, D. J., and Nakai, H. (1998). Versatile action of Escherichia coli ClpXP as protease or molecular chaperone for bacteriophage Mu transposition. *J. Biol. Chem.*, **273**, 459.

107. Arlt, H., Tauer, R., Feldmann, H., Neupert, W., and Langer, T. (1996). The YTA10–12 complex, an AAA protease with chaperone-like activity in the inner membrane of mitochondria. *Cell*, **85**, 875.

108. Rep, M., van Dijl, J. M., Suda, K., Schatz, G., Grivell, L. A., and Suzuki, C. K. (1996). Promotion of mitochondrial membrane complex assembly by a proteolytically inactive

yeast Lon [published erratum appears in *Science* 1997 Feb 7; 275(5301): 741]. *Science*, **274**, 103.

109. Hoskins, J. R., Pak, M., Maurizi, M. R., and Wickner, S. (1998). The role of the ClpA chaperone in proteolysis by ClpAP. *Proc. Natl. Acad. Sci. USA*, **95**, 12135.

110. Singh, S. K., Guo, F., and Maurizi, M. R. (1999). ClpA and ClpP remain associated during multiple rounds of ATP-dependent protein degradation by ClpAP protease. *Biochemistry*, **38**, 14906.

111. Maurizi, M. R., Thompson, M. W., Singh, S. K., and Kim, S. H. (1994). Endopeptidase Clp: ATP-dependent Clp protease from Escherichia coli. *Methods Enzymol.*, **244**, 314.

112. Johnson, E. S., Gonda, D. K., and Varshavsky, A. (1990). cis-trans recognition and subunit-specific degradation of short-lived proteins. *Nature*, **346**, 287.

113. Zhou, N. Y., Gottesman, S., Hoskins, J. R., Maurizi, M., and Wickner, S. (2001). The RssB response regulator directly targets σ^s for degradation by ClpXP. *Genes & Dev.* (in press.)

114. Polissi, A., Goffin, L. G., and Georgopoulos, C. (1995). The Escherichia coli heat shock response and bacteriophage lambda development. *FEMS Microbiol. Rev.*, **17**, 159.

115. Sherman, M. Y. and Goldberg, A. L. (1996). Involvement of molecular chaperones in intracellular protein breakdown. *EXS*, **77**, 57.

116. Sherman, M. Y. and Goldberg, A. L. (1991). Formation in vitro of complexes between an abnormal fusion protein and the heat shock proteins from *Escherichia coli* and yeast mitochondria. *J. Bacteriol.*, **173**, 7249.

117. Jubete, Y., Maurizi, M. R., and Gottesman, S. (1996). Role of the heat shock protein DnaJ in the lon-dependent degradation of naturally unstable proteins. *J. Biol. Chem.*, **271**, 30798.

118. Kandror, O., Busconi, L., Sherman, M., and Goldberg, A. L. (1994). Rapid degradation of an abnormal protein in Escherichia coli involves the chaperones GroEL and GroES. *J. Biol. Chem.*, **269**, 23575.

119. Keller, J. A. and Simon, L. D. (1988). Divergent effects of a *dnaK* mutation on abnormal protein degradation in *Escherichia coli*. *Mol. Microbiol.*, **2**, 31.

120. Sherman, M. and Goldberg, A. L. (1992). Involvement of the chaperonin dnaK in the rapid degradation of a mutant protein in Escherichia coli. *EMBO J.*, **11**, 71.

121. Straus, D. B., Walter, W. A., and Gross, C. (1988). *Escherichia coli* heat shock gene mutants are defective in proteolysis. *Genes Dev.*, **2**, 1851.

122. Inoue, I. and Rechsteiner, M. (1994). On the relationship between the metabolic and thermodynamic stabilities of T4 lysozymes. Measurements in Escherichia coli. *J. Biol. Chem.*, **269**, 29241.

123. Laskowska, E., Kuczynska-Wisnik, D., Skorko-Glonek, J., and Taylor, A. (1996). Degradation by proteases Lon, Clp and HtrA, of Escherichia coli proteins aggregated in vivo by heat shock; HtrA protease action in vivo and in vitro. *Mol. Microbiol*, **22**, 555.

124. Suzuki, C. K., Suda, K., Wang, N., and Schatz, G. (1994). Requirement for the yeast gene LON in intramitochondrial proteolysis and maintenance of respiration. *Science*, **264**, 891.

125. Wagner, I., Arlt, H., van Dyck, L., Langer, T., and Neupert, W. (1994). Molecular chaperones cooperate with PIM1 protease in the degradation of misfolded proteins in mitochondria. *EMBO J.*, **13**, 5135.

126. Yaglom, J. A., Goldberg, A. L., Finley, D., and Sherman, M. Y. (1996). The molecular chaperone Ydj1 is required for the p34CDC28-dependent phosphorylation of the cyclin Cln3 that signals its degradation. *Mol. Cell Biol.*, **16**, 3679.

127. Lee, D. H., Sherman, M. Y., and Goldberg, A. L. (1996). Involvement of the molecular chaperone Ydj1 in the ubiquitin-dependent degradation of short-lived and abnormal proteins in Saccharomyces cerevisiae. *Mol. Cell Biol.*, **16**, 4773.

128. Ellgaard, L., Molinari, M., and Helenius, A. (1999). Setting the standards: quality control in the secretory pathway. *Science*, **286**, 1882.

129. Glover, J. R. and Lindquist, S. (1998). Hsp104, Hsp70, and Hsp40: a novel chaperone system that rescues previously aggregated proteins. *Cell*, **94**, 73.

130. Motohashi, K., Watanabe, Y., Yohda, M., and Yoshida, M. (1999). Heat-inactivated proteins are rescued by the DnaK.J-GrpE set and ClpB chaperones. *Proc. Natl. Acad. Sci. USA*, **96**, 7184.

131. Mhammedi-Alaoui, A., Pato, M., Gama, M. J., and Toussaint, A. (1994). A new component of bacteriophage Mu replicative transposition machinery: the Escherichia coli ClpX protein. *Mol. Microbiol.*, **11**, 1109.

132. Kruklitis, R., Welty, D. J., and Nakai, H. (1996). ClpX protein of Escherichia coli activates bacteriophage Mu transposase in the strand transfer complex for initiation of Mu DNA synthesis. *EMBO J.*, **15**, 935.

133. Roudiak, S. G. and Shrader, T. E. (1998). Functional role of the N-terminal region of the Lon protease from Mycobacterium smegmatis. *Biochemistry*, **37**, 11255.

134. Woo, K. M., Kim, K. I., Goldberg, A. L., Ha, D. B., and Chung, C. H. (1992). The heat-shock protein ClpB in Escherichia coli is a protein-activated ATPase. *J. Biol. Chem.*, **267**, 20429.

135. Parsell, D. A., Kowal, A. S., and Lindquist, S. (1994). Saccharomyces cerevisiae Hsp104 protein: purification and characterization of ATP-induced structural changes. *J. Biol. Chem.*, **269**, 4480.

136. Squires, C. L., Pedersen, S., Ross, B. M., and Squires, C. (1991). ClpB is the *Escherichia coli* heat shock protein F84.1. *J. Bacteriol.*, **173**, 4254.

137. Parsell, D. A., Kowal, A. S., Singer, M. A., and Lindquist, S. (1994). Protein disaggregation mediated by heat-shock protein Hsp104. *Nature*, **372**, 475.

138. Sanchez, Y., Taulien, J., Borkovich, K. A., and Lindquist, S. (1992). Hsp104 is required for tolerance to many forms of stress. *EMBO J.*, **11**, 2357.

139. Chernoff, Y. O., Lindquist, S. L., Ono, B., Inge-Vechtomov, S. G., and Liebman, S. W. (1995). Role of the chaperone protein Hsp104 in propagation of the yeast prion-like factor [psi+]. *Science*, **268**, 881.

140. Paushkin, S. V., Kushnirov, V. V., Smirnov, V. N., and Ter-Avanesyan, M. D. (1996). Propagation of the yeast prion-like [psi+] determinant is mediated by oligomerization of the SUP35-encoded polypeptide chain release factor. *EMBO J.*, **15**, 3127.

141. DebBurman, S. K., Raymond, G. J., Caughey, B., and Lindquist, S. (1997). Chaperone-supervised conversion of prion protein to its protease-resistant form. *Proc. Natl. Acad. Sci. USA*, **94**, 13938.

142. Schmitt, M., Neupert, W., and Langer, T. (1995). Hsp78, a Clp homologue within mitochondria, can substitute for chaperone functions of mt-hsp70. *EMBO J.*, **14**, 3434.

143. Schmitt, M., Neupert, W., and Langer, T. (1996). The molecular chaperone Hsp78 confers compartment-specific thermotolerance to mitochondria. *J. Cell. Biol.*, **134**, 1375.

144. Turgay, K., Hahn, J., Burghoorn, J., and Dubnau, D. (1998). Competence in Bacillus subtilis is controlled by regulated proteolysis of a transcription factor. *EMBO J.*, **17**, 6730.

145. Turgay, K., Hamoen, L. W., Venema, G., and Dubnau, D. (1997). Biochemical characterization of a molecular switch involving the heat shock protein ClpC, which controls the activity of ComK, the competence transcription factor of Bacillus subtilis. *Genes Dev.*, **11**, 119.

146. Msadek, T., Dartois, V., Kunst, F., Herbaud, M. L., Denizot, F., and Rapoport, G. (1998). ClpP of Bacillus subtilis is required for competence development, motility, degradative

enzyme synthesis, growth at high temperature and sporulation. *Mol. Microbiol.*, **27**, 899.

147. Missiakas, D. and Raina, S. (1997). Protein folding in the bacterial periplasm. *J. Bacteriol.*, **179**, 2465.

148. Hu, S. I., Carozza, M., Klein, M., Nantermet, P., Luk, D., and Crowl, R. M. (1998). Human HtrA, an evolutionarily conserved serine protease identified as a differentially expressed gene product in osteoarthritic cartilage. *J. Biol. Chem.*, **273**, 34406.

149. Spiess, C., Bell, A., and Ehrmann, M. (1999). A temperature-dependent switch from chaperone to protease in a widely conserved heat shock protein. *Cell*, **97**, 339.

150. Silber, K. R., Keiler, K. C., and Sauer, R. T. (1992). Tsp: a tailspecific protease that selectively degrades proteins with nonpolar C termini. *Proc. Natl. Acad. Sci. USA*, **89**, 295.

151. Sassoon, N., Arie, J. P., and Betton, J. M. (1999). PDZ domains determine the native oligomeric structure of the degP (HtrA). protease. *Mol. Microbiol.*, **33**, 583.

10 | Regulation of expression of molecular chaperones

PETER A. LUND

1. Introduction

Many molecular chaperones are also heat shock proteins: their expression in cells is elevated when the cells are exposed to temperatures a few degrees above normal growth temperature. The regulation of the heat shock response has been the subject of intense study for some years, partly because, as with all expression studies, there is considerable interest in tracing the pathways that lead from an external effect (in this case, an increase in temperature) which generates some kind of inducing signal that can be read by components in the cell, through to a change in gene expression. Following the pathways from external signal to gene induction becomes particularly intriguing when the signal is read in a cellular compartment which is separated by one or more membranes from where the response takes place. As many heat shock proteins are also molecular chaperones, there is also considerable interest in relating the nature of the inducing signal(s) to the processes with which the chaperones are involved. In this chapter, I will discuss some of the many ways in which organisms can sense the need for elevated chaperone expression, predominantly by using examples drawn from prokaryotes where many of these pathways have been well characterized. I will endeavour in several cases to follow the pathways from the external event to the change in gene expression, and to discuss the experimental evidence for the nature of the intracellular inducing signals. Some aspects of the regulation of chaperone gene expression by stress in eukaryotes will also be briefly discussed in the same light.

2. Chaperone gene expression in prokaryotes

2.1 Expression of many cytoplasmic chaperones is constitutive and is up-regulated by heat shock and related stresses

The major cytoplasmic chaperones in *E. coli* (as always, the best studied example) are all expressed constitutively, some at quite high levels, which reflects their importance under all growth conditions (see Chapter 1). In addition, these proteins are also expressed at higher levels following a heat shock. This has been shown in numerous studies over the years, using both analysis of labelled proteins in pulse chase experiments, and studies of the expression of individual genes and operons. The availability of an ordered cosmid library covering the complete *E. coli* genome, and more recently the availability of the complete genome sequence of *E. coli*, has made the simultaneous analysis of expression of all the genes in the organism under different conditions possible, and such analysis has confirmed that nearly all the known cytoplasmic chaperones, along with many proteases, including those referred to as charonins in Chapter 9, are strongly induced on heat shock (1). Exceptions include the rather specialized cases such as trigger factor (which is also a peptidyl prolyl *cis-trans* isomerase, see Chapter 1) and SecB (which has a limited range of substrates, see Chapter 2). Relative levels of expression in terms of actual functional complexes of some of the main chaperones have been determined both at 30 °C and 42 °C (2), and they show that in all cases examined (ClpB, HtpG, DnaK, GroEL, and IbpA/B) the levels of active species increase after heat shock by between two- and eightfold depending on the particular protein under study. DnaK is the most abundant cytoplasmic chaperone (in terms of active species) at both temperatures, consistent with the important role of this protein in preventing and reversing protein aggregation at high temperature (2; see Chapter 1). The increases in levels of these proteins generally result from increased transcription of the genes at higher temperatures.

The stress most commonly used to study the heat shock response is of course heat shock itself, which is easily administered to a culture of growing cells. In this case, the response is always to a growth temperature a few degrees above the normal growth temperature of the organism; thus, psycrophiles that grow at 5 °C show induction of Hsps at 15 °C, while thermophiles that flourish at 84 °C still show a heat shock response when their growth temperature is elevated to 88 °C (3, 4). The precise kinetics of the heat shock response vary between different organisms: in *E. coli*, the synthesis of Hsps peaks rapidly and then falls back to a new level, greater than that seen in non-heat shocked cells. Other organisms show continuously elevated synthesis of some components of the response for as long as the stress is present (for example, GroEL in *Streptomyces albus*; 5). In addition, studies on *E. coli* and other prokaryotes have shown that a variety of other stresses can induce many or all of the same proteins as those that are induced by heat shock. For example, growth in the presence of ethanol (4%), puromycin, nalidixic acid, and cadmium chloride have all been shown to induce Hsps (6). The simplest explanation of these two phenomena is that a common inducing signal, which is produced as a consequence of both heat shock and a variety of other stressful treatments, is sensed by all cells, and this results in the induction of the proteins of the heat shock response. More complex explanations are also possible: for example, different stresses may produce different signals that are detected by alternate pathways but which still result in the induction of the heat shock response. What is known about the signals that can induce the response, and how do these relate to the role of the molecular chaperones and proteases which are produced?

2.2 The presence of unfolded protein is a signal for elevation of chaperone expression in the cytoplasm

Most proteins are only marginally more stable in their folded state than in their unfolded state (7), and are hence predicted to begin to unfold as soon as they are heated much beyond the optimal growth temperature of the organism within which they have evolved to function. Once unfolded, proteins have a greater propensity to aggregate, particularly within the crowded environment of the cell, as the hydrophobic forces that mainly drive protein folding will also lead to protein aggregation if the local concentration of protein molecules is high. Given that the primary role of molecular chaperones is to prevent protein aggregation and to promote correct protein folding, it is therefore not surprising that many of the molecular chaperones are expressed more highly in heat shocked cells. One of the simplest intracellular signals for induction of the heat shock response would clearly be the presence of unfolded proteins, and a number of lines of evidence indicate that this is indeed the case.

It was shown in 1985 (8) that the presence of abnormal proteins in *E. coli* led to induction of several proteins of the heat shock response. The protein most closely studied in this work was the Lon protease (a heat shock protein but not a molecular chaperone) but it was clear that several other Hsps were also induced, although none of them were specifically identified. Treatments that led to Hsp expression included growth on the amino acid canavanine (known to prevent correct folding when present in proteins), treatment with puromycin (which results in the release of prematurely terminated polypeptides from the ribosome), and over-production of a eukaryotic secreted protein that failed to fold correctly in the bacterial cytosol. Subsequent studies using mutant derivatives of a phage lambda repressor that was known to be predominantly unfolded under physiological conditions confirmed and extended these findings, and showed that it is the level of unfolded protein per se that determines the degree of Hsp induction (9). The proteins which induce the heat shock response in the above studies were all to an extent 'abnormal', and hence perhaps not perfect models of those that would in fact be important in a genuine heat shock, but it has also been shown that a heat shock response can be induced simply by preventing secretion of some periplasmic *E. coli* proteins, either by expressing a fusion protein that blocks normal secretion (10). or by using a null mutation of *secB* (11). Since proteins which have evolved to fold in the periplasm (an oxidizing environment) often fold poorly or not at all in the cytoplasm (a reducing environment), it is reasonable to assume that it is again the fact that these proteins had not folded correctly that induced the heat shock response in these strains.

There is, unfortunately, a paucity of experimental evidence on whether the presence of unfolded proteins is also a potent signal for the heat shock response in other prokaryotic organisms, although this seems likely to be the case. Only in *Bacillus subtilis* has it been shown that the agents that increase the levels of unfolded proteins and induce the heat shock response in *E. coli* also induce expression of the *groE* and *dnaK* operons in this organism (12). As will be discussed further, below, these operons are under a different regulatory system than that of some of the other chaperone genes in this organism, and a fascinating coda to this research is that although some of the chaperone genes in different regulons are induced by heat and ethanol, they show different responses to other effectors which are expected to increase the presence of unfolded protein in the cell, such as puromycin. Thus in the case of *B. subtilis* and other organisms where there are several regulatory circuits governing the expression of different chaperone genes, differences exist in how these various circuits respond to the presence of unfolded protein. This presumably relates to the differing physiological roles of the various chaperones, and is an area ripe for further study.

2.3 In *E. coli*, heat shock inducible molecular chaperones are regulated by the σ^{32} factor of RNA polymerase

RNA polymerase in *E. coli* is composed of five subunits and has the composition $\alpha_2\beta\beta'\sigma$. The $\alpha_2\beta\beta'$ subunits make up the apoenzyme, which contains all the machinery for polymerization of mRNA from a DNA template, but which binds non-specifically to DNA. It is the presence of the fifth subunit that directs the polymerase to particular sites (promoters) within the cell (13). For the housekeeping genes of *E. coli*, this subunit is a protein of approximately 70 kDa which is referred to as σ^{70}. However, a variety of experiments published in the late seventies and early eighties showed that the promoters of the heat shock genes, which include the major chaperones of the cell, are recognized in addition by a different sigma factor, σ^{32} (Fig. 1). Under heat shock conditions, the levels of σ^{32} rise in the cell, and this hence leads to greater expression of the genes whose promoters are now recognized. Strains of *E. coli* which are mutated in the gene for σ^{32} (*rpoH*) are deficient in the induction of the heat shock response (14), and in fact complete deletion of the *rpoH* gene leads to severe temperature sensitivity, with strains carrying such a mutation being unable to grow at temperatures above 20 °C. Two important conclusions follow from this latter result: first, at least some components of the *rpoH* regulon are important for normal growth even at 37 °C, and second, σ^{32} is directly or indirectly involved in determining their expression even at non-heat shock temperatures. Selection at varying temperatures for revertants of the strains carrying a deletion of the *rpoH* gene showed that up-regulation of the *groE* operon alone was sufficient to permit growth of *E. coli* lacking σ^{32} factor at up to 40 °C, conclusive evidence for the pre-eminent importance of the *groE* chaperones in normal *E. coli* growth (15). The converse experiment has also been carried out, showing that over-expression of σ^{32} caused induced expression of most of the heat shock genes in the absence of a heat shock (16), although intriguingly this did not confer any additional thermotolerance on these strains. It was also interesting in these experiments to note that not all the heat shock genes are induced by over-expression of σ^{32} alone; the small Hsps IbpA and IbpB, for example, are not seen, implying that additional signals are required for the elevation of expression of these proteins. These signals have not yet been identified. To reconstruct

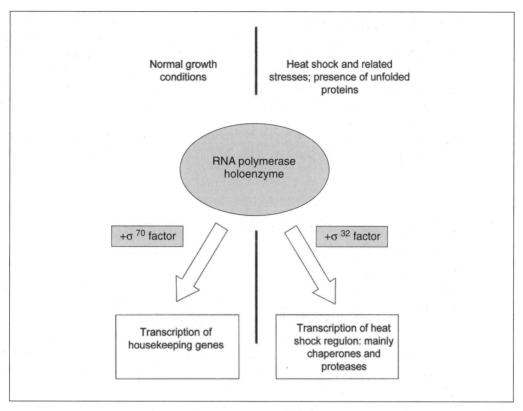

Fig. 1 Under normal growth conditions, the σ^{70} subunit directs RNA polymerase to *E. coli* promoters. The promoters of heat shock genes are recognized by the alternate σ^{32} subunit, levels of which are transiently elevated by heat shock.

the chain of events that leads from a heat shock (or other stress) via elevated σ^{32} levels to induced heat shock gene expression, and to investigate the role that the presence of unfolded proteins plays in this pathway, it was necessary to investigate the mechanism by which σ^{32} is elevated under heat shock conditions. This proved a complex task, which is still underway, as several distinct mechanisms appear to play a part. First, the transcription of σ^{32} can be increased by increases in temperature, although this is only significant at lethally high temperatures (17). Of more importance are two other effects which occur at heat shock temperatures: increased translation of *rpoH* mRNA, and transient increased stabilization of the σ^{32} protein itself. The latter appears to provide a mechanism whereby levels of σ^{32} can be regulated by the amount of unfolded protein present in the cell, and will be described in the next section. The regulation of σ^{32} levels by increased translatability of message may be more directly related to actual temperature, and thus may provide a separate route whereby the signal of heat shock can lead to a heat shock response. This is further considered in section 2.5.

2.4 In *E. coli*, σ^{32} levels are regulated in part by unfolded protein via the DnaK/DnaJ chaperones

σ^{32} is a remarkably unstable protein. In non-heat shocked *E. coli* cells, the half life of the protein as measured by pulse chase experiments is a mere 60 seconds. After heat shock, the half life increases significantly, to 8 minutes, although this increase is only transient (18, 19). This increased stabilization of σ^{32} is also seen in cells which are displaying an elevated heat shock response due to the presence of unfolded proteins at normal growth temperature (20). Several proteases are involved in the degradation of σ^{32}, including HslVU, ClpAP, Lon, and (most significantly) the membrane bound ATP-dependent protease FtsH (previously referred to as HflB)(21–23). All these proteases are themselves members of the σ^{32} regulon, and thus, all other things being equal, it would be expected that σ^{32} would be less, not more, stable after heat shock. Some other factor must be operating to determine the susceptibility of σ^{32} to these proteases.

The clue as to what this factor is comes from studies showing that mutations in any one of the members of the DnaK chaperone team (DnaK, DnaJ, and GrpE) leads to an elevation of expression of the heat shock response under normal growth conditions, and this in turn is due to stabilization of σ^{32} (24). Not only do strains carrying such mutations show elevated levels of Hsps at normal growth temperatures, but they also take longer to reach the new steady state for Hsp synthesis after the initial high induction. A plausible explanation for this result would be that σ^{32} has to be presented to the proteases that degrade it by the action of the DnaK chaperone team, and so if members of this team become less effective, the action of proteases on σ^{32} is reduced. An attractive corollary of this model is that it provides a mechanism whereby the levels of unfolded protein could regulate levels of σ^{32}, in that as the DnaK team is known to have a major chaperone role at high temperatures (see Chapter 1), it could be that there is a simple competition between σ^{32} and unfolded proteins for the DnaK team (25, 26). By this model, as levels of unfolded protein increase, the amount of DnaK and its co-chaperones which are available to bind σ^{32} and present it to proteases is reduced, and thus σ^{32} is stabilized, inducing the heat shock response. If this is true, the control of the heat shock response should be closely linked to the actual levels of the DnaK team members within the cell.

Recent studies have confirmed this prediction (27). Alteration of DnaK and DnaJ levels in the cell has been shown to have several effects on the heat shock response. First, alteration of DnaK and DnaJ levels (even by as little as 20%) produced the opposite change in the levels of σ^{32} itself, and also in levels of GrpE and GroEL. This is in accordance with the above model, as (for example) lower DnaK and DnaJ levels should decrease the proteolysis of σ^{32} and hence increase the level of expression of proteins in the σ^{32} regulon. Second, when DnaK and DnaJ levels are lowered, the shut-off phase of the heat shock response becomes much longer, again as predicted. Third, and most importantly for this model, the induction of the heat shock response by the presence of an unfolded substrate protein depended on the levels of DnaK and DnaJ: as the levels of these two proteins were raised artificially, the ability of an unfolded protein to induce the heat shock response was decreased. This is exactly as predicted by the titration model.

Precisely how the association with the DnaK system leads to degradation of σ^{32} by FtsH is not known. It has been shown, however, that σ^{32} bound to RNA polymerase is resistant to FtsH mediated degradation (27), so a distinct possibility is that DnaK binding to σ^{32} may prevent this association and permit FtsH a chance to degrade σ^{32}.

It thus seems likely that a large component of regulation of the heat shock response in *E. coli*, and the production of the chaperones and proteases which largely comprise the heat shock regulon, is due to the increased presence of unfolded proteins in heat shocked cells. These proteins bind to DnaK and DnaJ, and these two chaperones are then no longer available to bind and destabilize σ^{32}. Thus, increased expression of the σ^{32} regulon occurs. As the molecular chaperone and charonin networks, in particular the DnaK team and ClpB, cooperate to either hold, refold, or degrade the misfolded proteins, shut-off of the initial high induction of the heat shock response occurs as more free DnaK and DnaJ become available and σ^{32} is again destabilized. This is, however, only part of the story, as the altered translatability of the *rpoH* mRNA itself is also an important component in mediating the heat shock response.

2.5 σ^{32} levels are also modulated directly by temperature

As described above, σ^{32} levels are mediated not only by changes in the stability of the protein but also in the translatability of the mRNA. This means that in normal non-heat shocked cells, levels of *rpoH* mRNA are high but the translation of this mRNA is poor; upon heat shock, the pre-existing mRNA pool is now more translationally active. The mechanism for this appears to be related to the secondary structure of the *rpoH* mRNA (28). Two key regions of the *rpoH* mRNA have been shown to play an important role in this aspect of σ^{32} regulation (29). The first region (region A) shows strong homology to a known enhancer of translation in 16S rRNA called the 'downstream box'. A second part of the mRNA, 100 nucleotides upstream from region A, can base-pair with region A to form a secondary structure which inhibits translation by blocking the ribosome binding site and the start codon. It has been suggested that following a rise in temperature the formation of this secondary structure does not occur, and thus efficient translation proceeds. In support of this, a direct correlation between the melting curves of transcripts of this region and accessibility for 30S ribosomes with the levels of expression at different temperatures was observed (30). Therefore the σ^{32} mRNA appears to have a built in thermostat operating via the secondary structure which allows direct modulation of translation by different temperatures.

Thus there are at least two mechanisms whereby the heat shock regulon of *E. coli* can be induced at heat shock temperatures (Fig. 2): the presence of unfolded protein altering the stability of σ^{32}, mediated by DnaK and DnaJ, and the direct sensing of temperature, mediated by changes in *rpoH* mRNA secondary structure. How do these two effects interact? The most recent evidence is consistent with the idea that the primary effector for induction of the response in *E. coli* is the latter, that is, the

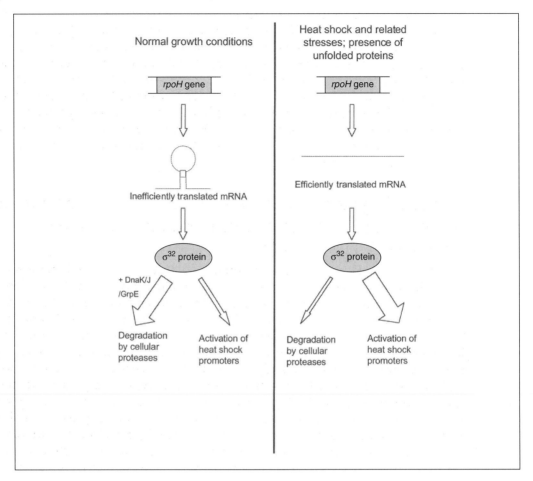

Fig. 2 Two main mechanisms operate to increase the cellular levels of σ^{32} subunit after heat shock: improved translation of *rpoH* mRNA, which encodes σ^{32}, and decreased degradation of σ^{32} by cellular proteases.

chaperone-independent change in mRNA translatability (19). A significant finding in this study was that the increase in stability of σ^{32} on heat shock is extremely transient, and in fact within 10 minutes of heat shock σ^{32} is even more unstable ($t_{1/2} = 20$ seconds) than before the onset of heat shock. This is likely to account for the transient nature of the heat shock response in *E. coli*, and to be mediated by DnaK and DnaJ by the mechanism described above. Presumably the increase in σ^{32} on heat shock is only transient because once the response is well established, DnaK and DnaJ levels are elevated and can again bind σ^{32}, and moreover, the levels of the proteases which degrade σ^{32} are also elevated. The regulation of the heat shock response is thus the result of a balance of at least two different effects. The way in which stresses other than heat shock induce the σ^{32} regulon has not been closely studied, but under these stresses, the presence of unfolded protein must be more significant for the induction phase as well as for the subsequent shut-off.

Direct sensing of temperature may also be occurring in other systems. For example, a heat shock dependent mRNA cleavage has been reported for the *groE* operon of *Agrobacterium tumefaciens*, occurring between the *groES* and *groEL* genes, leading to the presence of a transcript encoding GroEL only (31). Such an event could be an important part of the regulation of the GroE chaperones after heat shock, and would change the ratio of GroES and GroEL present in the cell under these conditions. Whether such events do indeed occur in other bacteria remains to be determined and has not been extensively researched. The RheA repressor, discussed in section 2.6 below, is the only case reported to date where a repressor protein may itself be directly sensitive to temperature.

2.6 The *E. coli* system is unusual: most bacteria appear to use repressor based mechanisms to regulate the major cytoplasmic chaperones

Paradigms in science are made to be broken, and the *E. coli* paradigm for regulation of the heat shock response is no exception. First, although *rpoH* homologues are found in numerous other bacteria, they are by no means universal. Some of the ways in which other bacteria regulate their heat shock responses are described below. Second, even in those bacteria which do contain *rpoH* genes, the way in which the levels of σ^{32} are regulated appears to vary significantly between different species. Some appear to use predominantly a translational mechanism, as described for *E. coli* (32). But in other cases, the increased synthesis is regulated predominantly at the transcriptional level (including *Caulobacter crescentus* (33, 34) and *Bradyrhizobium japonicum*; the latter is a particularly complex case as the organism contains three *rpoH* homologues which are differentially regulated (35)). How the different mechanisms for enhancing σ^{32} levels in response to stress in these different organisms actually operate—whether, for example, they also respond to the level of unfolded protein or to temperature—has yet to be determined.

Many bacteria (including some of those with the simplest genomes) either do not have an *rpoH* gene in their genomes or use other methods for regulating the expression of their heat shock genes. The most widespread and best understood of these methods is the CIRCE/HrcA system. This system was uncovered following analysis of promoter regions of the DnaK and GroE operons in *B. subtilis*, which identified a conserved inverted repeat (IR) as a possible operator site involved in regulation (36, 37). It was shown that mutations in this IR in the *dnaK* operon led to a higher level of expression of the operon under non-heat shock conditions. A variety of similar experiments in different bacteria which also contain this IR have subsequently given the same result. Following their experiments Zuber and Schumann named the IR 'CIRCE' (**C**ontrolling **I**nverted **R**epeat of **C**haperone **E**xpression). The CIRCE element is highly conserved, appearing in all classes of Eubacteria including highly divergent groups such as Cyanobacteria, Spirochetes, and low G-C gram positives (reviewed in 38). Generally speaking it is found only upstream of *groE* and *dnaK* operons, although exceptions are known to this.

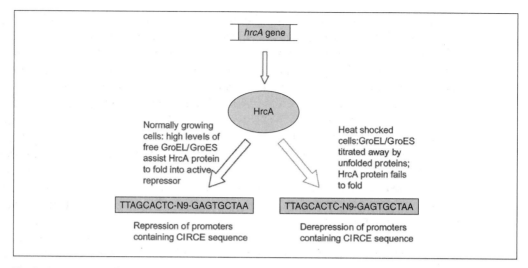

Fig. 3 HrcA protein binds at the CIRCE inverted repeat to repress transcription of many *dnaK* and *groEL* operons under normal growth conditions. According to current models, GroEL is required for the efficient folding of HrcA. When GroEL is sequestered by unfolded proteins following heat shock, HrcA fails to fold efficiently and HrcA-regulated operons are derepressed.

Having identified the role of the CIRCE region in chaperone and other heat shock gene regulation, it was necessary to define the regulator protein which was predicted to be involved in binding to this region. Indirect evidence indicating that a member of the *dnaK* operon of *B. subtilis* was responsible for the repression at the CIRCE element came when Schulz and colleagues mutated the first gene of *dnaK* operon (*orf39*). This eliminated expression of the downstream genes in this operon, and also led to elevated levels of expression of the unlinked *groE* operon, suggesting that *orf39* coded for a repressor which regulated both operons (39). Subsequently, it was confirmed in both *B. subtilis* and *C. crescentus* that this gene encoded a repressor that acted at the CIRCE element (40–42), and this was renamed HrcA (Fig. 3). The *hrcA* gene has been cloned from a number of bacteria. The proteins show rather low sequence identity even with closely related species (43) and nothing is known about the structure of HrcA. The way in which this repressor may respond to the presence of unfolded protein is discussed in section 2.7 below.

Unlike the situation with σ^{32}, where a large number of different chaperones and proteases are under the control of a single effector, HrcA in general seems to be involved mainly with the regulation of the GroE and DnaK chaperone machines. Other chaperones and proteases involved in stress resistance in *B. subtilis* and other gram positive bacteria, such as the ClpB homologue ClpC, are under the control of at least two separate systems: a novel sigma subunit called σ^{B}, and a repressor called CtsR (44). Although these two systems have been well characterized and appear to be widespread, the precise signals that they respond to inside the cell are not known at present; they do not appear to respond to non-native proteins (12). The CIRCE/HrcA system is not the only repressor based mechanism known. At least two other repressor based mechanisms have been described in prokaryotes which are involved with regulation of chaperone gene expression: HspR and RheA.

HspR was first identified as a negative regulator of the *dnaK* operon in *Streptomyces coelicolor* (45, 46). The gene has also been identified in *Streptomyces albus* and in both cases is part of the *dnaK* operon (47). An HspR homologue in *Helicobacter pylori* regulates at least three operons including one which contains itself, another which contains an *hrcA* homologue, *grpE* and *dnaK*, and the *groESL* operon (48). HspR repression is relieved by heat shock in *Streptomyces*, but not in *H. pylori*. Of the three HspR regulated operons in *H. pylori*, two are induced at the transcriptional level by osmotic shock; the environmental signal(s) which induces the third (the *hrcA-dnaK-grpE* operon) has yet to be established. Recent data suggest that DnaK is required for repression by HspR in *S. coelicolor* (49a). The fact that some organisms, which do not contain an *rpoH* gene, do contain genes for both *hspR* and *hrcA*, coupled with the fact that HrcA expression may itself be regulated by levels of HspR, points to the existence of a complex system for the control of chaperones and other Hsps which is likely to be finely tuned to respond to different stresses by expressing the appropriate combinations of proteins. The advent of whole-genome assays for transcription should greatly aid the understanding of these regulatory circuits.

Another repressor regulated heat shock gene which has been described in *S. albus*, namely the *hsp18* gene, is regulated by the RheA repressor protein. As a member of the small Hsp family, Hsp18 may well prove to have chaperone properties, although this has not yet been established. What makes the RheA protein unique among the repressors so far described is that good *in vitro* evidence exists to suggest that it may be directly sensitive to temperature, changing its structure and failing to bind its cognate operator under heat shock conditions (49). Thus this may be the first example of a chaperone whose expression is linked to heat shock through a direct effect of heat on a repressor protein. The mechanism for this process has not been determined.

Finally, mention may be made of the ROSE (**r**epression **o**f heat **s**hock gene **ex**pression) element, which provides a further degree of complexity to the regulation of the heat shock response in *B. japonicum* (50, 51). This element is found in the promoters of the genes for several small Hsps in this organism, but is also present in the regulatory region of *rpoH*. The element is large at approximately 100 bp, and has been shown to be retarded by an unknown protein from a crude cell extract; no retardation occurs if the extract is made from heat shocked cells. It is assumed that the ROSE element is also bound by a repressor which loses activity after heat shock, but no further details of this system are known.

2.7 Levels of the HrcA repressor are controlled by the GroE chaperone system

It seems likely, although it has only been formally demonstrated for the case of *B. subtilis* (12), that the HrcA regulatory system also acts in response to the level of unfolded protein in the cell. How does this work? Convincing evidence, again predominantly from work on *B. subtilis*, supports a titration model analogous to the

DnaK/σ^{32} system in *E. coli*. In this case, however, the key chaperone appears to be GroEL and not DnaK. Preliminary evidence that GroEL might be involved in the regulation of at least its own expression in those organisms that use a CIRCE-mediated system for regulation came initially from *B. japonicum*, where it was shown that deletion of one of the *groE* operons that was CIRCE-regulated led to an increase in expression from the promoter of this operon (52), implying direct or indirect negative autoregulation. More direct evidence came from experiments where the expression of DnaK and GroEL were measured in strains of *B. subtilis* in which the level of GroES and GroEL could be experimentally manipulated (53). It was found in these experiments that a decrease in the level of the GroE proteins led to an elevation of expression from the promoters for the *dnaK* and *groE* operons, whereas over-production of GroES and GroEL had the opposite effect. This result is as would be expected if the GroE proteins are required for folding of HrcA to its active form. To confirm that this was indeed due to an interaction between the HrcA repressor protein and GroE, the system was reconstituted in *E. coli*. Heat shock regulation of a *B. subtilis* CIRCE-containing promoter was seen as long as HrcA was also present. The level of repression before a heat shock was much reduced in genetic backgrounds containing mutated *groES* or *groEL* alleles, supporting the model that both GroES and GroEL are required to form active HrcA. Some in vitro evidence also supports this model, in that purified HrcA (from *Bacillus stearothermophilus*, as the protein purified from *B. subtilis* is insoluble) was shown to retard DNA which contained the CIRCE element in a gel shift assay, and this effect was markedly enhanced in the presence of GroEL. However, similar experiments with the HrcA protein from *B. japonicum* failed to show this effect (54); thus the evidence from *in vitro* studies cannot yet be regarded as conclusive.

HrcA, σ^{32}, and HspR thus appear to use the levels of unfolded protein in the cell, at least in part, as a way of monitoring the need for the expression of major cellular chaperones, and this detection is mediated by the chaperones themselves. This appealing model may also ultimately be extended to other components of the heat shock response that are mediated by other regulators and mechanisms. It would be of considerable interest to look at the global effects of the presence of unfolded protein on gene expression in organisms that have different regulators, particularly those where several different regulator proteins are present in the same organism. The advent of techniques allowing analysis of expression of complete genomes at the mRNA or protein level make such experiments technically feasible.

2.8 Extra-cytoplasmic unfolded proteins in *E. coli* also induce a chaperone response

As has been discussed in Chapter 2, there are several proteins in the periplasm that are required for correct folding of periplasmic proteins and also for correct folding and insertion of outer membrane proteins. These include the disulfide oxidases and isomerases DsbA and DsbC, and also a range of proteins with prolyl *cis-trans* isomerase activity, such as FkpA, SurA, and RotA. FkpA has recently been suggested to have chaperone activity which is independent of its isomerase activity (55, 56). The levels of expression of these proteins are affected by the presence of misfolded proteins in the periplasm or associated with the inner or outer membranes, as well as by other signals. Understanding this regulation presents an interesting problem since the signal must not only be detected but it must also be transduced across the inner membrane. How is this accomplished?

There appear to be at least two systems present in *E. coli* for achieving this. The Cpx pathway regulates (among others) the DsbA and RotA proteins, while the σ^E pathway regulates (among others) the FkpA protein. The two systems are not completely independent—both regulate the expression of the periplasmic protease DegP, for example, and loss of function mutants in the σ^E pathway can be suppressed by activation of the Cpx pathway (57). However, the nature of the inducing signals does differ somewhat between the two pathways and they achieve their targets of gene regulation by quite different mechanisms. The roles of the two systems also differ: evidence suggests that the σ^E regulated proteins are of particular importance in the correct folding of outer membrane proteins. The Cpx system, on the other hand, appears to be very important in the assembly of pilli.

The Cpx system is a classic two-component regulator system, with a membrane bound sensor molecule (CpxA) and a cytoplasmic response protein (CpxR). CpxA can act to phosphorylate (or dephosphorylate) CpxR, and when in its phosphorylated form, CpxR acts by binding upstream of a number of genes and activating their expression (58–60). The presence of misfolded proteins, particularly at the inner membrane, leads to the activation of the kinase activity of CpxA, but the mechanism for this has not been established, although it has been suggested that the periplasmic protein CpxP may play a role.

The pathway for activation of σ^E is different. σ^E (a cytoplasmic protein) is normally bound to a protein in the inner membrane called RseA, which acts as an anti-sigma factor, a class of proteins which act by interacting reversibly with sigma factors and preventing them from binding to RNA polymerase (61–63). A second protein factor, RseB, is also involved in this pathway, interacting with the periplasmic domain of RseA and further repressing σ^E activation. The signal in this case appears again to be the presence of unfolded protein, either in the periplasm or in the outer membrane, but again the way in which the presence of unfolded protein is detected by RseA and/or RseB has not been established. Activation of σ^E by loss of the interaction with RseA leads to activation of transcription at a small number of promoters, including those for *fkpA* and *degP* but also for *rpoH* (the σ^{32} gene) and *rpoE*, which codes for σ^E itself.

3. Chaperone gene expression in eukaryotes

Although this chapter focuses deliberately on the well-understood prokaryotic systems, it is interesting to compare them with some examples of chaperone gene regulation in eukaryotes, and to ask whether similar mechanisms have evolved to enable molecular chaperones to respond at the level of expression to signals such as the presence of unfolded protein. The discussion is limited to cytoplasmic events, as the unfolded protein response which signals the need for enhanced chaperone activity in the endoplasmic reticulum has been discussed extensively in Chapter 8. The regulation of CCT expression is also not discussed here, as this has been covered in Chapter 4.

3.1 Stress induced cytoplasmic chaperones are activated by the heat shock transcription factor (HSF)

As is the case in prokaryotes, many of the proteins of the heat shock response also act as molecular chaperones in eukaryotes, although there are exceptions. Expression of the CCT chaperone complex for example, as discussed in Chapter 4, is generally not heat shock induced. Some chaperones are represented by both heat inducible and non-inducible forms (such as Hsp70 and Hsc70), where the different forms show some degree of specialization of function related to the expression signals which regulate them (see Chapter 5). Patterns of expression of different chaperones also can vary between species, or even within tissues in the same species. For this section, I will discuss the way in which heat shock gene regulation is accomplished in eukaryotic cells, and then look at the signals that cause induction of the heat shock response.

Activation of heat shock genes in eukaryotes requires both the binding of several eukaryotic transcription factors such as TfIID (which binds to TATA elements in promoters) and GAGA (which binds to a GAG motif) (64, 65). In addition, a specific transcriptional activator factor called HSF must bind to promoters for their activation by heat shock. Transfer of heat shock inducibility to other promoters can be achieved by fusing promoter regions from heat shock genes to other genes and this approach (together with deletion and scanning mutagenesis of heat shock promoters) has allowed the definition of a heat shock element (HSE) to which HSF binds (66–68); this is formed of three inverted repeats of the sequence AGAAn. Studies on either cells or cell extracts where *de novo* protein synthesis is blocked show that binding of HSF to HSEs is enhanced following heat shock, thus implying that the important process in the induction of a heat shock response is the activation of pre-existing HSF protein rather than its *de novo* synthesis (69, 70). The nature of this activation is now moderately well understood; the way in which it relates to the conditions inside the cell rather less so.

The most important step in the conversion of HSF from an inactive to an active form is trimerization of the protein. This process has been demonstrated using a variety of techniques (reviewed in 71) and requires a conserved domain that is found in the C-terminal region of the part of the protein which is involved with DNA binding (72). Trimerization causes an increase in binding affinity of HSF for the HSE by around four orders of magnitude (71), the major consequence of this being the activation of transcription of genes which contain an HSE. The trimerization reaction can be reproduced *in vitro* with purified protein using some but not all of the signals that induce the eukaryotic heat shock response (73), so at least in part the induction of the heat shock response may be due to the ability of HSF to directly respond to alterations in temperature by changing from an inactive to an active conformation. However, experiments where HSF factors are expressed in heterologous species shows that the story must be more complex than this, as heat shock mediated by the novel HSF takes place at the typical heat shock temperature for the organism in which it is expressed, not that of the one from which it was originally derived (74). By analogy with prokaryotic systems, two mechanisms (not mutually exclusive) might be proposed to account for this: the HSF could be sensing the presence of unfolded or damaged proteins in some way, and could also be affected by the levels of molecular chaperones in the cell. What evidence is there for these two possibilities?

3.2 Unfolded proteins can induce a heat shock response in eukaryotes

As is the case with prokaryotes, growth of cells in the presence of amino acid analogues can induce the heat shock response in eukaryotic cells (75). A more direct demonstration of the potential role of unfolded protein in heat shock induction came from experiments where proteins which had been chemically denatured (by, for example, reduction and carboxymethylation) and then microinjected into *Xenopus* oocytes containing an *hsp70* promoter-*lacZ* fusion (76). Induction of β-galactosidase expression was clearly seen in preliminary experiments, and was not seen when the same proteins were microinjected in a native form. Subsequent experiments showed that the picture was more complex, as not all proteins had this effect even when in the denatured form, and that even for proteins that did induce the induction of the *hsp70* promoter, different preparations of the same protein varied greatly in their ability to do this. Careful investigation established that the key factor was the degree of aggregation of the microinjected protein, which correlated well with the amount of induction seen (77). Ubiquination was not required. Moreover, microinjection was much more successful at inducing a response if it was directly into the nuclei of the cells. Treatments that result in the formation of non-native disulfides, which could be shown to lead to denaturation of protein and exposure of hydrophobic regions, also caused HSF-dependent induction of Hsps(78). Conversely, treatment of cells with protein stabilizing reagents such as glycerol was shown to block the induction of stress proteins by heat shock (79). The precise nature of the inducing signal(s) remains unclear, but it has been proposed that a key factor in determining whether a given effector induces the heat shock response is its ability to cause oxidation of glutathione, which in turn causes the formation of inter-chain disulfides and of protein-glutathione mixed disulfides, both of which are likely to destabilize protein structure in the cytoplasm (80).

3.3 Molecular chaperones modulate HSF activity

Again by analogy with the prokaryotic system, an obvious way for unfolded proteins to cause the stress response would be by titration of molecular chaperones. Again there is evidence to support this, though the picture is complex. Reduction in cellular

Hsp70 levels results in elevated activity of HSF in yeast, even in the absence of heat shock, as would be predicted by a titration model (81). Also, in the microinjection experiments described above, the effects of denatured protein could be reduced if they were co-injected with purified Hsc70 (82); injection of Hsc70 alone in these studies also reduced the response to thermal stress. A direct association between Hsp70 protein and HSF has also been demonstrated with a variety of techniques (reviewed in 71). The association of Hsp70 is with the domain of HSF which is responsible for activation of gene expression, and it would appear that a major role of Hsp70 is to block the ability of HSF which has already trimerized and bound to DNA to induce gene expression (83). The eukaryotic DnaJ homologue Hdj1 is also involved with this process. Hsp70 is not the only chaperone involved with the regulation of HSF activity; Hsp90 has also been shown to play a role. As discussed more extensively in Chapter 7, Hsp90 can be shown to bind HSF directly, and to play a role in its oligomerization (84–87).

Building a model of how chaperones enable HSF to monitor the level of unfolded protein within the cell, if indeed this is what they do, is complicated by the fact that they probably also play a role in the disassembly of the HSF trimers after the heat shock has passed. If for simplicity we consider only the forward reaction, a basic model would be that in non-stressed cells, HSF is predominantly in the monomeric form and bound (perhaps transiently) in a complex with Hsp90 and other proteins which keeps it in this state. Depletion of Hsp90 (for example by titration with unfolded proteins) may shift the equilibrium to the trimeric, DNA binding form. However, activation of gene expression by this form also requires depletion of Hsp70. By this simple model, control of Hsp expression will be tight and precisely tuneable to the needs of the cell for molecular chaperone activity.

4. Summary of major points

- In both prokaryotes and eukaryotes, many molecular chaperones are also Hsps, induced by changes in temperature and a variety of other stressful treatments.

- A common factor in induction of these proteins is likely to be the presence of unfolded or partially folded proteins, and abundant evidence supports the hypothesis that such proteins are an important part of the induction signal for the heat shock response in prokaryotes and eukaryotes.

- A variety of mechanisms exist which regulate chaperone gene expression in response to cellular stress. The best characterized in prokaryotes is that mediated by the RNA polymerase σ^{32} factor, which allows expression of heat shock genes in *E. coli*. This factor is regulated at at least two levels. First the mRNA for this sigma factor is more efficiently translated at high temperature, due to loss of secondary structure. Second, the σ^{32} factor is transiently stabilized after heat shock through titration of the DnaK chaperones by unfolded proteins, which allows the σ^{32} factor to assemble into RNA polymerase and hence evade the proteolytic action of several cellular proteases.

- Regulation as described by this detailed model does not appear to be widespread among other prokaryotes. Repressor based systems are more widely used, of which the best studied is the HrcA repressor, typified by the one found in *Bacillus subtilis*. This repressor binds to and represses *groE* and *dnaK* operons in many species, and requires GroE for its activity. After heat shock, titration of the GroE chaperone machine by unfolded proteins leads to loss of HrcA repression. Other repressor based systems are being characterized, although the way in which they detect cellular stress has not always been described to date.

- Activation of chaperones and protein foldases in the bacterial periplasm in *E. coli* is via at least two separate but overlapping mechanisms. One employs an anti-sigma factor which prevents activity of the σ^E subunit of RNA polymerase unless abnormal proteins begin to accumulate in the periplasm. The other is via a two-component system with reversible phosphorylation of a cytoplasmic protein leading to expression of appropriate periplasmic proteins.

- In eukaryotic cells, induction of heat shock gene expression requires the activity of the HSF protein, which trimerizes and binds to DNA on heat shock. This may be partly mediated directly by temperature, but Hsp70 and Hsp90 and their cofactors also clearly play a role. Hsp70 acts to specifically block the activation of gene expression by HSF even when it is bound to DNA. Hsp90 appears to play a key role in the oligomerization of HSF into its active state. Both chaperones may help the cell to monitor the level of unfolded protein and hence respond to any need for enhanced chaperone activity.

References

1. Richmond, C. S., Glasner, J. D., Mau, R., Jin, H., and Blattner, F. R. (1999). Genome-wide expression profiling in *Escherichia coli* K-12 *Nucl. Acids Res.*, **27**, 3821.
2. Mogk, A., Tomoyasu, T., Goloubinoff, P., Rüdiger, S., Röder, D., Langen, H., and Bukau, B. (1999). Identification of thermolabile *Escherichia coli* proteins: prevention and reversion of aggregation by DnaK and ClpB. *EMBO J.*, **18**, 6934.
3. McCallum, K. L., Heikkila, J. J., and Inniss, W. E. (1986). Temperature dependent pattern of heat shock protein synthesis in psychrophilic and psychrotrophic micro-organisms. *Can. J. Microbiol.*, **32**, 516.
4. Takai, K., Nunoura, T., Sako, Y., and Uchida, A. (1998). Acquired thermotolerance and temperature-induced protein accumulation in the extremely thermophilic bacterium *Rhodothermus obamensis*. *J. Bacteriol.*, **180**, 2770.
5. Servant, P. and Mazodier, P. (1996). Heat induction of hsp18 gene expression in Streptomyces albus G: transcriptional and posttranscriptional regulation. *J. Bacteriol.*, **178**, 7031.
6. van Bogelen, R. A., Vaughn, V., and Neidhardt, F. C. (1987). Differential induction of heat shock, SOS, and oxidation stress proteins and accumulation of nucleotides in *Escherichia coli*. *J. Bacteriol.*, **153**, 26.
7. Gill, K. A. (1990). Dominant forces in protein folding. *Biochem.*, **29**, 7133.
8. Goff, S. A. and Goldberg, A. L. (1985). Production of abnormal proteins in *E. coli* stimulates transcription of *lon* and other heat shock genes. *Cell*, **41**, 587.

9. Parsell, D. A. and Sauer, R. T. (1989). Induction of a heat shock like response by unfolded protein in *Escherichia coli*: dependence on protein level not protein degradation. *Genes Dev.*, **3**, 1226.

10. Ito, K., Akiyama, Y., Yura, T., and Shiba, K. (1986). Diverse effects of the MalE-LacZ hybrid protein in *Escherichia coli* cell physiology. *J Bacteriol.*, **167**, 201.

11. Wild, J., Walter, W. A., Gross, C. A., and Altman, C. (1993). Accumulation of secretory protein precursors in *Escherichia coli* induces the heat shock response. *J. Bacteriol.*, **175**, 3992.

12. Mogk, A., Völker, A., Engelmann, S., Hecker, M., Schumann, W., and Völker, U. (1998). Nonnative proteins induce expression of the *Bacillus subtilis* CIRCE regulon. *J. Bacteriol.*, **180**, 2895.

13. Wosten, M. M. S. M. (1998). Eubacterial sigma-factors. *FEMS Micro. Rev.*, **22**, 127.

14. Yamamori, T. and Yura, T. (1981). Genetic control of heat-shock protein synthesis and its bearing on growth and thermal regulation in *Escherichia coli*. *Proc. Natl. Acad. Sci. USA.*, **79**, 860.

15. Kusukawa, N. and Yura, T. (1988). Heat shock protein GroE of *Escherichia coli*: key protective roles against thermal stress. *Genes & Dev.*, **2**, 874.

16. van Bogelen, R. A., Acton, M. A., and Neidhardt, F. C. (1987). Induction of the heat shock regulon does not produce thermotolerance in *Escherichia coli*. *Genes & Dev.*, **1**, 525.

17. Erickson, J. W., Vaughn, V., Walter, W. A., Neidhardt, F. C., and Gross, C. A. (1987). Regulation of the promoters and transcripts of *rpoH*, the *Escherichia coli* heat-shock regulatory gene. *Genes & Dev.*, **1**, 419.

18. Strauss, D. B., Walter, W. A., and Gross, C. A. (1989). The heat shock response of *E. coli* is regulated by changes in the concentration of σ^{32}. *Nature*, **329**, 341.

19. Morita, M. T., Kanemori, M., Yanagi, H., and Yura, T. (2000). Dynamic interplay between antagonistic pathways controlling the σ^{32} level in *Escherichia coli*. *Proc. Natl. Acad. Sci. USA.*, **97**, 5860.

20. Kanemori, M., Mori, H., and Yura, T. (1994). Induction of heat shock proteins by abnormal proteins results from stabilisation and not increased synthesis of σ^{32} in *Escherichia coli*. *J. Bacteriol.*, **176**, 5648.

21. Herman, C., Thénevet, R. D., and Bouloc, P. (1995). Degradation of σ^{32}, the heat shock regulator in *Escherichia coli*, is governed by HflB. *Proc. Natl. Acad. Sci. USA.*, **92**, 3516.

22. Tomoyasu, T., Gamer, J., Bukau, B., Kanemori, M., Mori, H., Rutman, A. J., *et al.* (1995). *Escherichia coli* FtsH is a membrane-bound, ATP-dependent protease which degrades the heat-shock transcription factor, σ^{32}. *EMBO J.*, **14**, 2551.

23. Kanemori, M., Nishihara, K., Yanagi, H., and Yura, T. (1997). Synergistic roles of HslVU and other ATP-dependent proteases in controlling *in vivo* turnover of abnormal proteins in *Escherichia coli*. *J. Bacteriol.*, **179**, 7219.

24. Strauss, D. B., Walter, W., and Gross, C. A. (1990). DnaK, DnaJ, and GrpE negatively regulate heat shock gene expression by controlling the synthesis of σ^{32}. *Genes & Dev.*, **4**, 2202.

25. Gross, C. A. (1996). Function and regulation of the heat shock proteins. In *Escherichia coli and Salmonella: cellular and molecular biology* (Editor in chief: F. C. Neidhardt), pp. 1382–99. ASM Press.

26. Bukau, B. (1993). Regulation of the *Escherichia coli* heat shock response. *Mol. Microbiol.*, **9**, 671.

27. Tomoyasu, T., Ogura, T., Tatsuta, T., and Bukau, B. (1998). Levels of DnaK and DnaJ provide tight control of heat shock gene expression and protein repair in *Escherichia coli*. *Mol. Microbiol.*, **30**, 567.

28. Strauss, D. B., Walter, W., and Gross, C. A. (1989). The activity of is σ^{32} is reduced under conditions of excess heat shock protein production in *Escherichia coli*. *Genes Dev.*, **3**, 2003.

29. Nagai, H., Yuzawa, H, and Yura, T. (1991). Interplay of two *cis*-acting elements in translation control of synthesis during the heat shock response of *Escherichia coli*. *Proc. Natl. Acad. Sci. USA*, **88**, 10515.

30. Morita, M. T., Tanaka, Y., Kodama, T. S., Kyogoku, Y., Yanagi, H., and Yura, T. (1999). Translational induction of heat shock transcription factor sigma(32): evidence for a built-in RNA thermosensor. *Genes & Dev.*, **13**, 655.

31. Segal, G. and Ron, E. (1995). The *groESL* operon of *Agrobacterium tumefaciens*: evidence for heat shock dependent mRNA cleavage. *J. Bacteriol.*, **177**, 750.

32. Nakahigashi, K., Yanagi, H., and Yura, T. (1998). Regulatory conservation and divergence of σ^{32} homologues from Gram-negative bacteria. *J. Bacteriol.*, **180**, 2402.

33. Reisenauer, A., Mohr, C. D., and Shapiro, L. (1996). Regulation of a heat shock σ^{32} homolog in *Caulobacter crescentus*. *J. Bacteriol.*, **178**, 1919.

34. Wu, J. G. and Newton, A. (1997). The *Caulobacter* heat shock sigma factor gene *rpoH* is positively autoregulated from a σ^{32}-dependent promoter. *J. Bacteriol.*, **179**, 514.

35. Narberhaus, F., Krummenacher, P., Fischer, H. M., and Hennecke, H. (1997). Three disparately regulated genes for σ^{32}-like transcription factors in *Bradyrhizobium japonicum*. *Mol. Microbiol.*, **24**, 93.

36. Zuber, U. and Schumann, W. (1994). CIRCE, a novel heat-shock element involved in regulation of heat-shock operon dnaK or *Bacillus subtilis*. *J. Bacteriol.* **176**, 1359.

37. Yuan, G. and Wong, S. L. (1995). Regulation of GroE expression in *Bacillus subtilis*: the involvement of the σ^{A} like promoter and the roles of the inverted repeat sequence (CIRCE). *J. Bacteriol.*, **177**, 5427.

38. Segal, G. and Ron, E. Z. (1996). Regulation and organisation of the *dnaK* and *groE* operons in Eubacteria. *FEMS Micro. Letters*, **138**, 1.

39. Schulz, A., Tzschaschel, B., and Schumann, W. (1995). Isolation and analysis of mutants of the *dnaK* operon of *Bacillus subtilis*. *Mol. Microbiol.* **15**, 421.

40. Yuan, G. and Wong, S. L. (1995). Isolation and characterisation of *Bacillus subtilis groE* regulatory mutants: evidence for orf39 in the *dnaK* operon as a repressor gene in regulating both the expression of both *groE* and *dnaK*. *J. Bacteriol.* **177**, 6462.

41. Schulz, A. and Schumann, W. (1996). hrcA, the first gene of the *Bacillus subtilis dnaK* operon, encodes a negative regulator of class I heat shock genes *J. Bacteriol.*, **178**, 1088.

42. Roberts, R. C., Toochinda, C., Avedissian, M., Baldini, R. L., Gomes, S. L., and Shapiro, L. (1996). Identification of a *Caulobacter crescentus* operon encoding *hrcA*, involved in negatively regulating heat-inducible transcription, and the chaperone gene *grpE*. *J. Bacteriol.*, **178**, 1829.

43. Narberhaus, F. (1999). Negative regulation of bacterial heat shock genes. *Mol. Microbiol.*, **31**, 1.

44. Hecker, M. and Völker, U. (1998). Non-specific, general and multiple stress resistance of growth-restricted *Bacillus subtilis* cells by the expression of the σ^{B} regulon. *Mol. Microbiol.*, **29**, 1129.

45. Bucca, G., Ferina, G., Puglia, A. M., and Smith, C. P. (1995). The *dnaK* operon of *Streptomyces coelicolor* encodes a novel heat shock protein which binds to the promoter region of the operon. *Mol. Microbiol.*, **17**, 663.

46. Bucca, G., Hindle, Z., and Smith, C. P. (1997). Regulation of the *dnaK* operon of *Streptomyces coelicolor* A3(2) is governed by HspR, an autoregulatory repressor protein. *J. Bacteriol.*, **179**, 5999.

47. Grandvalet, C., Servant, P., and Mazodier, P. (1997). Disruption of *hspR*, the repressor gene of the *dnaK* operon in *Streptomyces albus* G. *Mol. Microbiol.*, **23**, 77.

48. Spohn, G. and Scarlato, V. (1999). The autoregulatory HspR repressor protein governs chaperone gene transcription in *Helicobacter pylori*. *Mol. Microbiol.*, **34**, 663.

49. Servant, P., Grandvalet, C., and Mazodier, P. (2000). The RheA repressor is the thermosensor of the HSP18 heat shock response in *Streptomyces albus*. *Proc. Natl. Acad. Sci USA*, **97**, 3538.

49a. Bucca, G., Brassington, A. M. E., Schönfeld, H.-J., and Smith, C. P. (2000). The HspR regulon of *Streptomyces coelicolor*: a role for the DnaK chaperone as a transcriptional co-repressor. *Mol. Microbiol.*, **38**, 1093.

50. Narberhaus, F., Kaser, R., Nocker, A., and Hennecke, H. (1998). A novel DNA element that controls bacterial heat shock gene expression. *Mol. Microbiol.*, **28**, 315.

51. Munchbach, M., Nocker, A., and Narberhaus, F. (1999). Multiple small heat shock proteins in rhizobia. *J. Bacteriol.*, **181**, 83.

52. Babst, M., Hennecke, H., and Fischer, H. M. (1996). Two different mechanisms are involved in the heat-shock regulation of chaperonin gene expression in *Bradyrhizobium japonicum*. *Mol. Microbiol.*, **19**, 827.

53. Mogk, A., Homuth, G., Scholz, C., Kim, L., Schmid, F. X., and Schumann, W. (1997). The GroE chaperonin machine is a major modulator of the CIRCE heat shock regulon of *Bacillus subtilis*. *EMBO J.*, **16**, 4579.

54. Minder, A. C., Fischer, H. M., Hennecke, H., and Narberhaus, F. (2000). Role of HrcA and CIRCE in the heat shock regulatory network of *Bradyrhizobium japonicum*. *J. Bacteriol.*, **182**, 14.

55. Bothmann, H. and Pluckthun, A. (2000). The periplasmic *Escherichia coli* peptidylprolyl cis,trans-isomerase FkpA—I. Increased functional expression of antibody fragments with and without cis-prolines. *J. Biol. Chem.*, **275**, 17100.

56. Ramm, K. and Pluckthun, A. (2000). The periplasmic *Escherichia coli* peptidylprolyl cis,trans-isomerase FkpA—II. Isomerase-independent chaperone activity *in vitro*. *J. Biol. Chem.*, **275**, 17106.

57. Connolly, L., de las Penas, A., Alba, B. M., and Gross, C. A. (1997). The response to extra-cytoplasmic stress in *E. coli* is controlled by partially overlapping pathways. *Genes Dev.*, **11**, 2012.

58. Raivio, T. L. and Silhavy, T. J. (1997). Transduction of envelope stress in *Escherichia coli* by the Cpx two-component system. *J. Bacteriol.*, **179**, 7724.

59. Pogliano, J., Lynch, A. S., Belin, D., Lin, E. C.C., and Beckwith, J. (1997). Regulation of the *Escherichia coli* cell envelope proteins involved in protein folding and degradation by the Cpx two-component system. *Genes & Dev.*, **11**, 1169.

60. Danese, P. N. and Silhavy, T. J. (1997). The σ^E and the Cpx signal transduction systems control the synthesis of periplasmic protein-folding enzymes in *Escherichia coli*. *Genes & Dev.* **11** 1183.

61. Helmann, J. D. (1999). Anti-sigma factors. *Current Opin. Micro.*, **2**, 135.

62. de las Penas, A., Connolly, L., and Gross, C. A. (1997). The σ^E mediated response to extracytoplasmic stress in *Escherichia coli* is transduced by RseA and RseB, two negative regulators of σ^E. *Mol. Microbiol.*, **24**, 373.

63. Missiakis, D., Mayer, M. P., Lemaire, M., Georgopoulos, C., and Raina, S. (1997). Modulation of the *Escherichia coli* σ^E (RpoE) heat-shock transcription factor activity by the RseA, RseB and RseC proteins. *Mol. Microbiol.*, **24**, 355.

64. Wu, C. (1985). An exonuclease protection assay reveals heat shock elements and TATA-box binding proteins in crude nuclear extracts. *Nature*, **317**, 84.

65. Tsukiyama, T., Becker, P. B., and Wu, C. (1994). ATP-dependent nucleosome disruption at a heat shock promoter mediated by DNA binding of GAGA transcription factor. *Nature*, **367**, 525.

66. Pelham, H. R. B. (1982). A regulatory upstream promoter element in the Drosophila hsp70 heat shock gene. *Cell*, **30**, 517.

67. Cuniff, N. F. A. and Morgan, W. D. (1993). Analysis of heat shock element recognition by saturation mutagenesis of the human HSP70.1 gene promoter. *J. Biol. Chem.*, **268**, 8317.

68. Fernandes, M., Xiao, H., and Lis, J. T. (1994). Fine structure analysis of the *Drosophila* and *Saccharomyces* heat shock factor-heat shock element interactions. *Nucl. Acids Res.*, **22**, 167.

69. Zimarino, V. and Wu, C. (1987). Induction of sequence specific binding of *Drosophila* heat shock activator protein without protein synthesis. *Nature*, **327**, 727.

70. Larson, J. S., Schütz, T. J., and Kingston, R. E. (1988). Activation in vitro of sequence specific DNA-binding by a human regulatory factor. *Nature*, **355**, 372.

71. Wu, C. (1995). Heat shock transcription factors: structure and regulation. *Ann. Rev. Cell Dev. Biol.*, **11**, 441.

72. Sorger, P. K. and Nelson, H. C. M. (1989). Trimerisation of a yeast transcriptional activator via a coiled coil motif. *Cell*, **59**, 807.

73. Zhong, M., Orosz, A., and Wu, C. (1998). Direct sensing of heat and oxidation by *Drosophila* heat shock transcription factor. *Molec. Cell*, **2**, 101.

74. Clos, J., Rabindran, S., Wisniewski, J., and Wu, C. (1993). Induction temperature of a human heat shock factor is reprogrammed in a *Drosophila* cell environment. *Nature*, **364**, 252.

75. DiDomenco, B. J., Bugaisky, G. E., and Lindquist, S. (1982). The heat shock response is self-regulated at both the transcriptional and translational levels. *Cell*, **31**, 593.

76. Ananthan, J., Goldberg, A. L., and Voellmy, R. (1986). Abnormal proteins serve as eukaryotic stress signals and trigger the activation of heat shock genes. *Science*, **232**, 522.

77. Miflin, L. C. and Cohen, R. E. (1994). Characterisation of denatured protein inducers of the heat shock (stress) response in *Xenopus laevis* oocytes. *J. Biol. Chem.*, **269**, 15710.

78. McDuffee, A. T., Senisterra, G., Huntley, S., Lepock, J. R., Sekhar, K. J., Meredith, M. J., *et al.* (1997). Proteins containing non-native disulfide bonds generated by oxidative stress can act as signals for the induction of the heat shock response. *J. Cell. Physiol.*, **171**, 143.

79. Edington, B. V., Whelan, S. A., and Hightower, L. E. (1989). Inhibition of heat shock (stress) protein induction by deuterium oxide and glycerol: additional support for the abnormal protein hypothesis of induction. *J. Cell. Physiol.*, **139**, 219.

80. Zou, J. Y., Salminen, W. F., Roberts, S. M., and Voellmy, R. (1998). Correlation between glutathione oxidation and trimerization of heat shock factor 1, an early step in stress induction of the Hsp response. *Cell Stress and Chaps.*, **3**, 141.

81. Boorstein W. R. and Craig, E. A. (1990). Transcriptional regulation of SSA3, an HSP70 gene from *Saccharomyces cerevisiae*. *Mol. Cell. Biol.*, **10**, 3262.

82. Miflin, L. C. and Cohen, R. E. (1994). Hsc70 moderates the heat shock (stress) response in *Xenopus laevis* oocytes and binds to denatured protein inducers. *J. Biol. Chem.*, **269**, 15718.

83. Shi, Y., Mosser, D. D., and Morimoto, R. I. (1998). Molecular chaperones as HSF1-specific transcriptional repressors. *Genes & Dev.*, **12**, 654.

84. Nadeau, K., Das, A., and Walsh, C. T. (1993). Hsp90 chaperonins possess ATPase activity and bind heat-shock transcription factors and peptidyl-prolyl isomerases. *J. Biol. Chem.*, **268**, 1479.

85. Ali, A., Bharadwaj, S., O'Carroll, R., and Ovsenek, N. (1998). HSP90 interacts with and regulates the activity of heat shock factor 1 in *Xenopus* oocytes. *Mol. Cell. Biol.*, **18**, 4949.

86. Zou, J. Y., Guo, Y. L., Guettouche, T., Smith, D. F., and Voellmy, R. (1998). Repression of heat shock transcription factor HSF1 activation by HSP90 (HSP90 complex) that forms a stress-sensitive complex with HSF1. *Cell*, **94**, 471.

87. Bharadwaj, S., Ali, A., and Ovsenek, N. (1999). Multiple components of the HSP90 chaperone complex function in regulation of heat shock factor 1 in vivo. *Mol. Cell. Biol.*, **19**, 8033.

11 | Partial unfolding as a precursor to amyloidosis: a discussion of the occurrence, role, and implications

NEIL M. KAD and SHEENA E. RADFORD

1. Introduction

An astounding feature of proteins is their ability to acquire a unique three-dimensional structure despite the vast sequence space accessible to the polypeptide chain. However, on rare occasions a polypeptide chain finds an energy minimum that is not the native state. These are often in the form of alternately folded structures, or insoluble aggregates. Usually regarded as an experimental nuisance, protein aggregation is now emerging as an exciting area of research with both biological and biotechnological importance. For example, heterogeneous recombinant proteins expressed in bacterial cells can be deposited as inclusion bodies, or the aggregation of proteins *in vivo* may cause deposition disorders such as Alzheimer's or Creutzfeldt–Jacob disease. This chapter will not consider the intricacies and implications of the former (this is dealt with in (1)). Instead, the discussion here will focus on the significance of ordered aggregation of proteins into fibrous structures, which lead to a complaint known as amyloid disease. This ordered aggregation involves the association of the protein or peptide subunits into long fibrils with an overall cross-β structure involving a long array of hydrogen-bonded β-strands (2). Such fibrils are seen in a number of different human diseases e.g. Alzheimer's disease, senile systemic amyloidosis, and dialysis related amyloidosis but it is not yet known whether the fibrils or their precursors are the pathogenic agents (see (3)). In this chapter we review the structure of amyloid fibres and describe current views on the possible mechanisms underlying the conversion of soluble protein into these insoluble aggregates. In particular the importance of partial unfolding in the disease process will be considered, focusing on three well-studied examples. Potential drug therapies will then be discussed, which draw upon a molecular understanding of the fibrillogenic process.

1.1 Amyloid fibril structure

One of the most fascinating features of amyloid diseases is that, despite the range of proteins that are involved, all fibres exhibit similar structural characteristics. This has been long established, from early electron microscopy examination (4) and dye binding assays (5), to more recent X-ray diffraction studies (6). These examinations established that amyloid consists of long unbranching protein fibres, approximately

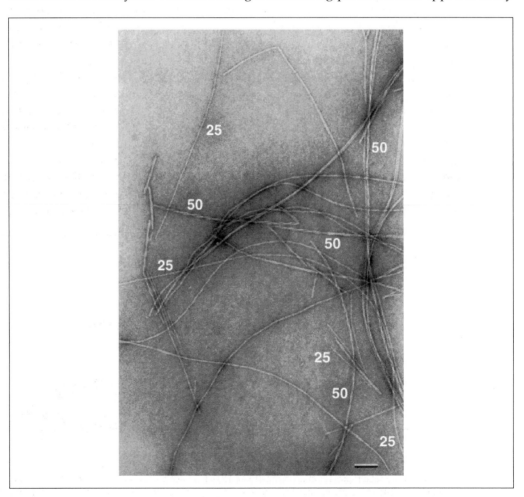

Fig. 1 Electron microscope images of human amylin fibrils. These fibrils show typical characteristics of amyloid fibrils, indeterminate length, unbranching and with a twisting morphology. The latter is consistent with different numbers of protofibrils comprising the fibrils. The numbers indicate the repeat distance in nanometres. Bar, 100 nm. Taken from (7) with permission.

10 nm in diameter. The binding of the dye Congo red to fibrils and their consequent red-green birefringence when viewed through cross polarizers on a light microscope, was (and still is) regarded as a key diagnostic test for fibrils (5). Birefringence was suggested to occur by the linear arrangement of dye molecules on the fibril, reinforcing the EM observations of long repetitive structures.

Electron microscopy images show the overall structure of a fibril; these appear as long unbranched structures with repeats that reflect the twisting of the component filaments around one another ((7); Fig. 1). A variety of fibril forms can be seen in Fig. 1, which arise from the number of intertwining protofilaments that make up the fibrils. The morphologies may correspond to the maturity of the fibrils, with those of greater diameter being seen at later time points in fibril growth. Indeed, a hierarchy for fibre assembly has recently been proposed using atomic force microscopy (AFM) imaging (8). This method, outlined in Fig. 2, allows a more detailed image of the fibril structure to be made than by negative stain EM. More importantly, this method enables imaging in solution thus allowing the measurement of fibril growth in real time (9). In this method, a tip is oscillated over the sample and the change in amplitude as it encounters objects on the atomically flat mica surface is translated into an image (10). A major drawback of using such a method is that the sample is directly handled and, given the fragile nature of biological samples, this can lead to the loss of fine structural features. Pioneered for amyloid proteins by the Lansbury group (11, 12) this method has led to a greater understanding of amyloid structure and growth, even at the level of a single molecule. Such experiments have shown that 10 nm diameter mature amyloid fibrils assemble from smaller diameter units known as protofibrils. In turn, these protofibrils consist of even smaller diameter units known as filaments.

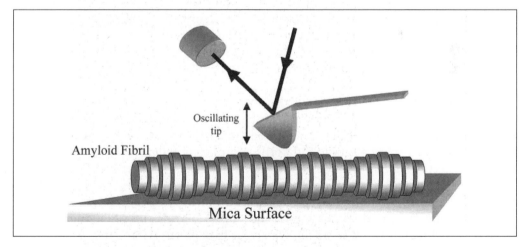

Fig. 2 This schematic representation is presented to give a brief overview of atomic force microscopy imaging of amyloid fibrils. The fibrils are immobilized on an atomically flat mica surface and the cantilever tip is then passed over the sample. The tip is oscillated at its resonant frequency and therefore only intermittently contacts the surfaces. A laser is reflected off the cantilever into a photodiode, hence measuring tip displacement detects the amplitude of oscillation. Thus, details of the surface topology are revealed by changes in the amplitude of oscillation.

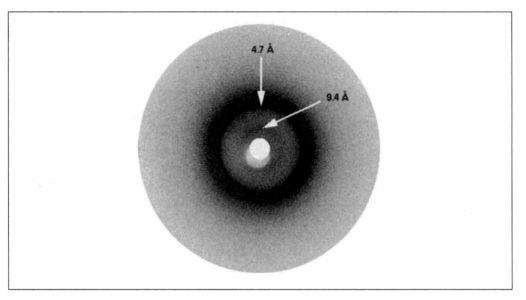

Fig. 3 A typical X-ray diffraction plate of phosphatidylinosito1 3-kinase SH3 domain (21) showing the equatorial (4.7 Å – interstrand) and meridional (9.4 Å – intersheet) reflections as described in the text. Due to the inherent difficulty in aligning fibrils the reflections are circular, and establishment of the meridional and equatorial axes is therefore difficult. Taken from (21) with permission.

As well as differing in diameter, the helical twist of fibrils differs from that of proto-fibrils and filaments, consequently introducing an easy method of identification (8, 11, 12). Using the immunoglobulin light chain domain SMA involved in light-chain amyloidosis, Fink's group observed the formation of another type of fibril (8). This fibril type comprises three filaments twisted together without the involvement of the protofibrillar precursor described above.

Although EM and AFM can provide very detailed topological information, how fibrils are internally organized from their component protofibrils and filaments re-quires the use of other methods such as X-ray diffraction and cryo-EM. X-ray diffraction images allow one to penetrate the exterior of the fibrils to visualize their constituents. By aligning the fibrils and placing them into an X-ray beam, a series of reflections along the meridional and equatorial planes are seen (Fig. 3). The former provides information about repeating structures perpendicular to the fibril axis, whereas the latter gives information on structures parallel to the fibril axis. The position of the dominant reflection on the meridional plane indicates a repeating structure with a 4.7 Å repeat, which corresponds exactly to the interstrand spacing in a β-sheet (Fig. 4, top). Similarly, the main reflection on the equatorial plane lies at ~10 Å and corresponds to the intersheet spacing in a protofibril (Fig. 4, bottom). More distant reflections suggest longer range repeats providing information on the diameter of protofibrils and the intrinsic 15° twist of the β-sheet that results in a 115 Å helical repeat along the axis of the fibril (6). On the basis of such information a model

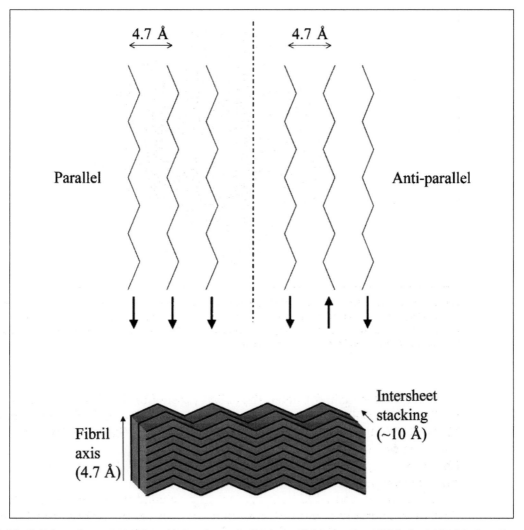

Fig. 4 Schematic representations of strand organization in parallel and anti-parallel β-sheets (top) to form the overall architecture of a fibril (bottom). Whether in parallel or anti-parallel β-sheet structure, the spacing between strands is almost identical. The arrangement of sheets into amyloid fibrils requires packing of sheets laterally to yield the filamentous substructure of a fibril (bottom).

can be constructed for the ultrastructure of a fibril (13), which consists of a continuous β-sheet in the direction of the fibril axis, with similar sheets packing laterally against it with an intersheet spacing of around 10 Å. These sheets twist at the normal pitch of a β-sheet (Fig. 5). FTIR experiments (14) and CD measurement (15) have confirmed the presence of such large amounts of β-sheet.

The use of X-ray diffraction was not incidental for these studies. By virtue of their insolubility and indeterminate length, amyloid fibrils have not as yet been crystallized. Also, their size does not allow examination by solution NMR methods,

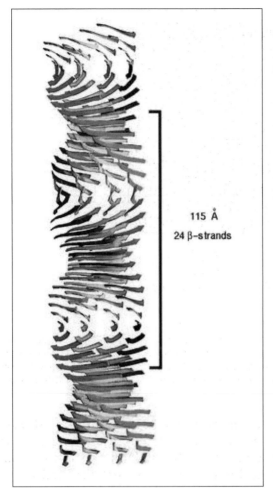

115 Å

24 β-strands

Fig. 5 This image is a reconstruction of the structure of a fibril modelled using X-ray fibre diffraction studies. The lateral stacking of sheets to yield a fibril is clearly seen. Also, the longer range repeats (115 Å) are seen as the β-sheets twist with a 15° pitch between strands, a full 360° rotation involves 24 β-strands. Taken from (6) with permission.

although awe-inspiring steps are being made in this direction (16). Solid-state NMR measurements have already been successful in examining the organization of Aβ fibril peptides (17–19). More recently, cryo-EM has also been used to examine the fibril ultrastructure (20). This technique does not require heavy metal staining and therefore can give a more detailed view of fibril structure. The much higher resolution pictures available using this method have allowed speculation on the constitution of the fibrils. Studies using phosphatidylinositol 3-kinase SH3 domain, a protein that has no known pathology but has been induced to form long and straight amyloid fibrils (21), have shown that the cross-section of a fibril reveals a ring of density (Fig. 6; (20)). By model building it was shown that this ring can accommodate four non-native SH3 domains, potentially by domain-swapping (for an excellent review of domain-swapping see (22)). In this model each individual SH3 subunit in the fibril is composed of two β-sheets arranged with anti-parallel strands (this involves an

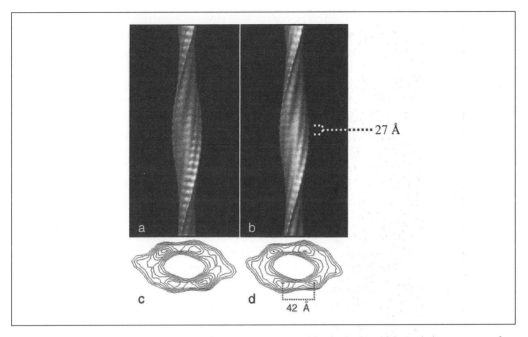

Fig. 6 Using cryo-electron microscopy (cryo-EM) it has been possible to obtain a high-resolution structure of an amyloid fibril. Clear topological features can be discerned, a regular 27 Å subunit repeat and a difference in pitch (610 Å (a, c), 580 Å (b, d)). The cross-section images (c, d) show four main regions of density corresponding to four protofilaments. It was proposed in this study that a non-native SH3 subunit occupies each of these densities. Taken from (20) with permission.

approximate 90° rotation of one sheet in the monomeric precursor). Hydrogen bonds involving backbone amides between each subunit then create the continuous β-sheet structure typical of a fibril. The corollary from this model is that each of the filaments is comprised of two β-sheets. At the time of writing, such detailed modelling of cryo-EM data has not been applied to other forms of amyloid. Using cross-sectional analyses, negative stain EM has established that fibrils from the amyloidogenic protein transthyretin (TTR) are composed of four units of density (23), whilst those of Aβ are composed of five or six (24).

1.2 Amyloid fibril precursors

The central dogma of protein folding is that the sequence of a polypeptide chain will find its lowest energy conformation, in the ambient conditions. Amyloid proteins fulfil this agenda by forming three-dimensional structures that are prone to aggregation. Therefore, the proteins (or peptides) involved in these disorders find a further energy minimum that is the aggregated fibrillar state. The discussion in the previous sections has indicated that amyloid fibrils are predominantly β-sheet in nature, and that in the case of SH3 domain and SMA light-chain the native state cannot be modelled into the density seen in cryo-EM or AFM images (8, 20). How, then, can a soluble, sometimes mixed, α/β protein aggregate into an insoluble predominantly β-sheet fibril, with electron density that does not correspond to the native state? Two possibilities exist, (i) the unfolded state aggregates, (ii) a partially unfolded state aggregates. At least for both variant lysozyme (25) and L55P TTR (26) it is possible to attain native-like structures, suggesting that the protein *unfolds* in perturbed solvent conditions from the native state to a species with high amyloid forming propensity. Or, in the case of peptides, this structural destabilization can be achieved by proteolytic cleavage (27).

The overriding view now emerging from a wealth of studies *in vitro* is that partial unfolding, at least, is required to form fibrils (21, 25, 28–32). A particularly clear example of this postulate has emerged from studies of TTR. This β-sheet tetrameric protein is involved in sporadic and familial amyloid diseases. Led mainly by the Kelly group, a number of studies have determined that upon acidification the tetramer dissociates to form native-like monomers, which are able to form amyloid fibrils (28, 33). Other studies on light-chain associated amyloid (34, 35), lysozyme (25), and prions (36) have also shown that destabilization of the native protein is necessary for amyloid fibril growth. However, in contrast to these examples gelsolin has been proposed to form fibrils from the denatured state ensemble (37).

Whilst these amyloid forming proteins exhibit pathology *in vivo*, two proteins (SH3 and acyl phosphatase (ACP)) have recently been identified that have no known pathology, but still form amyloid fibrils (21, 32). The population of partially unfolded state(s) through destabilization of the native state also appears to be a prerequisite for amyloid fibril formation with these proteins. Interestingly for ACP, the structure of the partially unfolded form is strongly α-helical, converting to a β-sheet form over many hours. This species may aggregate or indeed the aggregation process in its initial stages may be responsible for this β-sheet signal. The implications of being able to form fibrils from proteins that do not have a known pathology are manifold. It may well be that the population of partially unfolded species of many proteins in conditions that do not disrupt hydrogen bond formation will lead to the formation of amyloid fibrils (38). The only other factors that would then be involved are the complementary nature of the associating faces (35).

Elucidation of the molecular mechanism of amyloid formation requires a detailed analysis of the structure of partially folded amyloidogenic precursors. This task is not trivial, because it is necessary to elucidate the structure of an ensemble of species in equilibrium with both the native and denatured states. It is fortuitous that over the past ten or so years a number of biophysical techniques have been developed that permit detailed analysis of partially unfolded states (39) . Perhaps the most potent of these is hydrogen exchange. Using NMR methods coupled with hydrogen exchange, the residue specific location of stable hydrogen bonds can be determined. In addition, by using electrospray ionization mass spectrometry (ESI-MS) in conjunction with NMR studies the populations of species with distinct hydrogen-exchange can be established, thus giving both the overall structure and population of the intermediate. Both TTR (40) and lysozyme (25) have been investigated using hydrogen exchange ESI-MS. These examinations have shown that for both proteins the amyloid precursor is weakly protected from hydrogen exchange, consistent with its partially folded character. Another application is native state hydrogen exchange. This method has been used to examine the conformational dynamics of a protein in its native state. By the addition of denaturant the stability of specific regions of the protein is explicated, potentially mapping the structure of rarely and/or transiently populated species during unfolding. The utility of such a method has been demonstrated for cytochrome c and RNase H, where resemblance was struck between rarely populated partially unfolded states and partially unfolded species populated at low pH (41). Indeed, this method has been used for the human prion protein, revealing that a small nucleus of structure may be the amyloidogenic determinant (30).

It is not only possible to examine the amyloidogenic precursor by excursions from the native state. Stabilization of the precursor allows direct examination of its conformational properties in conditions where it is maximally populated. In order to achieve this, acidification, reduction of disulfide bonds, altered temperatures, or sequence mutation must be employed. Acidification is thought to destabilize the native state by charge repulsions between groups that ionize. Further acidification or addition of salt will attenuate this charge repulsion effect, leading to a re-collapse or refolding of the polypeptide chain (42). The partially unfolded state thus formed has been named the acid- or A-state. The properties of this state have been characterized extensively (see (42)); proteins in this state typically exhibit slightly attenuated secondary structure but almost complete loss of fixed tertiary structure. Hydrophobic surfaces are exposed that will bind the probe 1-anilinonapthalene-8-sulfonic acid (ANS) (43), and the molecules are compact and prone to aggregation (42). It is intuitively attractive to speculate that the self-association of the A-state could lead to amyloid formation. Indeed, it has been shown that ionic strength and acidification is critical to the formation of the light-chain amyloid variant SMA (8), Aβ (44), and β_2-microglobulin (44a). However, studies on TTR have show that in this case the A-state will not form fibrils; instead only lower order oligomers form (33).

The evidence presented here points towards the view that fibrillogenesis involves the transient unfolding of soluble native proteins and population of a partially unfolded state (this case is distinct from peptides that require other factors e.g. proteolysis). To achieve this, alteration of the ambient conditions is a prerequisite. *In vivo*, such alterations in the environment are mainly observed in subcellular compartments, such as the lysosomes where the pH is decreased and proteases are abundant.

In the following section we review three specific well-studied examples in more detail, with a view to examining the methods, trends, and themes that currently exist in the field.

2. Transthyretin

Transthyretin (TTR), formerly known as prealbumin, is a thyroxin (T_4) carrying protein in the cerebrospinal fluid (45). Although essential as a T_4 transporter in the brain, its role in the serum is reduced by the presence of other T_4 binding proteins (thyroxin binding globulins (46)). In up to 25% of aged people systemic TTR polymerizes into amyloid fibrils, leading to senile systemic amyloidosis. In other forms mutations of the protein lead to a condition known as familial amyloid polyneuropathy (FAP) (27). The *in vitro* conversion of TTR to amyloid has been studied extensively using many techniques (47). These molecular insights have pointed the way towards the development of a potential cure to amyloid disease (48).

In native conditions TTR is a very stable tetramer, each subunit of which is a β-sandwich containing three and four strands in each β-sheet. The T_4 binding face is formed at the dimer–dimer interface, which is stabilized by the AB loop from each dimer partner interacting with the H-strand across the interface ((45); Fig. 7). Early studies of TTR fibril formation *in vitro* examined the pH dependence of fibrillization using turbidity and Congo red binding (28). Fibrils were formed upon reduction of the pH from 7 to 5.1–3.9. Additionally, refolding of the protein from monomeric acid-denatured TTR (pH 2) led to the formation of fibrils in the same region (pH 3.9–5.8). Interestingly at the optimal pH (pH 4.4 (33)) value, refolded TTR forms fibrils faster from the acid-denatured state relative to those formed from native tetramer by reduction of pH (33). Since amyloid formation is at least a bimolecular process, and therefore concentration dependent, it was assumed that the precursor is maximally populated at this pH. The discrepancy between the rate of fibril formation when denaturing the tetramer and refolding the acid denatured monomer, reflects the fact that monomerization of the original tetramer precedes aggregation when denaturing. During refolding monomerization is not a factor. Conformational analysis at pH 4.4 by far- and near-UV CD spectroscopy indicates that TTR retains native-like secondary and tertiary structure, and SDS-PAGE analysis indicated the tetramer had dissociated to monomer under these conditions. Based on these data a pathway for amyloid formation (Fig. 8) was suggested. Essentially, reducing the pH induces the tetramer to dissociate, possibly through a dimer, into a monomer, which can then self-associate into amyloid fibrils. Upon further acidification, the monomer is no longer able to form

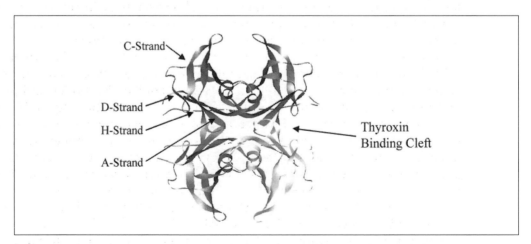

Fig. 7 The three-dimensional structure of transthyretin reveals a distinct ligand-binding cleft between the top and bottom dimers as depicted in this orientation. The tetramer is stabilized by interactions through this region and further upon ligand binding, providing the basis for inhibitor design strategies. This image was produced from pdb coordinates (85) using Weblab viewer v.2.01.

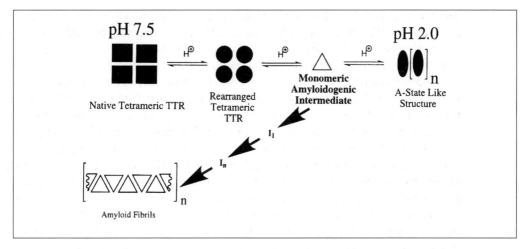

Fig. 8 This represents the structural transitions that TTR undergoes during pH denaturation/refolding. Starting from the native tetramer, a reduction in pH causes a structural rearrangement followed by dissociation into monomers, which can then form amyloid. Further reduction in pH results in a second conformational transition to populate a species no longer able to form amyloid fibrils. Instead of forming long fibrils this species associates into much lower order arrays. Reproduced from (58) with permission.

fibrils. Instead, by the addition of salt the acid-unfolded monomers can be coerced into forming much shorter non-amyloid fibrils. Hence, two pH-induced transitions are observed by intrinsic tryptophan fluorescence. The first reflects the rearrangement of the monomer after tetramer dissociation to yield the amyloidogenic intermediate, whilst the second reflects the transition from the amyloidogenic intermediate to the acid-unfolded state (33, 49). The complicating side reaction of amyloid formation was initially obviated by use of a detergent. Later, however, reducing the concentration of the protein to tenfold below physiological concentrations impeded extensive fibril formation. This was confirmed by analytical ultracentrifugation (50, 51).

Having successfully established an overall pathway for amyloid formation from TTR, the nature of the specific structural elements that are exposed in the amyloidogenic precursor comes into question. An analysis of the distribution and properties of 40 or so naturally occurring TTR mutants (52) found the region 45–58 (edge strands) to be a 'hot spot' for FAP mutations. All but one of these mutations removes interactions within the core of the protein, suggesting that a common feature of the mutations is that they destabilize the protein. However, the one residue that is not associated with core interactions is residue 55; when mutated to a proline this is the most potent of all TTR variants. This variant has a native-like X-ray structure (26) but is less stable as a tetramer (40, 53). As a consequence, tetramer dissociation and amyloid formation both occur at higher pH values (5.5–5.0) than found for the wild-type protein (51, 53). Interestingly, mutations involved in FAP, such as the most common mutant, Val30Met, also destabilize TTR, although this mutant is more stable as a tetramer than the Leu55Pro mutant (40, 50, 53). Another mutation (Thr119Met) not associated with amyloidosis is located close to residue 55 in the three-dimensional structure of TTR. However, this mutant is more acid stable than wild-type TTR, with the result that long incubation times and very high protein concentrations (10 times higher than that found physiologically) are required for it to form fibrils (49). This confirms the hypothesis (52) that this region of the protein is extremely important in amyloid formation. In support of this, Va130Met is found proximal to these residues. From these studies a model for the structural conversion of native TTR to its amyloidogenic precursor has been suggested (33). This involves unfolding of the C- and D-strands to form a long loop linking strands B and E. Crystallization of the Leu55Pro mutant held promise for the elucidation of the structure of this intermediate. Indeed, this structure confirmed some of the Serpell hypothesis (52), that the region 48–55 is important. The D-strand unfolded as proposed; however, the C-strand remained intact (26).

2.1 Inhibition studies

The most hopeful strategies to inhibit amyloid diseases are centred on the reduction of the amount of amyloid material deposited in tissues. This can be achieved by two means, (i) prevention or alteration of subunit interactions within a fibril, and (ii) prevention of the structural conversion to the amyloid precursor. The former of the two approaches is based upon the interruption of the specific interstrand interactions involved in amyloid formation. Such methods have been explored for Aβ inhibition (54–56). The mainstay of transthyretin inhibitor research has been the second of the two strategies. Given that tetramer dissociation is a crucial first step towards fibril formation, stabilization of the tetramer is an ideal strategy for inhibitor design. Moreover, since hysteresis is observed in the folding/unfolding of TTR, interactions within the tetramer lead to stabilization of the monomer (57). One possible route to tetramer stabilization is through the addition of the ligand T_4 (40, 58). In support of this strategy, FAP is found less in the brain than the serum (59). This suggests that stabilization by T_4 binding leads to protection from the disease. Hence, the approach of looking for a T_4 analogue that will not cross the blood–brain barrier, and with a high affinity for TTR, is an ideal therapeutic strategy. One such molecule has been found; flufenamic acid is a bisarylamine non-steroidal anti-inflammatory drug that binds to the active site of TTR (48). At pH 7.6 flufenamic acid has a high affinity for TTR (K_d = 30 nM and 255 nM (48)), comparable to T_4 at pH 7.4 (K_d = 10 nM and 96 μM (58)). Accordingly, use of excess flufenamic acid at pH 4.4 was shown to prevent fibril formation *in vitro* through stabilization of the native tetramer (48). This confirmed the validity of the original rationale and has opened the door to the use of this clinically acceptable drug for a new use. Moreover, and perhaps most excitingly, this study provides a more general strategy for the prevention of other amyloid diseases that involve structural perturbations from the native state.

3. Light chain amyloid

Immunoglobulin light chains were the first known major proteins to be associated with amyloid (60). Light chain amyloid disease involves the deposition of these immuno-globulin-derived proteins either alone or with fragments of the immuno-globulin constant domain (61–63). It is also possible to form amyloid from the constant domain alone (64). Studies using these proteins have also provided inroads into the structural basis of amyloidosis. Given that these proteins are generated from a protean genetic source, the variability in the sequences of these domains is vast (see (65) for an overview). This has lent itself to the study of the occurrence and amyloidogenicity of these sequences. Certain 'hotspots' important for amyloidosis have been identified using such sequence comparisons (29). One of these is the salt bridge formed between the residues Arg61 and Asp82. To test the involvement of these residues in amyloidosis, mutations were made singly and together, and then the stability of the protein and its propensity to form amyloid examined. Mutation of Arg61 to Asn made the normally non-amyloidogenic light-chain variant REI aggregate into ordered fibrils *in vitro* (29, 35). By contrast, mutation of residue 82 to Ile results in the formation of insoluble non-amyloid aggregates. Interestingly, this mutant is associated with light chain deposition disease (66), an aggregation disease resulting from the deposition of insoluble amorphous aggregates. This very important result demonstrates that destabilization of the native protein through destruction of a stabilizing salt bridge is in itself not sufficient to explain amyloid formation. Instead, it suggests that sequence alterations in a polypeptide chain not only dictate that an intermediate be populated but also *how* the subunits of the aggregate associate.

Domain stability appears to play a key role in the propensity of light chain to form amyloid. Previous studies (34) have suggested that reduction of the disulfide bond that lies between the two β-sheets leads to oligomerization. This disulfide bond has been shown to contribute approximately 4.5 kcal/mol to the stability of the protein (67). A number of primary sequence variants (both naturally occurring and rationally designed) have been studied with respect to their stability and amyloidogenicity. Using Congo red binding as a measure of the amount of fibril formation, a direct correlation was observed between stability and amyloidogenicity (29, 68). In another study, simply destabilizing the native state using guanidinium hydrochloride showed that amyloid fibrils form over a discrete range of denaturant concentration (68). The most interesting aspect of this work is that the light chain LEN was used, which shows no amyloid disease morphology *in vivo*. Therefore, although in some cases the sequence may dictate whether non-specific or ordered aggregation occurs, the data suggest that the overriding force behind amyloid formation is the partial unfolding of the domains.

4. β₂-microglobulin

Another protein with an immunoglobulin fold has been noted to form amyloid deposits, in a condition known as dialysis related amyloidosis (69). This disorder involves the deposition of full-length wild-type β₂m aggregates into typical (although morphologically distinct (70)) amyloid fibrils. β₂-microglobulin (β₂m) is the non-covalently linked light chain component of the HLA class I major histocompatibility complex (Fig. 9). Since this complex is found on the cell surface, an equilibrium exists

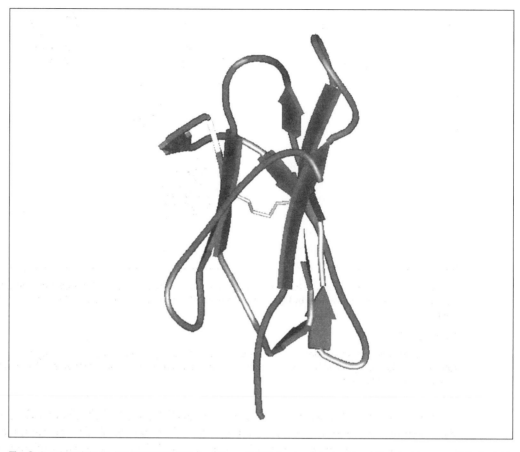

Fig. 9 β_2-microglobulin is composed of two anti-parallel β-sheets organized in a classical immunoglobulin fold. There is a single conserved disulfide formed between the only Cys residues (25 and 80). As discussed in the text, this protein normally associates with the major histocompatibility complex (class I) on certain cell surfaces. Upon renal failure, circulating levels of β_2m rise and this protein is then seen to deposit as amyloid fibrils, primarily at the joints. A conformational change upon pH reduction is suggested to be a critical first step in this deposition disease (44a). This figure was drawn using Weblab viewer v.2.01 using pdb coordinates taken from co-crystallixation with the major histocompatibility complex (86).

between HLA-bound β_2m and free monomeric serum β_2m, thus maintaining a level of approximately 0.2 mg/litre in the serum (69). β_2m is normally degraded in the proximal tube of the kidney; however, upon renal failure degradation is compromised and serum levels rise up to 50 times that of normal. Indeed, even during dialysis with highly permeable membranes the body produces more β_2m than can be removed (69). Such an increase in β_2m concentration facilitates amyloid deposition, since fibrillization is a multi-molecular process (Fig. 10). However, merely increasing the concentration of β_2m *in vitro* to above that *in vivo* is not sufficient to form fibrils (71, 72). Attempts to form fibrils *in vitro* have included the incubation of β_2m with other factors normally associated with fibrillar deposits, such as serum amyloid

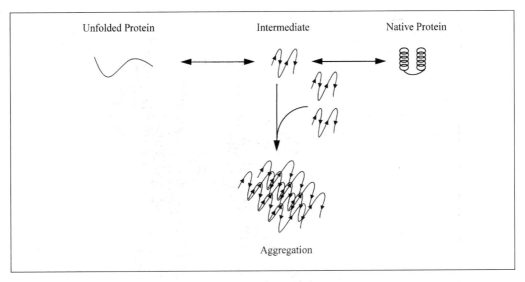

Fig. 10 This depicts the effect of stabilization of an intermediate in the context of amyloid formation. Aggregation from the intermediate is favoured by high concentrations of this species due to the multi-molecular nature of the association event.

$$v_{agg} = k_{agg} [I]^n$$

Therefore an increase in [I] will lead to an increase in the rate of aggregation by $[I]^n$. Where v_{agg} is the aggregation rate and k_{agg} is the aggregation rate constant. [I] is the concentration of the intermediate and n refers to the molecularity of aggregation.

component-P (SAP) and glucosaminoglycans (GAGs) (71, 72). This study concluded that β₂m is fibrillogenic only in the presence of at least SAP. By contrast, Connors *et al.* (73) showed that it was possible to form amyloid fibrils *in vitro* by reducing the ionic strength at pH 7.2 in the absence of other factors. Unfortunately no biological comparisons can be made from this approach, only that β₂m alone is sufficient to form fibrils. Given the intrinsic difficulty of forming β₂m fibrils *in vitro*, another study sought an alternative method of studying β₂m fibril growth, by extending fibrils purified from extracts *ex vivo* (71). This study found that the reduction of pH was critical to the extension process, implying that a conformational change may be required prior to the formation of an amyloid precursor, commensurate with the studies on TTR and light chain amyloid described above. Indeed, antibody mapping studies have suggested that the C-terminal region of β₂m is involved in fibril formation, since a monoclonal antibody raised against this region (residues 98–99) inhibits fibrillization *in vitro* (74). However, whether this arises from the inhibition of a partial unfolding event or by blocking interacting faces is unclear.

What are the factors that drive β₂m towards amyloid formation? A study of the epidemiology of the disease may provide important clues as to its causes. Fibrils formed *in vivo* are localized in the joints; tissues rich in collagen (75). β₂m has been shown to have an affinity towards collagen, explaining its deposition in joints (76). Also, macrophages cluster in joints where β₂m is present (77, 78), suggesting that these may play a part in fibrillization. Uptake of β_2m into macrophage lysosomes does not lead to proteolytic cleavage (79). However, the low pH in these compartments may act to initiate the fibrillization process. Indeed, β_2m fibrils have been observed directly in macrophages *in vivo* (see (69)). Although protection from proteolytic cleavage has been observed, other groups have noted the presence of truncated and acidic forms of β_2m from *in vivo* extracts of dialysis-related amyloid (80, 81), suggesting that these may nucleate fibrillization. Nevertheless, a body of data points toward the significance of acidification in β_2m amyloidosis. Studies performed in our own laboratory using a variety of biophysical techniques have shown that upon acidification wild-type, full length β_2m undergoes a conformational rearrangement to a partially unfolded form. Specifically in these conditions, β_2m has been seen to form fibrils. This indicates the importance of an acid-induced partial unfolding event prior to the formation of β_2m amyloid fibrils (44a). This is in accord with the mechanisms implied for TTR derived and light chain amyloidosis (see above), suggesting that a common mechanism of partial unfolding prior to amyloid fibril formation may prevail. As mentioned above, it is possible to convert soluble non-amyloid proteins such as SH3 and ACP into fibrillar structures (21, 32); further reinforcing the idea that partial unfolding is a prerequisite to amyloid fibril formation.

5. Discussion

Amyloid disease represents a unique pathology in which a normally soluble and functional protein (or peptide) aggregates to form long fibrils. The overriding theme that appears to underpin the formation of amyloid is that the protein constituent experiences conditions that are not conducive to maintenance of the native state. Changes in the solvent or the primary sequence of the polypeptide can bring about such conditions, either by acidification, mutation, or proteolytic cleavage. By thus reducing the native state stability, it is possible to populate one or more non-native states that then self-associate. A reduction in the overall free energy of the protein then results, by the burial of the exposed hydrophobic surfaces and the formation of an extensive hydrogen-bonding network.

One method of preventing the occurrence and propagation of such diseases could be to reduce the concentration of the amyloidogenic precursor. In the case of TTR this has been achieved through the use of a small molecule that binds to the active site and stabilizes the non-amyloidogenic tetramer. In other cases, however, this may not be so simple to achieve, since there is not always a ligand that can bind and stabilize the native state. Here, molecular chaperones could play a role, by binding to the amyloidogenic precursor. The exposure of hydrophobic faces provides a major determinant in the interaction between chaperonin and substrate (for reviews see (82, 83)). Such a partially unfolded state, exposing hydrophobic surfaces that are a characteristic feature of at least some amyloidogenic precursors, provides a rationale for such an approach. The effect of reducing the concentration of aggregation prone proteins has been previously termed as the 'bulk solvent effect' (84), and can in principle be achieved through solvent perturbation as well as protein–protein interactions. Clearly, a mutation in the chaperoning apparatus of a cell that compromises its ability to bind amyloid precursors could be extremely detrimental. Given the sporadic aetiology of senile systemic amyloidosis, a factor other than the primary sequence of wild-type TTR may be involved. In such cases it may well be that molecular chaperones are involved, although currently there is no experimental evidence for this. In any case, most amyloid deposition diseases occur extracellularly where molecular chaperones cannot be directly involved. Much more remains to be discovered about the functions and toxicity of amyloid fibrils *in vivo*, including how fibrils with a similar morphology give rise to different diseases and the identity of the toxic agent. The culprits are still unclear. Nevertheless, much is being learned by the application of biophysical methods to the study of amyloidosis; these offer hope for the design of therapeutic strategies in the future, and new and fascinating insights into the features of self-association events both *in vitro* and *in vivo*.

Acknowledgements

We thank David Smith and the rest of the members of the Radford group for helpful discussions. N. M. K. is supported by the BBSRC. S. E. R. thanks the BBSRC, MRC, EPSRC, The University of Leeds, The Wellcome Trust, and British Biotech Pharmaceuticals Ltd. for financial support.

References

1. Speed, M. A., Wang, D. I. C., and King, J. (1996). Specific aggregation of partially folded polypeptide chains: the molecular basis of inclusion body composition, *Nature Biotechnology*, **14**, 1283.
2. Pauling, L. and Corey, R. (1951). Configuration of polypeptide chain with favoured orientation around single bonds: two pleated sheets. *Proc. Natl. Acad. Sci. USA*, **37**, 729.
3. Lansbury, P. T. (1999). Evolution of amyloid: what normal protein folding may tell us about fibrillogenesis and disease. *Proc. Natl. Acad. Sci. USA*, **96**, 3342.
4. Shirahama, T. and Cohen, A. S. (1965). Structure of amyloid fibrils after negative staining and high-resolution electron microscopy. *Nature*, **206**, 737.
5. Puchtler, H., Sweat, F., and Levine, M. (1962). On the binding of Congo Red by amyloid. *J. Histochem. Cytochem.*, **10**, 355.
6. Sunde, M., Serpell, L. C., Bartlam, M., Fraser, P. E., Pepys, M. B., and Blake, C. C. F. (1997). Common core structure of amyloid fibrils by synchrotron X-ray diffraction. *J. Mol. Biol.*, **273**, 729.
7. Goldsbury, C. S., Cooper, G. J. S., Goldie, K. N., Muller, S. A., Saafi, E. L., Gruijters, W. T. M., and Misur, M. P. (1997). Polymorphic fibrillar assembly of human amylin, *J. Struct. Biol.*, **119**, 17.
8. Ionescu-Zanetti, C., Khurana, R., Gillespie, J. R., Petrick, J. S., Trabachino, L. C., Minert, L. J., *et al.* (1999). Monitoring the assembly of Ig light-chain amyloid fibrils by atomic force microscopy, *Proc. Natl. Acad. Sci. USA*, **96**, 13175.
9. Goldsbury, C., Kistler, J., Aebi, U., Arvinte, T., and Cooper, G. J. S. (1999). Watching amyloid fibrils grow by time-lapse atomic force microscopy. *J. Mol. Biol.*, **285**, 33.

10. Engel, A., Gaub, H. E., and Muller, D. J. (1999). Atomic force microscopy: a forceful way with single molecules. *Curr. Biol.*, **9**, R133.

11. Harper, J. D., Lieber, C. M., and Lansbury, P. T. (1997). Atomic force microscopic imaging of seeded fibril formation and fibril branching by the Alzheimer's disease amyloid-beta protein. *Chem. Biol.*, **4**, 951.

12. Harper, J. D., Wong, S. S., Lieber, C. M., and Lansbury, P. T. (1997). Observation of metastable A beta amyloid protofibrils by atomic force microscopy. *Chem. Biol.*, **4**, 119.

13. Blake, C. and Serpell, L. (1996). Synchrotron X-ray studies suggest that the core of the transthyretin amyloid fibril is a continuous beta-sheet helix. *Structure*, **4**, 989.

14. Lansbury, P. T. (1992). In pursuit of the molecular-structure of amyloid plaque – new technology provides unexpected and critical information. *Biochemistry*, **31**, 6865.

15. Walsh, D. M., Hartley, D. M., Kusumoto, Y., Fezoui, Y., Condron, M. M., Lomakin, A., *et al.* (1999). Amyloid beta-protein fibrillogenesis – structure and biological activity of protofibrillar intermediates. *J. Biol. Chem.*, **274**, 25945.

16. Wider, G. and Wuthrich, K. (1999). NMR spectroscopy of large molecules and multi-molecular assemblies in solution. *Curr. Opin. Struct. Biol.*, **9**, 594.

17. Lansbury, P. T., Costa, P. R., Griffiths, J. M., Simon, E. J., Auger, M., Halverson, K. J., *et al.*, (1995). structural model for the beta-amyloid fibril based on interstrand alignment of an antiparallel-sheet comprising a C-terminal peptide. *Nat. Struct. Biol.*, **2**, 990.

18. Costa, P. R., Kocisko, D. A., Sun, B. Q., Lansbury, P. T., and Griffin, R. G. (1997). Determination of peptide amide configuration in a model amyloid fibril by solid-state NMR. *J. Am. Chem. Soc.*, **119**, 10487.

19. Benzinger, T. L. S., Gregory, D. M., Burkoth, T. S., MillerAuer, H., Lynn, D. G., Botto, R. E., and Meredith, S. C. (1998). Propagating structure of Alzheimer's beta-amyloid((10–35)) is parallel beta-sheet with residues in exact register, *Proc. Natl. Acad. Sci. USA*, **95**, 13407.

20. Jimenez, J. L., Guijarro, J. L., Orlova, E., Zurdo, J., Dobson, C. M., Sunde, M., and Saibil, H. R. (1999). Cryo-electron microscopy structure of an SH3 amyloid fibril and model of the molecular packing *EMBO J.*, **18**, 815.

21. Guijarro, J. I., Sunde, M., Jones, J. A., Campbell, I. D., and Dobson, C. M. (1998). Amyloid fibril formation by an SH3 domain. *Proc. Natl. Acad. Sci. USA*, **95**, 4224.

22. Schlunegger, M. P., Bennett, M. J., and Eisenberg, D. (1997). Oligomer formation by 3D domain swapping: a model for protein assembly and misassembly. *Adv. Prot. Chem.*, **50**, 61.

23. Serpell, L. C., Sunde, M., Fraser, P. E., Luther, P. K., Morris, E. P., Sangren, O., *et al.* (1995). Examination of the structure of the transthyretin amyloid fibril by image-reconstruction from electron-micrographs. *J. Mol. Biol.*, **254**, 113.

24. Fraser, P. E., Duffy, L. K., Omalley, M. B., Nguyen, J., Inouye, H., and Kirschner, D. A. (1991). Morphology and antibody recognition of synthetic beta-amyloid peptides. *J. Neuro. Res.*, **28**, 474.

25. Booth, D. R., Sunde, M., Bellotti, V., Robinson, C. V., Hutchinson, W. L., Fraser, P. E., *et al.* (1997). Instability, unfolding and aggregation of human lysozyme variants underlying amyloid fibrillogenesis. *Nature*, **385**, 787.

26. Sebastiao, M. P., Saraiva, M. J., and Damas, A. M. (1998). The crystal structure of amyloidogenic Leu(55)→ Pro transthyretin variant reveals a possible pathway for transthyretin polymerization into amyloid fibrils. *J. Biol. Chem.*, **273**, 24715.

27. Sipe, J. D. (1992). Amyloidosis. *Annu. Rev. Biochem.*, **61**, 947.

28. Colon, W. and Kelly, J. W. (1992). Partial denaturation of transthyretin is sufficient for amyloid fibril formation in vitro. *Biochemistry*, **31**, 8654.

29. Hurle, M. R., Helms, L. R., Li, L., Chan, W. N., and Wetzel, R. (1994). A role for de-stabilizing amino-acid replacements in light-chain amyloidosis. *Proc. Natl. Acad. Sci. USA*, **91**, 5446.

30. Hosszu, L. L. P., Baxter, N. J., Jackson, G. S., Power, A., Clarke, A. R., Waltho, J. P., *et al.* (1999). Structural mobility of the human prion protein probed by backbone hydrogen exchange. *Nat. Struct. Biol.*, **6**, 740.

31. Litvinovich, S. V., Brew, S. A., Aota, S., Akiyama, S. K., Haudenschild, C., and Ingham, K. C. (1998). Formation of amyloid-like fibrils by self-association of a partially unfolded fibronectin type III module. *J. Mol. Biol.*, **280**, 245.

32. Chiti, F., Webster, P., Taddei, N., Clark, A., Stefani, M., Ramponi, G., and Dobson, C. M. (1999). Designing conditions for in vitro formation of amyloid protofilaments and fibrils. *Proc. Natl. Acad. Sci. USA*, **96**, 3590.

33. Lai, Z. H., Colon, W., and Kelly, J. W. (1996). The acid-mediated denaturation pathway of transthyretin yields a conformational intermediate that can self-assemble into amyloid. *Biochemistry*, **35**, 6470.

34. Klafki, H. W., Pick, A. I., Pardowitz, I., Cole, T., Awni, L. A., Barnikol, H. U., *et al.* (1993). Reduction of disulfide bonds in an amyloidogenic BenceJones protein leads to formation of amyloid-like fibrils in-vitro. *Biol. Chem. Hoppe-Seyler*, **374**, 1117.

35. Helms, L. R. and Wetzel, R. (1996). Specificity of abnormal assembly in immunoglobulin light chain deposition disease and amyloidosis. *J. Mol. Biol.*, **257**, 77.

36. Jackson, G. S., Hosszu, L. L. P., Power, A., Hill, A. F., Kenney, J., Saibil, H., *et al.* (1999). Reversible conversion of monomeric human prion protein between native and fibrilogenic conformations. *Science*, **283**, 1935.

37. Ratnaswamy, G., Koepf, E., Bekele, H., Yin, H., and Kelly, J. (1999). The amyloidogenicity of gelsolin is controlled by proteolysis and pH. *Chem. Biol.*, **6**, 293.

38. Dobson, C. M. (1999). Protein misfolding, evolution and disease. *TIBS*, **24**, 329.

39. Brockwell, D., Smith, D. A., and Radford, S. E. (2000). Protein folding mechanisms: new methods and emerging ideas. *Curr. Opin. Struct. Biol.*, **10**, 16.

40. Nettleton, E. J., Sunde, M., Lai, Z. H., Kelly, J. W., Dobson, C. M., and Robinson, C. V. (1998). Protein subunit interactions and structural integrity of amyloidogenic trans-thyretins: evidence from electrospray mass spectrometry. *J. Mol. Biol.*, **281**, 553.

41. Raschke, T. M. and Marqusee, S. (1997). The kinetic folding intermediate of ribonuclease H resembles the acid molten globule and partially unfolded molecules detected under native conditions. *Nat. Struct. Biol.*, **4**, 298.

42. Fink, A. L., Calciano, L. J., Goto, Y., Kurotsu, T., and Palleros, D. R. (1994). Classification of acid denaturation of proteins – intermediates and unfolded states. *Biochemistry*, **33**, 12504.

43. Semisotnov, G. V., Rodionova, N. A., Razgulyaev, O. I., Uversky, V. N., Gripas, A. F., and Gilmanshin, R. I. (1991). Study of the molten globule intermediate state in protein folding by a hydrophobic fluorescent-probe. *Biopolymers*, **31**, 119.

44. Harper, J. D., Wong, S. S., Lieber, C. M., and Lansbury, P. T. (1999). Assembly of A beta amyloid protofibrils: an in vitro model for a possible early event in Alzheimer's disease. *Biochemistry*, **38**, 8972.

44a. McParland, V. J., Kad, N. M., Kalverda, A. P., Brown, A., Kerwin-Jones, P., Hunter, M. G., Sunde, M., and Radford, S. E. (2000). Partially unfolded states of β_2-microglobulin and amyloidosis *in vitro*. *Biochemistry*, **39**, 8735.

45. Blake, C. C. F., Geisow, M. J., Swan, I. D., Rerat, C., and Rerat, B., Structure of human plasma prealbumin at 2–5 A resolution. A preliminary report on the polypeptide chain conformation, quaternary structure and thyroxine binding., *J. Mol. Biol.*, **88**, 1 (1974).

46. Nilsson, S. F., Rask, L., and Peterson, P. A. (1975). Studies on thyroid hormone-binding

proteins. II. Binding of thyroid hormones, retinol-binding protein, and fluorescent probes to prealbumin and effects of thyroxine on prealbumin subunit self association. *J. Biol. Chem.*, **250**, 8554.

47. Kelly, J. W., Colon, W., Lai, Z. H., Lashuel, H. A., McCulloch, J., McCutchen, S. L., et al. (1997). Transthyretin quaternary and tertiary structural changes facilitate misassembly into amyloid *Adv. Prot. Chem.*, **50**, 161.
48. Peterson, S. A., Klabunde, T., Lashuel, H. A., Purkey, H., Sacchettini, J. C., and Kelly, J. W. (1998). Inhibiting transthyretin conformational changes that lead to amyloid fibril formation. *Proc. Natl. Acad. Sci. USA*, **95**, 12956.
49. McCutchen, S. L., Lai, Z. H., Miroy, G. J., Kelly, J. W., and Colon, W. (1995). Comparison of lethal and nonlethal transthyretin variants and their relationship to amyloid disease. *Biochemistry*, **34**, 13527.
50. Lashuel, H. A., Lai, Z. H., and Kelly, J. W. (1998). Characterization of the transthyretin acid denaturation pathways by analytical ultracentrifugation: implications for wild-type, V30M, and L55P amyloid fibril formation. *Biochemistry*, **37**, 17851.
51. Lashuel, H. A., Wurth, C., Woo, L., and Kelly, J. W. (1999). The most pathogenic transthyretin variant, L55P, forms amyloid fibrils under acidic conditions and protofilaments under physiological conditions. *Biochemistry*, **38**, 13560.
52. Serpell, L. C., Goldsteins, G., Dacklin, I., Lundgren, E., and Blake, C. C. F. (1996). The 'edge strand' hypothesis: prediction and test of a mutational 'hot-spot' on the transthyretin molecule associated with FAP amyloidogenesis. *Amyloid-Int. J. Exp. Clin. Inv.*, **3**, 75.
53. McCutchen, S. L., Colon, W., and Kelly, J. W. (1993). Transthyretin Mutation Leu-55-Pro significantly alters tetramer stability and increases amyloidogenicity. *Biochemistry*, **32**, 12119.
54. Tjernberg, L. O., Lilliehook, C., Callaway, D. J. E., Naslund, J., Hahne, S., Thyberg, J., et al. (1997). Controlling amyloid beta-peptide fibril formation with protease-stable ligands. *J. Biol. Chem.*, **272**, 12601.
55. Tjernberg, L. O., Naslund, J., Lindqvist, F., Johansson, J., Karlstrom, A. R., Thyberg, J., et al. (1996). Arrest of beta-amyloid fibril formation by a pentapeptide ligand. *J. Biol. Chem.*, **271**, 8545.
56. Pallitto, M. M., Ghanta, J., Heinzelman, P., Kiessling, L. L., and Murphy, R. M. (1999). Recognition sequence design for peptidyl modulators of beta-amyloid aggregation and toxicity. *Biochemistry*, **38**, 3570.
57. Lai, Z. H., McCulloch, J., Lashuel, H. A., and Kelly, J. W. (1997). Guanidine hydrochloride-induced denaturation and refolding of transthyretin exhibits a marked hysteresis: equilibria with high kinetic barriers. *Biochemistry*, **36**, 10230.
58. Miroy, G. J., Lai, Z. H., Lashuel, H. A., Peterson, S. A., Strang, C., and Kelly, J. W. (1996). Inhibiting transthyretin amyloid fibril formation via protein stabilization. *Proc. Natl. Acad. Sci. USA*, **93**, 15051.
59. Sipe, J. D. (1994). Amyloidosis. *Crit. Rev. Clin. Lab. Sci.*, **31**, 325.
60. Glenner, G. G., Terry, W., Harada, M., Isersky, C., and Page, D. (1971). Amyloid fibril proteins: proof of homology with immunoglobulin light chains by sequence analyses. *Science*, **172**, 1150.
61. Glenner, G. G., Ein, D., Eanes, E. D., Bladen, H. A., Terry, W., and Page, D. L. (1971). Creation of amyloid fibrils from Bence Jones proteins in vitro. *Science*, **174**, 712.
62. Glenner, G. G. (1980). Amyloid deposits and amyloidosis. The beta-fibrilloses (first of two parts), *N. Engl. J. Med.*, **302**, 1283.
63. Glenner, G. G. (1980). Amyloid deposits and amyloidosis: the beta-fibrilloses (second of two parts). *N. Engl. J. Med.*, **302**, 1333.

64. Solomon, A., Weiss, D. T., Murphy, C. L., Hrncic, R., Wall, J. S., and Schell, M. (1998). Light chain-associated amyloid deposits comprised of a novel kappa constant domain. *Proc. Natl. Acad. Sci. USA*, **95**, 9547.

65. Milstein, C. and Neuberger, M. S. (1996). Maturation of the immune response. *Adv. Prot. Chem.*, **49**, 451.

66. Gallo, G., Goni, F., Boctor, F., Vidal, R., Kumar, A., Stevens, F. J., *et al.* (1996). Light chain cardiomyopathy. Structural analysis of the light chain tissue deposits. *Am. J. Pathol.*, **148**, 1397.

67. Frisch, C., Kolmar, H., Schmidt, A., Kleemann, G., Reinhardt, A., Pohl, E., *et al.* (1996). Contribution of the intramolecular disulfide bridge to the folding stability of REI(v), the variable domain of a human immunoglobulin kappa light chain. *Fold. Des.*, **1**, 431.

68. Raffen, R., Dieckman, L. J., Szpunar, M., Wunschl, C., Pokkuluri, P. R., Dave, P., *et al.* (1999). Physicochemical consequences of amino acid variations that contribute to fibril formation by immunoglobulin light chains. *Prot. Sci.*, **8**, 509.

69. Van Ypersele, C. and Drucke, T. B. (1996). *Dialysis Amyloid*. Oxford University Press, Oxford, UK.

70. Nishi, S., Ogino, S., Maruyama, Y., Honma, N., Gejyo, F., Morita, T., and Arakawa, M., (1990). Electron-microscopic and immunohistochemical study of beta-2-microglobulin-related amyloidosis. *Nephron*, **56**, 357.

71. Naiki, H., Hashimoto, N., Suzuki, S., Kimura, H., Nakakuki, K., and Gejyo, F. (1997). Establishment of a kinetic model of dialysis-related amyloid fibril extension in vitro. *Amyloid-Int. J. Exp. Clin. Inv.*, **4**, 223.

72. Ono, K. and Uchino, F. (1994). Formation of amyloid-like substance from beta-2-microglobulin in vitro. *Nephron*, **66**, 404.

73. Connors, L. H., Shirahama, T., Skinner, M., Fenves, A., and Cohen, A. S., (1985). In vitro formation of amyloid fibrils from intact beta-2-microglobulin. *Biochem. Biophys. Res. Comm.*, **131**, 1063.

74. Stoppini, M., Bellotti, V., Mangione, P., Merlini, G., and Ferri, G. (1997). Use of anti-(beta 2 microglobulin) mAb to study formation of amyloid fibrils. *Eur. J. Biochem.*, **249**, 21.

75. Gejyo, F., Kazama, J. J., Hasegawa, S., Nishi, S., Arakawa, M., and Odano, I. (1995). 131I-beta 2-microglobulin scintigraphy in patients with dialysis amyloidosis. *Clin. Nephrol.*, **44** (Suppl. 1), S14.

76. Homma, N., Gejyo, F., Isemura, M., and Arakawa, M. (1989). Collagen-binding affinity of beta-2-microglobulin, a preprotein of hemodialysis-associated amyloidosis. *Nephron*, **53**, 37.

77. Ohashi, K., Hara, M., Kawai, R., Ogura, Y., Honda, K., Nihei, H., and Mimura, N. (1992). Cervical disks are most susceptible to beta-2-microglobulin amyloid deposition in the vertebral column. *Kidney Int.*, **41**, 1646.

78. Ayers, D. C., Athanasou, N. A., Woods, C. G., and Duthie, R. B. (1993). Dialysis arthropathy of the hip. *Clinical Orthopaedics and Related Research*, **290**, 216.

79. GarciaGarcia, M., Argiles, A., GouinCharnet, A., Durfort, M., GarciaValero, J., and Mourad, G. (1999). Impaired lysosomal processing of beta 2-microglobulin by infiltrating macrophages in dialysis amyloidosis. *Kidney Int.*, **55**, 899.

80. Linke, R. P., Kerling, A., and Rail, A. (1993). Hemodialysis – demonstration of truncated beta-2-microglobulin in Ab-amyloid insitu. *Kidney Int.*, **43**, S100.

81. Bellotti, V., Stoppini, M., Mangione, P., Sunde, M., Robinson, C., Asti, L., *et al.* (1998). Beta 2-microglobulin can be refolded into a native state from ex vivo amyloid fibrils. *Eur. J. Biochem.*, **258**, 61.

82. Coyle, J. E., Jaeger, J., Gross, M., Robinson, C. V., and Radford, S. E. (1997). Structural and mechanistic consequences of polypeptide binding by GroEL. *Fold Des.*, **2**, R93.
83. Ranson, N. A., White, H. E., and Saibil, H. R. (1998). Chaperonins., *Biochem. J.*, **333** (Pt 2), 233.
84. Buchner, J., Schmidt, M., Fuchs, M., Jaenicke, R., Rudolph, R., Schmid, F. X., and Kiefhaber, T. (1991). GroE facilitates refolding of citrate synthase by suppressing aggregation. *Biochemistry*, **30**, 1586.
85. Wojtczak, A. (1997). Crystal structure of rat transthyretin at 2.5 A resolution: first report on a unique tetrameric structure. *Acta Biochim. Pol.*, **44**, 505.
86. Saper, M. A., Bjorkman, P. J., and Wiley, D. C. (1991). Refined structure of the human histocompatibility antigen HLA-A2 ar 2.6 A resolution. *J. Mol. Biol.*, **219**, 277.

Index